INFORMATION TECHNOLOGY FOR MANAGERS

INFORMATION
TECHNOLOGY FOR
MANAGERS

George W. Reynolds
University of Cincinnati

COURSE TECHNOLOGY
CENGAGE Learning™

Australia • Brazil • Japan • Korea • Mexico • Singapore • Spain • United Kingdom • United States

COURSE TECHNOLOGY
CENGAGE Learning™

Information Technology for Managers

George W. Reynolds

VP/Editorial Director: Jack Calhoun

Senior Acquisitions Editor: Charles McCormick, Jr.

Product Manager: Kate Hennessy Mason

Development Editor: Dan Seiter, Mary Pat Shaffer

Editorial Assistant: Bryn Lathrop

Marketing Manager: Bryant Chrzan

Marketing Coordinator: Suellen Ruttkay

Content Product Manager: Heather Furrow

Senior Art Director: Stacy Jenkins Shirley

Cover Designer: Lou Ann Thesing

Cover Image: ©Getty Images/Photodisc

Technology Project Manager: Chris Valentine

Manufacturing Coordinator: Julio Esperas

Composition: GEX Publishing Services

For product information and technology assistance, contact us at
Cengage Learning Academic Resource Center, 1-800-354-9706

For permission to use material from this text or product,
submit all requests online at **www.cengage.com/permissions**
Further permission questions can be emailed to
permissionrequest@cengage.com

ISBN-13: 978-1-4239-0169-3
ISBN-10: 1-4239-0169-X

Course Technology
20 Channel Center Street
Boston, MA 02210
USA

Cengage Learning is a leading provider of customized learning solutions with office locations around the globe, including Singapore, the United Kingdom, Australia, Mexico, Brazil, and Japan. Locate your local office at: **international.cengage.com/region**

Cengage Learning products are represented in Canada by Nelson Education, Ltd.

For your lifelong learning solutions, visit **www.cengage.com/coursetechnology**

Visit our corporate website at **www.cengage.com**

Microsoft, Windows 95, Windows 98, Windows 2000, Windows XP, and Windows Vista are registered trademarks of Microsoft® Corporation. Some of the product names and company names used in this book have been used for identification purposes only and may be trademarks or registered trademarks of their manufacturers and sellers.

Printed in the United States of America
1 2 3 4 5 6 7 13 12 11 10 09

To my grandchildren: Michael, Jacob, Jared, Fievel, Aubrey, Elijah, Abrielle, Sofia, Elliot
—GWR

TABLE OF CONTENTS

Why This Text?

The undergraduate capstone course on information technology and the MBA level information technology course required of College of Business graduates are two of the most challenging courses in the business curriculum to teach. Students in both courses often start the term skeptical of the value of such a course. Indeed, "Why do I need to take this course?" is frequently their attitude. Unfortunately, this attitude is only perpetuated by most texts, which take the approach of "Here is a lot of technical stuff you have to understand." As a result, students complete the course without getting as much from it as they could. The instructors of such courses are disappointed, receive poor student evaluations, and wonder what went wrong. An opportunity to deliver an outstanding and meaningful course has been missed.

Information Technology for Managers takes a fundamentally different approach to this subject in three ways. First, it is targeted squarely at future managers, making it clear why IT does indeed matter to them and the organization. Second, it enables future business managers to understand how information technology can be applied to improve the organization. Third, it provides a framework for business managers to understand their important role vis-à-vis information technology. Said another way, *Information Technology for Managers* answers three basic questions—Why do I need to understand IT? What good is IT? What is my role in delivering good results through the use of IT?

Approach of this Text

Information Technology for Managers is intended for future managers who are expected to understand the implications of IT, identify and evaluate potential opportunities to employ IT, and take an active role in ensuring the successful use of IT within the organization. The text is also valuable for future IT managers who must understand how IT is viewed from the business perspective and how to work effectively with all members of the organization to achieve IT results.

Organization and Coverage

Chapter 1: Managers: Key to Information Technology Results presents a clear rationale for why managers must get involved in information technology strategic planning and project implementation. The chapter helps managers identify what they must do to advance the effective use of IT within their organizations. It also helps them understand how to get involved with IT at the appropriate times and on the appropriate issues.

Chapter 2: Strategic Planning describes how to develop effective strategic planning by defining key business objectives and goals, which are used to identify a portfolio of potential business projects that are clearly aligned with business needs. Further refinement is required to narrow the portfolio to the projects that should be executed and for which sufficient resources are available. This process is illustrated by the example of the United Parcel Service, a major global organization respected for its highly effective use of IT to support business objectives.

Chapter 3: Project Management provides a helpful overview of the project management process. The presentation is consistent with the Project Management Institute's Body of Knowledge, an American National Standard. The chapter describes the nine project management knowledge areas of scope, time, cost, quality, human resources, communications, risk, procurement, and integration. A business manager can take many roles throughout the project life cycle, including champion, sponsor, project manager, change agent, and end user. The chapter identifies frequent causes of project failure and offers invaluable suggestions for how to avoid these problems.

Chapter 4: Business Process and IT Outsourcing discusses the major business reasons for outsourcing as well as many of its potential pitfalls. It also outlines and describes an effective process for selecting an outsourcing firm and successfully transitioning work to the new organization. The chapter covers the importance of establishing service-level agreements and monitoring performance.

Chapter 5: Corporate Governance and IT describes the responsibilities and practices that a company's executive management uses to ensure delivery of real value from IT and to ensure that related risks are managed appropriately. The chapter covers two frameworks for meeting these objectives: the IT Infrastructure Library (ITIL) and Control Objectives for Information and Related Technology (COBIT). The discussion includes related issues such as the Sarbanes-Oxley Act, business continuity planning, and oversight of outsourcing arrangements.

Chapter 6: Collaboration Tools and Wireless Networks covers the fundamentals of electronic communications systems, with a focus on wireless and mobile communications. The chapter presents the benefits and disadvantages of various wide area and local area wireless networks, and how managers can understand and deal with related business issues.

Chapter 7: E-Business discusses the use of electronic business methods to buy and sell goods and services, interact with customers, and collaborate with business partners and government agencies. Several forms of e-business are covered, including business-to-business, business-to-consumer, consumer-to-consumer, and government-to-consumer. The chapter also covers m-commerce, an approach to conduct e-commerce in a wireless environment. The chapter prepares managers to understand and deal with many of the business, legal, and ethical issues associated with e-business.

Chapter 8: Enterprise Resource Planning explains what an ERP system is, identifies several of the benefits associated with ERP implementation, outlines a "best practices" approach to implementing an ERP system, and discusses future trends of ERP systems. The chapter also explains the key role that business managers play in successfully implementing ERP systems.

Chapter 9: Business Intelligence discusses a wide range of applications that help businesses gather and analyze data to improve decision making: data extraction and data cleaning, data warehousing and data mining, online analytical processing (OLAP), business activity monitoring, key performance indicators, dashboards, and balanced scorecards. The chapter discusses the complications and issues associated with each business intelligence system, and discusses the role of the business manager in developing and using these systems.

Chapter 10: Knowledge Management describes explicit and tacit information and how organizations use knowledge management to identify, select, organize, and disseminate that information. In this chapter, you will learn about communities of practice, social network analysis, Web 2.0 Technologies, business rules management systems, and enterprise search. The chapter also covers how to identify and overcome knowledge management challenges, as well as a set of best practices for selling and implementing a knowledge management project.

Chapter 11: Enterprise Architecture describes the use of enterprise architecture to establish a series of reference frameworks that define necessary business and IT changes. The chapter also describes the business manager's role in defining the architecture and business needs of an organization. You will learn about current architecture styles, including centralized, distributed, client/server, and service-oriented architectures. The chapter focuses on the differences between these models and how are they used in practice. You will also be exposed to the Open Group Architecture Framework, a proven process for developing enterprise architecture.

Chapter 12: Ethical, Privacy, and Security Issues provides a brief overview of ethics and identifies key privacy and security issues that managers need to consider in their use of IT to achieve organizational benefits. Ethics, privacy, and computer security are discussed from the perspective of what managers need to know about these topics.

Chapter Features

Opening vignette: Business majors and MBA students often have difficulty appreciating why they need to comprehend IT or what their role (if any) is vis-à-vis IT. In recognition of this, each chapter begins with an opening vignette that raises many of the issues that will be covered in the chapter. The vignette touches on these topics in such a way as to provide a strong incentive to the student to read further in order to gain clarity regarding the potential impact of IT on the business as well as management's responsibility in relation to IT.

Learning Objectives: A set of learning objectives follow the opening vignette and provide a preview of the major themes to be covered in the chapter.

Real-world examples: In an effort to maintain the interest and motivation of the reader, each chapter includes many real-world examples of business managers struggling with the issues covered in the chapter—some successfully, some unsuccessfully. The goal is to help the reader understand the manager's role in relation to information technology and to discover key learnings they can apply within their organizations.

A Manager Takes Charge: This special feature presents a real-world example of a manager taking the initiative to ensure the successful use of IT within his/her organization.

A Manager's Checklist: Each chapter contains a valuable set of guidelines for future business managers to consider as they weigh IT-related topics, including how they might use IT in the future within their organization.

Chapter Summary: Each chapter includes a helpful summary that highlights the managerial implications and key technical issues of the material presented.

Discussion Questions: A set of thought-provoking questions to stimulate a deeper understanding of the topics covered in the chapter.

Action Memos: Each chapter includes two action memos, which are mini-cases written in the style of an e-mail or text message, demanding a response, usually in the form of a decision or recommendation. The action memos provide realistic scenarios and test the student's knowledge, insight and problem-solving capability.

Web-Based Case: Each chapter includes an "open-ended" case that requires students to gather their own research information and do some critical thinking.

Case Study: Each chapter ends with a challenging real-world case of managers struggling with the issues covered in the chapter. These cases are unique because they look at IT from a manager's perspective, not from an IT technologist's point of view.

INSTRUCTOR RESOURCES

The teaching tools that accompany this text offer many options for enhancing a course. As always, we are committed to providing one of the best teaching resource packages available in this market.

Instructor's Manual

An *Instructor's Manual* provides valuable chapter overviews; chapter learning objectives, teaching tips, quick quizzes, class discussion topics, additional projects, additional resources and key terms. It also includes solutions to all end-of-chapter discussion questions, exercises, and case studies.

Test Bank and Test Generator

ExamView® is a powerful objective-based test generator that enables instructors to create paper-, LAN- or Web-based tests from test banks designed specifically for their Course Technology text. Instructors can utilize the ultra-efficient QuickTest Wizard to create tests in less than five minutes by taking advantage of Course Technology's question banks or customizing their own exams from scratch.

PowerPoint Presentations

A set of 50 or more Microsoft PowerPoint slides is available for each chapter. These slides are included to serve as a teaching aid for classroom presentation, to be made available to students on the network for chapter review, or to be printed for classroom distribution. The presentations help students focus on the main topics of each chapter, take better notes, and prepare for examinations. The slides are fully customizable. Instructors can either add their own slides for additional topics they introduce to the class or delete slides they won't be covering.

Figure Files

Figure files allow instructors to create their own presentations using figures taken directly from the text.

Blackboard and WebCT Level 1 Online Content.

If you use Blackboard or WebCT, the test bank for this textbook is available at no cost in a simple, ready-to-use format. Go to *www.cengage.com/coursetechnology* and search for this textbook to download the test bank.

ACKNOWLEDGMENTS

I want to thank all of the folks at Course Technology for their role in bringing this text to market. I offer many thanks to Mary Pat Shaffer and Dan Seiter, my wonderful development editors, who deserve special recognition for their tireless efforts and encouragement.

Heather Furrow, the Content Product Manager, guided the text through the production process. Thanks also to all the many people who worked behind the scenes to bring this effort to fruition including Charles McCormick, the Senior Acquisitions Editor. Special thanks to Kate Hennessy Mason, the Product Manager, for coordinating the efforts of these many people and keeping things moving forward.

I want to thank two contributors to the text: Ralph Brueggemann for his excellent help in providing material on enterprise architecture as well as his insightful feedback on the early chapters of the text, and Naomi Friedman, who wrote several of the opening vignettes and cases.

Last, but not least, I want to thank my wife, Ginnie, for her patience and support in this major project.

TO MY REVIEWERS

I greatly appreciate the following reviewers for their perceptive feedback on early drafts of this text:

Larry Booth, Clayton State University
Nicole Brainard, Principal, Archbishop Alter High School, Dayton, Ohio.
Ralph Brueggemann, University of Cincinnati
Rochelle A. Cadogan, Viterbo University
Wm. Arthur Conklin, University of Houston
Barbara Hewitt, Texas A&M Kingsville
William Hochstettler, Franklin University
Jerry Isaacs, Carroll College
Marcos Sivitanides, Texas State University
Gladys Swindler, Fort Hays State University
Jonathan Whitaker, University of Richmond

MY COMMITMENT

I welcome your input and feedback. If you have any questions or comments regarding *Information Technology for Managers,* please contact me through Course Technology at *www.cengage.com/coursetechnology*.

George Reynolds

CHAPTER 1

MANAGERS: KEY TO INFORMATION TECHNOLOGY RESULTS

HOW CAN YOU ENSURE YOUR FUTURE VALUE AND SUCCESS AS A LEADER AND MANAGER?

"The most valuable and successful leaders and managers today are those who consistently deliver the promised results of strategic initiatives. While visionary thinking and break-through innovation are still important ingredients, organizations are increasingly recogniz-ing that the real competitive differentiator is the ability to execute strategic initiatives reliably and deliver expected results—every time."

—Daryl Conner, founder of Conner Partners, a consulting firm that helps companies address the human side of organizational change

BELARUSBANK JSSB

Why Managers Must Get Involved in Information Technology (IT)

Belarusbank Joint Stock Savings Bank is one of the top 50 financial institutions in Europe. Its headquar-

ters is in Minsk, the capital of Belarus, a country in Eastern Europe with a population of 10 million. The

bank operates with six regional branches, 111 local branches, and 1812 outlets; it has 24,000 employees

and approximately $4 billion in assets. It offers customers a wide range of banking products and services,

including cash settlements, lending, deposit banking, leasing, foreign currency exchange and conversion,

and depository services.

Belarusbank began operations when Belarus was part of the Soviet Union. Its legacy systems and work processes were based on the Soviet style of banking, which emphasized tight control over efficiency. Each bank unit had its own separate accounting, reporting, and administrative systems and processes. The bank's operations were highly inefficient and overly complex.

Senior executives decided to compete globally, but recognized that the bank's convoluted systems were hindering its growth. For Belarusbank to achieve its objectives, it needed to change its decentralized systems and antiquated work processes. It needed a single system that would support streamlined, standard work processes and enable employees to share business and customer data stored anywhere in the company.

After studying the situation and evaluating many alternatives, management initiated a $20 million project to implement software from German vendor SAP AG and modernize the bank's operations. The software will replace many systems deployed over decades by various banking units. The new centralized banking operations will process some 2 million transactions daily and support 5000 employees. All credit, deposit, and payment processes will be standardized. The bank will be able to meet international accounting standards, international financial reporting standards, and Basel II financial reporting requirements, which are critical for expansion into international money markets. Management is working to identify the new roles, rewards, and expectations that employees must adopt to use the new information systems and work processes effectively.

The bank's new systems also will improve the efficiency of all the bank's business processes and provide a quicker "time to market" for new products and product enhancements. Customer satisfaction also is expected to improve through increased customer responsiveness and consistency of business processes. Business unit managers are preparing workers to recognize these new capabilities and take advantage of them in their everyday work.

A major portion of the project's benefits will come from a 10 percent reduction in employees, with a significant reduction in the number of accountants and IT workers. Further benefits will come from cost savings in computer software, hardware, and maintenance. These savings will recover the total cost of the SAP implementation over several years.

"IT solutions in the banking industry today are the essential part of a bank's strategy," said Vladimir Novik, deputy to the chairman of the board. "Belarusbank is the leader of the Belarus banking market and should become the example for other Belarus-based banks in every respect, including the modern banking business processes and solutions." [1,2,3]

LEARNING OBJECTIVES

As you read this chapter, ask yourself:
- What must managers do to advance the effective use of IT within their organizations?
- Am I prepared to get involved with IT at the appropriate times and on the appropriate issues?

This chapter provides a working definition of information technology, discusses the essential role of managers in ensuring good results from various types of IT systems, and warns of the dire consequences that can follow when managers fail to meet these responsibilities. But first let's answer the question—why should managers understand IT?

WHY MANAGERS MUST UNDERSTAND IT

Why learn about information technology? Isn't this area of the business best left to the IT professionals, and not managers? The answer is a simple, emphatic *No!* This section provides several reasons why managers must understand IT, and why they must lead the effort to decide what IT to invest in and how to use it most effectively.

New IT business opportunities, as well as threats, are coming at a faster and faster rate. Managers play a key role—they must frame these opportunities and threats so others can understand them, and then evaluate and prioritize problems and solutions. Finally, managers must lead the effort to pursue IT policies that best meet organizational needs.

Even if organizations invest in the same IT systems from the same vendors, they will not necessarily end up with identical solutions or use the systems in the same ways. As a result, one firm may profit greatly from an IT deployment while another struggles with unsatisfactory results. Managers, working in conjunction with IT specialists, must make many choices about the scope of the IT solution, what data to capture, how databases and applications should be tailored, what information will flow from the systems and to whom, and, most importantly, how people will use the data to make a difference.

True productivity improvements seldom come simply from automating work processes. Real gains in productivity require innovations to business practices and then automating these improved processes to take advantage of IT capabilities. Companies that merely insert IT into their operations without making changes that exploit the new IT capabilities will not capture significant benefits. Managers are the key to ensuring that IT innovations pay off; they must lead a holistic approach that includes encouraging the acceptance of change, addressing changes in business processes and organizational structure, addressing new employee roles and expectations, and establishing new measurement and reward systems.

To gain a sustainable competitive advantage, companies consistently must deliver increasing value to customers. Doing so requires essential information gained through the effective use of IT that better defines customers and their needs. This information can help companies improve products and develop better customer service, leading to sustained increases in revenue and profits. Managers must recognize the value of this information, know how to communicate their needs for it, and be able to work with IT staff to build effective IT systems that make useful information available.

In a rapidly changing global business environment, managers require life-long learning and flexibility in determining their business roles and career opportunities. Given the strong shift toward the use of IT, managers must be able to understand how technology affects their industry and the world at large.

WHAT IS INFORMATION TECHNOLOGY?

Information technology (IT) includes all tools that capture, store, process, exchange, and use information. The field of IT includes computer hardware, such as mainframe computers, servers, laptops, and PDAs; software, such as operating systems and applications for performing various functions; networks and related equipment, such as modems, routers, and switches; and databases for storing important data.

An organization's defined set of IT hardware, software, and networks is called its **IT infrastructure**. An organization also requires a staff of people called the **IT support organization** to plan, implement, operate, and support IT. In many firms, some or all technology support may be outsourced to another firm.

An organization's IT infrastructure must be integrated with employees and procedures to build, operate, and support **information systems**. These systems enable a firm to meet fundamental objectives, such as increasing revenue, reducing costs, improving decision making, enhancing customer relationships, and speeding up their products' time to market. For example, the new systems at Belarusbank will streamline work processes, provide access to customer data, and enable the bank to compete globally by offering new services to new customers. The bank's information system has many IT components: the mainframe computer and database that store business and customer information, the desktop and laptop computers used by employees, and network components that capture data at various branches and update the central database. A streamlined work process enables bank tellers, IT support staff, and other system users to operate efficiently and reliably.

Most organizations have a number of different information systems. When considering the role of business managers for working with IT, it is useful to divide information systems into three types: function IT, network IT, and enterprise IT.[4] Figure 1-1 shows the relationship among IT support staff, IT infrastructure, and the various types of information systems. These systems are explained in the following sections.

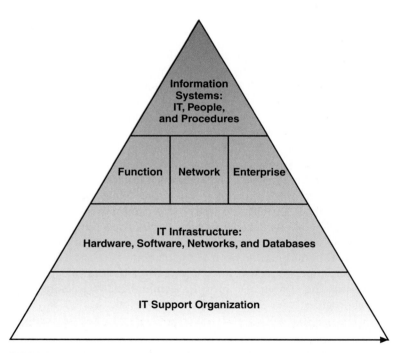

FIGURE 1-1 IT infrastructure supports function, network, and enterprise information systems

Function IT

Function IT includes information systems that improve the productivity of individual users in performing stand-alone tasks. Examples include using computer-aided design (CAD) software, word processors, spreadsheet software, decision support systems, and e-learning systems. One company that makes good use of function IT is Care Rehab, a small manufacturer of traction, electrotherapy, and biofeedback products that patients use for physical therapy and rehabilitation. Care Rehab engineers and scientists generate a constant stream of product innovations—they come out with a new medical device every six months and release upgrades for older devices every couple of months. Product designers use CAD software to create virtual product prototypes that are good enough to eliminate the need for physical prototypes. The firm can move directly into production using its CAD designs, cutting weeks off the time it takes to put new devices on the market.[5]

A **decision support system (DSS)** employs models and analytic tools to help users gain insights into data, draw conclusions from the data, and make recommendations. For example, a buyer might use a DSS to analyze supplier bids and select the least expensive provider for raw materials. The system must account for the suppliers' raw material costs and the cost of transporting the material from the supplier to the manufacturer. The DSS then quickly selects one or more suppliers that can provide the needed volume of raw materials for the best price. Without the DSS, such a task could otherwise take a buyer days or weeks. The DSS can even let a decision maker evaluate alternatives to the least expensive solution. Such "what if" analysis can provide the buyer with valuable options; for example, it can reduce the risk of a disruption in production by suggesting multiple suppliers instead of just one.

E-learning systems encompass a number of computer-enhanced learning techniques, including computer-based simulations, multimedia CD-ROMs, Web-based learning materials, hypermedia, podcasts, and Webcasts. Such use of information systems qualifies as an example of function IT. With the rapid changes in today's business environment, managers and employees must be continual learners to keep pace. For example, organizations like Whirlpool use e-learning systems to accelerate the development of new skills and aid employee performance. With manufacturing facilities in 15 countries that produce a variety of home appliances, Whirlpool was having trouble providing effective, cost-efficient training to its 20,000 salaried employees. Its traditional approach of face-to-face training was costly, and disrupted employees' work schedules. Therefore, Whirlpool implemented a new e-learning system called "Virtual University" that combines Web-enabled courseware and self-paced online modules with traditional instructor teaching. The electronic courses are available globally through the company's corporate portal, a Web site that provides a gateway to corporate information from a single point of access. There, students can review material in real time, get test results, and register both for electronic learning and face-to-face classes. The system also provides comprehensive tracking and reporting to monitor training progress.[6]

Network IT

In today's fast-moving, global work environment, success depends on the ability to communicate and collaborate with others, including co-workers, colleagues, clients, and customers. **Network IT** includes information systems that improve communications and support collaboration among members of a workgroup. Examples include the use of Web conferencing, wikis, and electronic corporate directories.

Web conferencing uses IT to conduct meetings or presentations in which participants are connected via the Internet. Screen sharing is the most basic form of Web conference—each participant sees whatever is on the presenter's screen, be it a spreadsheet, legal document, artwork, blueprint, or MRI image. Conference participants can communicate via voice or text. Another form of Web conferencing is Webcasting, in which audio and video information is broadcast from the presenter to participants. Still another type of Web conference, a Webinar, is a live Internet presentation that supports interactive communications between the presenter and the audience.

One company that uses Web conferencing is Cerner Corporation, which provides software solutions for hospitals and medical facilities. Cerner's 7300 employees are spread over 36 offices in 12 countries. Sales, marketing, and technology workers must be able to collaborate on customer activities, software demos, and internal meetings. Cerner implemented an online Web conferencing service to increase the frequency of meetings with clients and improve communications. Such Web conferencing is a good example of network IT; it shortens the time required to plan and conduct meetings, and greatly reduces travel expenses.[7]

Wiki (Hawaiian for *fast*) is a Web site that allows users to edit and change its content easily and rapidly. The wiki may be either a hosted Internet site or a site on the company intranet. A wiki enables individual members of a workgroup or project team to collaborate on a document, spreadsheet, or software application without having to send the materials back and forth. One company that uses wikis is MWW Group, a public relations and marketing firm of 200 employees in 11 locations. MWW Group serves such corporate clients as Bally Total Fitness, McDonald's, Nikon, Sarah Lee, and Verizon worldwide. A new director of media strategies introduced wikis to the company's copywriters and designers, who quickly caught on and began using wikis to produce ad campaigns and a client's new logo. Wikis have reduced e-mails, meetings, and conference calls so much that the creative teams at MWW Group claim their productivity has doubled.[8]

Electronic corporate directories are used in large organizations to find the right person with whom to collaborate on an issue or opportunity. Increasingly, organizations are creating online electronic corporate directories to solve this problem. IBM added many new features and capabilities when it recently reworked its online employee directory, called BluePages. This network IT application consists of three components—a database of information about employees' skills, knowledge areas, and experience; a search engine; and collaboration features that connect employees and facilitate the sharing of information. Employee profiles contain a photo and an audio file that provides the correct pronunciation of their name. Each profile is updated continually to show the local time at the person's location and his availability for immediate contact. The application is extremely popular with workers, who claim it saves them an average of one hour per month.[9]

Enterprise IT

Enterprise IT includes information systems that organizations use to define interactions among their own employees and/or with external customers, suppliers, and other business partners. These systems often require the radical redesign of fundamental work processes and the automation of new processes. Target processes may include purely internal activities within the organization (such as payroll) and those that support activities with external customers and suppliers. Three examples of enterprise systems are transaction processing, enterprise resource planning, and interorganizational systems. All three systems are explained in this section.

A **transaction processing system (TPS)** captures data for company transactions and other key events and updates the firm's records, which are maintained in electronic files or databases. Each TPS supports a specific activity of the firm, and several may work together to support an entire business process. For example, some organizations use many TPSs to support their order processing, which includes order entry, shipment planning, shipment execution, inventory control, and accounts receivable, as shown in Figure 1-2. The systems work together in the sense that data captured by an "upstream" system is passed "downstream" and made available to other systems later in the order processing cycle. Data captured using the order entry TPS is used to update a file of open orders—orders received but not yet shipped. The open order file, in turn, is used as input to the shipment planning TPS, which determines the orders to be filled, the shipping date, and the location from which each order will be shipped. The result is the planned order file, which is passed downstream to the shipment execution TPS, and so on.

Many organizations are moving from a collection of loosely linked transaction processing systems to an **enterprise resource planning system (ERP)**—a group of computer programs with a common database that a firm uses to plan, manage, and control its routine business operations (see Figure 1-3). This system enables information to be shared across business functions and all levels of management. The shared database eliminates such problems as lack of information and inconsistent information, which are common in multiple transaction processing systems that support only one business function or one department in an organization. Depending on an organization's needs, it may implement ERP software to support its finance, human capital management, manufacturing, or distribution operations. An ERP system can replace two or more independent TPSs, eliminating the need to pass files between the systems.

For example, Beall's Inc. is a Florida-based retailer with more than 600 department stores and outlet stores, primarily in Arizona, Florida, and Georgia. Recent annual sales of its apparel, footwear, gifts, and housewares were $1 billion. Beall's uses a hodgepodge of software packages and internally developed programs to run its business, but it has a three-year plan to replace all the programs with ERP software. The initial phase focuses on human resources, payroll, and point-of-sale operations. Later phases will address finance, merchandise management, and supply chains. Beall's is making this conversion for several reasons. The older software can no longer keep pace as the number of customers, inventory items, stores, and resulting business transactions continues to increase. The old software also is difficult to modify and support; few IT workers are familiar with the 20-year legacy systems. The new ERP system will improve many areas of the business, including inventory management, merchandise planning, and stock allocation and replenishment. These improvements should generate significant revenue increases and cost reductions.[10]

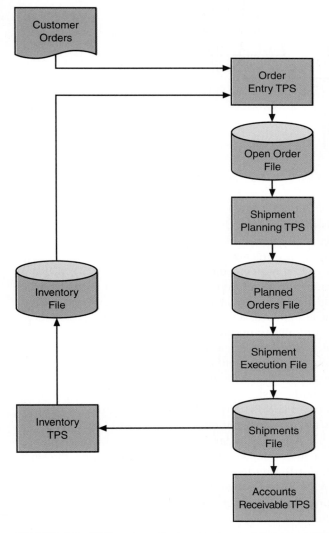

FIGURE 1-2   TPS systems that support order capture and fulfillment

Interorganizational information systems support the flow of data among organizations to achieve shared goals. For example, some organizations need to share data for purchase orders, invoices, and payments, along with information about common suppliers and financial institutions. Such a system speeds up the flow of material, payments, and information, while allowing companies to reduce the effort and costs of processing such transactions.

To ensure efficient and effective sharing of information, these organizations must agree in advance on the nature and format of information to be exchanged, and must use compatible technologies. Companies must resolve such technical issues as data definitions and

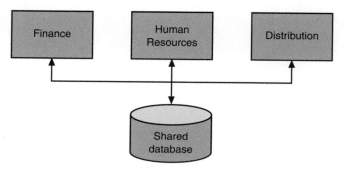

FIGURE 1-3   Common database used to plan, manage, and control an organization's routine business operations

formats, database designs, standards to ensure high data quality, and compatible network infrastructures. The full integration of interorganizational information systems often requires new work processes and significant organizational change.

One type of interorganizational system is **electronic data interchange (EDI)**. EDI supports the direct, computer-to-computer transfer of information in the form of predefined electronic documents. EDI standards dictate which data is required for each type of document and which data is optional. These standards also specify the sequence in which data must be presented and its length and format (numeric, alphabetic, or alphanumeric). For example, an EDI advanced shipment notification is sent from a shipper to a receiver with detailed information about the contents of the shipment and how it is packaged. Other information may be included if the parties agree to include it.

Two widely used sets of EDI standards exist. One standard, the United Nations/EDI for Administration, Commerce, and Transport (UN/EDIFACT) is defined and refined under the auspices of the United Nations. It is the only international standard, and is predominant outside North America. The other popular standard, ANSI ASC X12, includes more than 300 transaction sets that have been developed, integrated, and tested since the early 1980s. The ANSI standard is used in North America.

For example, Toys "R" Us employs EDI to perform effective cross docking, which reduces logistical costs and speeds products to market. Cross docking means that inbound materials are received and immediately loaded for shipping to clients, completely eliminating the intermediate step of warehouse storage. This process also eliminates the need to schedule and stage the shipment later, which saves on labor costs. To facilitate cross docking, suppliers transmit an advance shipment notice to Toys "R" Us for each inbound shipment. This notice provides a purchase order number, vendor identification number, product identification number, and carton counts for each item in the shipment. The EDI data is not handled manually, but flows directly from the suppliers' computers into the Toys "R" Us warehouse management system. With this information, Toys "R" Us distribution centers can anticipate inbound volume, prepare for receiving, and schedule the appropriate number of warehouse workers.[11]

TABLE 1-1 Examples of information systems in each of the Three Worlds of IT

Function IT	Network IT	Enterprise IT
Personal productivity software	Web conferencing	Transaction processing system
Decision support system	Wikis	ERP
E-learning system	Electronic corporate directories	EDI

THE ROLE OF MANAGERS VIS-À-VIS IT

When new IT is introduced in some organizations, managers adopt the technology first, then try to figure out what to do with the new information and cope with its implications. Such an approach is wrong, and can trigger major business disruptions. New IT is more powerful and diverse than the old systems, and is increasingly entwined with the organization's critical business practices.

Companies that successfully adopt new technology recognize that managers have a crucial role in leading the successful introduction and adoption of IT. Managers have three critical responsibilities when it comes to capturing real benefits from IT: identifying appropriate opportunities to apply IT, smoothing the way for its successful introduction and adoption, and mitigating its associated risks. These responsibilities are discussed in the following sections.

Identifying Appropriate IT Opportunities

The sheer magnitude of dollars spent on IT demands that management must ensure a good return on the investment. IT-related expenses in many organizations can account for 50 percent or more of capital spending.[12] Organizations typically spend one to six percent of their total revenues on IT; this spending is generally higher for industries in which IT is more critical to success, such as financial services (see Table 1-2).[13] IT spending as a percentage of revenue is also higher within small organizations than large organizations (see Table 1-3).[14]

These numbers represent rough averages. Great variation exists in IT-related spending, even among similar-sized companies within the same industry. While one company may outspend a competitor on IT, it is not necessarily making more effective use of IT. The most important consideration is what organizations are *getting out of their investments* in IT, not how much *they are investing* in IT. The most effective users of IT maximize value from IT investments that are aligned with their organization's strategic needs and that are well managed and executed. In today's global economy, new technologies, business opportunities, and business threats are coming at a faster and faster rate. According to industry observers Keri Pearlson and Carol Sanders, "Management's role is to frame these opportunities so others can understand them, to evaluate them against existing business needs, and finally to pursue any that fit with an articulated business strategy."[15] The next chapter will outline the strategic planning process and explain how managers can ensure that IT investments align with business strategies and support key objectives.

TABLE 1-2 IT spending based on industry

Industry	Percent of Revenue Spent on IT
Overall average	3-6%
Financial services	5-7%
Telecommunications	5-7%
Professional services	5-7%
Health care	4-5%
Insurance	3-5%
Utilities	3-4%
Retail	1-3%
Construction	1-2%

TABLE 1-3 IT spending based on size of firm

Firm Size	Annual Revenue	Percent of Revenue Spent on IT
Small	Less than $50 million	6.9%
Medium	$50 million to $2 billion	4.1%
Large	Over $2 billion	3.2%

Smooth Introduction and Adoption of IT

To implement an IT system successfully, a company might need to change its business processes, worker roles and responsibilities, reward systems, and decision making. For some IT systems, the amount of change may be trivial; for others, it may be monumental. It is human nature to resist change; researchers J.P. Kotter and L.A. Schlesinger identified four reasons for this resistance (see Table 1-4).[16] Many organizations have tried to implement a promising new IT system, only to have employees never use it or not use the system to its full potential. Managers must be able to overcome this resistance so that the new IT system is accepted and used throughout the organization.

TABLE 1-4 Four reasons people resist change

Reason to Resist Change	Explanation
Parochial self-interest	Some people are more concerned with the impact of the change on themselves than with how it might improve the organization
Misunderstanding	Some people have misconceptions or lack information about the change
Low tolerance to change	Some people require security and stability in their work
Different assessments of the situation	Some people disagree about the reasons for the change or do not support the process

Several theories on organizational change management can help smooth the introduction and adoption of IT. The following sections present three such theories: the Change Management Continuum Model, the Unified Theory of Acceptance and Use of Technology, and the Three Worlds of IT model.

Change Management Continuum Model

D. R. Conner developed the Change Management Continuum Model, which describes key activities that are needed to build commitment for change.[17] This model can identify actions to help an organization successfully introduce and adopt a specific IT system. Figure 1-4 depicts the change management continuum, which illustrates seven stages of commitment grouped into three major phases: inform, educate, and commit.[18] Table 1-5 briefly describes each phase and stage. An organization must execute each of the seven stages to get employees to commit to a new IT system. People will resist adoption of the new system if a stage is skipped or not successfully completed. For example, if a company fails to make employees understand the new IT system, they will not comprehend how they are expected to use it, and the company will be unable to achieve the system's benefits.

TABLE 1-5 The phases and stages of the Change Management Continuum Model

Phase	Goal	Stage	Description
Inform	Make people aware of the change and why it is occurring	Contact	Person first becomes aware that change is to take place
		Awareness	Person has basic knowledge of the change
		Understanding	Person comprehends nature and intent of change and how he/she will be affected
Educate	People recognize impact of change on them and their way of working	Positive Perception	Person develops positive disposition toward the change
		Adoption	Change has demonstrated a positive impact on the organization
		Institutionalization	Change is durable and has been formally incorporated into routine operating procedures of organization
Commit	The change is fully accepted and has become part of everyday life	Internalization	People are highly committed to change because it matches their interests, goals, and values

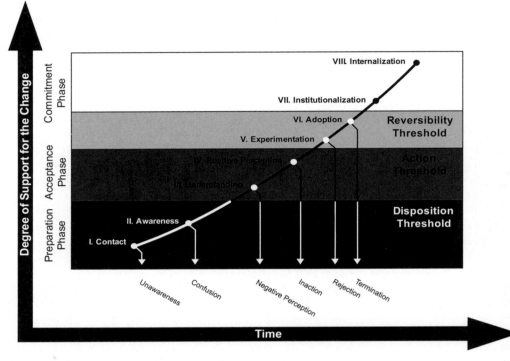

FIGURE 1-4   Change management continuum

Unified Theory of Acceptance and Use of Technology

The Unified Theory of Acceptance and Use of Technology identifies four key factors that directly determine a user's acceptance and usage of IT. These factors are listed in Table 1-6.[19] Companies can use the theory to help workers accept new IT.

TABLE 1-6   Key factors of IT acceptance and usage

Factor	Definition
Performance expectancy	Belief that using the system will help job performance
Effort expectancy	Degree of ease associated with the use of the system
Social influence	Degree of belief that important company officials want employees to use the system
Facilitating conditions	Belief that an organizational and technical infrastructure exists to support the system

For example, Payless ShoeSource is a retailer that implemented software from Kronos to automate the scheduling of its 20,000 sales associates in its 4600 stores.[20] The goal was to better balance the needs of its customers and employees while increasing sales and decreasing labor costs. Initially, some young, part-time employees were concerned that they could not negotiate their work hours to meet personal needs. Store managers were skeptical of the system's value—they believed they did not need a computer to develop a schedule. Fortunately, managers were sensitive to these concerns and addressed them effectively. They insisted on running a test of the software in 23 California stores before rolling it out nationally. The test demonstrated that managers easily could use the software to create schedules in 15-minute increments, accounting for minimum staffing requirements, employees' positions and pay rates, and any budgetary constraints.

Testing showed that the system also had the flexibility to account for individual work preferences and availability. The real payoff came when the improved schedules led to increased sales by ensuring that employees could complete assigned tasks without losing "face time" with customers. Managers used these successful test results of increased performance to overcome initial resistance to the system.[21] They also made it clear that store managers needed to use the system and that an organizational and technical infrastructure existed to support system use. Payless had successfully addressed the four factors that increase acceptance and usage of IT systems.

Three Worlds of IT

Successfully implementing the three types of IT (function, network, and enterprise) requires different types of organizational change, as summarized in Table 1-7. Management researcher Andrew McAfee suggests that companies need to change the way they get work done to enable improved performance with IT.[22] Four organizational complements allow these improvements with IT: better-skilled workers, higher levels of teamwork, redesigned processes, and new decision rights (which specify who can make certain types of decisions). Function IT can deliver results without the complements being in place, network IT allows the complements to emerge over time, and enterprise IT requires the complements to be deployed with the new technologies. Read the following feature, "A Manager Takes Charge," to learn how one manager identified the need for key complements and put them in place to ensure the smooth introduction and adoption of IT. These complements were overlooked in the initial implementation of the enterprise IT system.

TABLE 1-7 Manager's role in the Three Worlds of IT model

	Function IT	Network IT	Enterprise IT
Benefits	Improved productivity	Increased collaboration	Increased standardization and ability to monitor work
Examples	Personal productivity software, decision support system	E-mail, instant messaging, project management software	TPS, ERP, IOS
Organizational complements (better-skilled workers, better teamwork, redesigned processes, new decision rights)	Does not bring complements with it Partial benefits can be achieved without all complements being in place	Brings complements with it Allows users to implement and modify complements over time	Full complements must be in place when IT "goes live"

TABLE 1-7   Manager's role in the Three Worlds of IT model (continued)

	Function IT	Network IT	Enterprise IT
Manager's role	Encourage use, challenge workers to find new uses	Demonstrate how technology can be used, set norms for participation	Must identify and put into place the full set of organizational complements prior to adoption Intervene forcefully and continually to ensure adoption

A MANAGER TAKES CHARGE

Toys "R" Us EDI System Rescued

Toys "R" Us is a leading retailer of toys and baby products worldwide. The company sells merchandise through its stores and its Internet site. Toys "R" Us operated as a public company from 1978 until July 2005, when a private investment group acquired the company for $6.6 billion.

Tim Meester came to Toys "R" Us in 2003, as the director of vendor partnerships. His primary responsibility was to develop compliance and supply-chain performance programs. Before he could start any new initiatives, however, he saw a need for significant improvements in the accounts payable and electronic data interchange (EDI) groups. A recently implemented EDI system was creating problems in payment processes and information flows between the company and its vendors. Meester and his team learned that errors in EDI data exchanged between the company and its vendors was making it difficult for the vendors to get paid. The errors also were creating extra work and inefficiencies both for Toys "R" Us and its suppliers.

"We had to clean our own house before we could even address vendor performance," Meester said. "We estimated that 60 to 80 percent of the issues came from a lack of internal consistency and discipline, so we worked together with our vendors and internal partners to understand the process issues, identify gaps, and ensure that processes were integrated, product kept flowing, and vendors were paid accurately and on time."[23]

Perhaps the biggest challenge was to change the Toys "R" Us culture and make employees realize that less than 100 percent accuracy was unacceptable. Resolving the accuracy problems required cleaning up the data that Toys "R" Us transmitted to its vendors and the information they sent back. Meester and his team had to learn the detailed technical aspects of the EDI system and the business needs it was intended to address. Then the team had to develop corrective actions, address recurring problems both at Toys "R" Us and the vendor sites, and provide consistent advice to each vendor. To aid in this process, Toys "R" Us implemented an extranet to link itself with all 1300 of its vendors. The extranet enabled the company to share information, identify problems with the EDI system, and resolve them.[24]

continued

Discussion Questions

1. What essential complements and managerial actions were missed during the initial implementation of the EDI system? Why do you think they were missed?
2. Was there really any negative impact from the initial "poor implementation" of this system? After all, Toys "R" Us was able to correct the situation.
3. How do you think Meester got employees to improve the accuracy of EDI data?

Introducing an enterprise IT system requires large amounts of resources and significant changes in procedures, roles and responsibilities, reward systems, and decision making. In other words, it represents a major organizational change. Managers have their work cut out to gain acceptance of all these changes. A successful enterprise IT system requires the top-down imposition of standards and procedures that spell out exactly how transactions must be conducted and how the supporting information must be captured, stored, and shared. As a result, senior management sometimes encourages adoption of enterprise IT, by threatening penalties for nonconformance.

For example, the U.S. Health Insurance Portability and Accountability Act (HIPAA) specifies standards for the capture, storage, and sharing of electronic healthcare transactions, such as medical claims, electronic remittances, and claim status inquiries among healthcare providers, health insurance plans, and employers. As organizations scrambled to meet the May 2007 implementation deadline, there were complaints that several years of management time and millions of dollars in programming expenses were required to implement the full HIPAA standards for patient registration, medical records, billing, and claims processing. To encourage organizations to conform to the standards and meet the deadline, the Department of Health and Human Services (DHHS) required that the HIPAA standards be followed for organizations to receive payment for the services they provide patients. DHHS also established investigation procedures and set civil monetary penalties for violating HIPAA rules.[25]

Ensuring that IT Risks Are Mitigated

IT resources are used to capture, store, process, update, and exchange information that controls valuable organizational assets. As a result, special measures are needed to ensure that the information and its control mechanisms stand up to intense scrutiny. In the United States and many other countries, laws mandate stringent control requirements and accurate record keeping for publicly held corporations. For example, the U.S. Public Company Accounting Reform and Investor Protection Act of 2002—better known as the Sarbanes-Oxley Act, or simply SOX—was enacted in response to public outcry over several major accounting scandals, including those at Enron, WorldCom, Tyco, Adelphia, Global Crossing, and Qwest. Section 404 of the Sarbanes-Oxley Act states that all reports filed with the SEC must contain a signed statement by the CEO and CFO attesting that the information is accurate. The company also must submit to an audit to prove that it has controls in place to ensure accurate information. The penalties for false attestation are quite severe—up to 20 years in federal prison and significant monetary fines for senior executives. Because most, if not all, information in an SEC filing comes from information systems, managers have an additional incentive to ensure that appropriate controls exist for information systems, their associated processes, and the people who enter and process data.

Managers also are responsible for ensuring that physical IT assets, such as applications, databases, networks, and hardware, are protected against loss or damage. If assets are lost or destroyed as the result of a disaster, business continuity plans must be in place to ensure the ongoing operation of critical business functions. Management's responsibility also includes ensuring that data assets are secure from unwanted intrusion, loss, and alteration, and ensuring that personal data is secured to protect individual privacy rights. Table 1-8 identifies several examples of IT-related risks that concern managers.

TABLE 1-8 Examples of IT risks

IT Risks	Example
Inability to continue operations due to a natural disaster or accident	Fire destroys IT resources at corporate headquarters
Inability to continue operations due to a deliberate attack on IT assets	Hackers carry out denial-of-service attack on organization's Web site
Compromise of confidential data about organizational plans, products, or services	Senior executive loses laptop containing critical data
Compromise of personal, private data about employees or customers	Hackers access and download customer data, including account numbers
Violation of legally mandated procedures for controlling IT assets	IT system controls are inadequate to meet specific SOX guidelines for maintaining the integrity of financial data
Violation of established, generally accepted accounting principles	IT system controls are violated so that the same person can both initiate a purchase order and approve the invoice for that purchase order
Violation of the organization's defined procedures and/or accounting practices	IT system controls are circumvented by granting access to inappropriate people to adjust finished product inventory counts
Loss of physical IT assets	Theft of computers from corporate training facility
Inappropriate use of IT resources that places firm in a compromising position	Employees use corporate e-mail to disseminate sexually explicit material; firm is subjected to a sexual harassment lawsuit
Inappropriate use of IT resources that reduces worker productivity	Employees waste time at work visiting Web sites that are not related to their work

Managers cannot afford to ignore IT, because failed IT projects can lead to lowered returns on investment and missed opportunities. For example, A.G. Edwards had a long track record of unsuccessful project completions—nearly half of all projects exceeded cost and schedule limits. Worse yet, IT applications were often developed that did not fit the organization's overall strategic plan. As a result, projects failed to meet business objectives or deliver anticipated results. Some projects ended up being complete write-offs with no benefits delivered, including one that cost $46 million.[26] In recent times, however, the firm has taken strong actions to turn the situation around.

CFO Research Services and Deloitte Consulting recently completed a study that concluded "the billions invested in information technology still aren't producing great information."[27] The results showed that managers at 80 percent of surveyed firms believe that their operational and financial data is not as effective as it should be for developing strategies and planning. In addition, the data is not available when it is needed, nor is it sufficiently accurate. Too much time is spent reconciling data from different internal sources, creating a drain on profits and causing managers to miss opportunities.

Failure to ensure that IT risks are mitigated can lead to serious problems, such as unwanted oversight from federal regulators, IT-related fraud, and costly business disruptions. BearingPoint provides an example of what can go wrong. Previously known as KPMG Consulting, BearingPoint was spun off from KPMG LLC in 2001 when the "Big Five" accounting firms became concerned over potential conflicts of interest—many accounting clients were being encouraged to use associated consulting services to fix problems uncovered during audits.

In 2003, BearingPoint took on a challenging project to implement a new internal accounting system and associated standard business practices. The system was needed to make the firm compliant with Section 404 of the Sarbanes-Oxley regulations.[28] Because of delays in completing the project, BearingPoint had to negotiate several extensions to its SEC filing deadlines—quite embarrassing for a firm that promoted its ability to help clients comply with Sarbanes-Oxley.

Following several missed deadlines, the New York Stock Exchange announced that it would start suspension and delisting procedures if the company's 2005 Form 10-K annual report was not filed with the SEC by April 2007.[29] If this occurred, BearingPoint stock would be suspended from trading and it would be almost impossible to raise additional capital in the equity or debt markets. These well-publicized problems scared off potential customers, generated class-action shareholder lawsuits, diverted more than $100 million of company funds to resolve the accounting problems, and temporarily increased employee attrition to more than 25 percent. Fortunately, BearingPoint filed the report before the deadline and avoided delisting. BearingPoint managers realize that they should have managed the project more carefully and that they missed opportunities to make better choices.

This chapter has presented a powerful argument for why proper involvement by business managers at the right time is essential to obtain real value from the use of IT. This involvement is needed throughout the project, not just at certain key moments. The checklist in Table 1-9 recommends a set of actions that business managers can take to support and ensure the successful use of IT.

TABLE 1-9 A manager's checklist

Recommended Management Actions (appropriate response is "Yes")	Yes	No
Do you get involved in identifying and evaluating potential opportunities to apply IT?		
Do you work to smooth the introduction and adoption of IT in your area of the business?		
Do you work with appropriate resources to identify and mitigate IT-related risks?		
Do you understand that the successful implementation of each type of IT (function, network, and enterprise) requires different degrees and types of organization change?		

OVERVIEW OF REMAINING TEXT

This section provides a brief summary of the rest of the book.

Chapter 2: Strategic Planning describes how to develop effective strategic planning by defining key business objectives and goals and then applying them to multifunctional teams to identify a portfolio of potential business projects. The resulting set of projects is clearly aligned with business needs. Further refinement is required to narrow the portfolio to the projects that should be executed and for which sufficient resources are available. This process is illustrated by the example of the United Parcel Service, a major global organization respected for its highly effective use of IT to support business objectives.

Chapter 3: Project Management provides a helpful overview of the project management process. The presentation is consistent with the Project Management Institute's Body of Knowledge, an American National Standard. The chapter describes the nine project management knowledge areas of scope, time, cost, quality, human resources, communications, risk, procurement, and integration. A business manager can take many roles throughout the project life cycle, including champion, sponsor, project manager, change agent, and end user. The chapter identifies frequent causes of project failure and offers invaluable suggestions for how to avoid these problems.

Chapter 4: Business Process and IT Outsourcing discusses the major business reasons for outsourcing and identifies many of its issues and potential pitfalls. It also outlines and describes an effective process for selecting an outsourcing firm and successfully transitioning work to the new organization. The chapter covers the importance of establishing service-level agreements and monitoring performance.

Chapter 5: Corporate Governance and IT describes the responsibilities and practices that a company's executive management uses to ensure delivery of real value from IT and to ensure that related risks are managed appropriately. The chapter covers two frameworks for meeting these objectives: the IT Infrastructure Library (ITIL) and Control Objectives for Information and Related Technology (COBIT). The discussion includes related issues such as the Sarbanes-Oxley Act, business continuity planning, and oversight of outsourcing arrangements.

Chapter 6: Collaboration Tools and Wireless Networks covers the fundamentals of electronic communications systems, with a focus on wireless and mobile communications. You will learn about the benefits and disadvantages of various wide area and local area wireless networks, and how managers can understand and deal with related business issues.

Chapter 7: E-Business discusses the use of electronic business methods to buy and sell goods and services, interact with customers, and collaborate with business partners and government agencies. Several forms of e-business are covered, including business-to-business, business-to-consumer, consumer-to-consumer, and government-to-consumer. The chapter also covers m-commerce, an approach to conduct e-commerce in a wireless environment. The chapter prepares managers to understand and deal with many of the business, legal, and ethical issues associated with the use of e-business.

Chapter 8: Enterprise Resource Planning explains what an ERP system is, identifies several of the benefits associated with ERP implementation, outlines a "best practices" approach to implementing an ERP system, and discusses future trends of ERP systems. The chapter also explains the key role that business managers play in successfully implementing ERP systems.

Chapter 9: Business Intelligence discusses a wide range of applications that help businesses gather and analyze data to improve decision making: data extraction and data cleaning, data warehousing and data mining, online analytical processing (OLAP), information visualization, business activity monitoring, dashboards, and scorecards. The chapter discusses the complications and issues associated with each business intelligence system, and discusses the role of the business manager in developing and using these systems.

Chapter 10: Knowledge Management describes how organizations use knowledge management to identify, select, organize, and disseminate important information that is part of the "organization's memory." Unfortunately, much of this information and expertise is highly unstructured and informally communicated. In this chapter, you will learn about communities of practice, social network analysis, Web 2.0 Technologies, business rules management systems, and enterprise search. The chapter also covers how to identify and overcome knowledge management challenges, as well as a set of best practices for selling and implementing a knowledge management project.

Chapter 11: Enterprise Architecture describes the use of enterprise architecture to establish a series of reference frameworks that define necessary business and IT changes. The chapter also describes the business manager's role in defining the architecture and business needs of an organization. You will learn about current architecture topics, including common object request broker architecture (CORBA), the distributed component object model (DCOM), service-oriented architecture (SOA), and Web 2.0. The chapter focuses on the differences between these models and how are they used in practice.

Chapter 12: Ethical, Privacy, and Security Issues provides a brief overview of ethics and identifies key privacy and security issues that managers need to consider in their use of IT to achieve organizational benefits. Ethics, privacy, and computer security are discussed from the perspective of what managers need to know about these topics.

Chapter Summary

- An organization's IT infrastructure is integrated with people and procedures to build, operate, and support information systems that enable a firm to meet fundamental objectives.

- Managers have three critical responsibilities when it comes to IT: identifying appropriate opportunities to apply IT, smoothing the way for its successful introduction and adoption, and mitigating its associated risks.

- The most effective users of IT obtain maximum value from IT investments that align with the organization's strategic needs, and that are well managed and executed.

- Management's role is to frame changes, opportunities, and threats so that others can understand them, evaluate them against business needs, and address any opportunities that fit within an articulated business strategy.

- The Change Management Continuum Model describes key activities that are needed to build commitment for an organizational change, such as the introduction and adoption of IT.

- The Unified Theory of Acceptance and Use of Technology identifies four key factors that directly determine IT user acceptance and IT usage behavior: performance expectancy, effort expectancy, social influence, and facilitating conditions.

- Successfully implementing the three types of IT (function, network, and enterprise) requires different types of organizational change. Four organizational compliments—better-skilled workers, higher levels of teamwork, redesigned processes, and new decision rights—allow IT to improve performance. Function IT can deliver results without the complements being in place, network IT allows the complements to emerge over time, and enterprise IT requires the complements to be deployed with the new technologies.

- Managers must be able to vouch for the effectiveness of the organization's internal controls for financial reporting, protect the security and privacy of customer data, implement workable continuity plans that cover IT assets, and mitigate IT risks.

Discussion Questions

1. Do you agree that it is important for managers to understand IT? Write a paragraph to support your opinion.

2. Briefly define IT infrastructure, IT support organizations, and information systems. Describe how these entities are related.

3. Identify and briefly discuss two examples of function IT.

4. Why do some organizations have little to show for their investments in IT?

5. What is an enterprise resource planning system? How does it differ from a transaction processing system?

6. What percentage of revenue should a retail organization spend on IT? Discuss.

7. What are the basic reasons that people resist change? How can this resistance be overcome?

8. What is meant by social influence, and how can it affect the acceptance of new IT?

9. Which kind of system (from the three worlds of IT) requires the most management attention to ensure successful acceptance and adoption? Why?

10. What unfavorable results could occur if management is not appropriately involved with IT?

11. Should it be the responsibility of IT or management to identify and define tasks for the successful introduction and adoption of a new IT system?

Action Memos

1. You are the IT division controller for a company, and you just received the following e-mail from your manager, the organization's CFO. How do you respond?

   Preliminary results from the competitive analysis we commissioned show we spend 4 percent of annual revenue on IT-related expenses, while our top competitors spend just 3 percent. I'm meeting with the CEO in 10 minutes and he's bound to ask about this. What should I say?

2. You are a member of the Human Resources Department of Belarusbank, which was described in the opening vignette of this chapter. You have experience in expediting organizational change. The vice president for bank operations wants you to give him a list of actions he should take to ensure the smooth adoption and acceptance of the new IT system. He then wants to meet with you on this subject. What actions would you identify? What are your key talking points for this meeting?

Web-Based Case

Do research on the Web to find an example of a major IT-related project in which the actions of business managers made a major difference (either favorable or unfavorable) in the outcome. Document what you think were the key actions taken by business managers, and the key missed opportunities to take action.

Case Study

The Progressive Group of Insurance Companies

Managers Leverage Ongoing IT Investments to Achieve Competitive Advantages

The Progressive Group of Insurance Companies literally started from a garage. Today, it is the third largest U.S. auto insurance group in terms of net premiums written. The company sells auto insurance, other specialty property-casualty insurance, and related services through a network of 30,000 independent insurance agents, via direct sales over the phone, and through its Web site. Its premiums increased from $2 billion to $14 billion between 1993 and 2005, as shown in Figure 1-5. Much of Progressive's success is due to the leadership of its managers in investing in IT to support its many business innovations and fundamental business objectives for over 25 years. Progressive provides an excellent example for how a business can generate a sustainable advantage through continuous IT innovation.

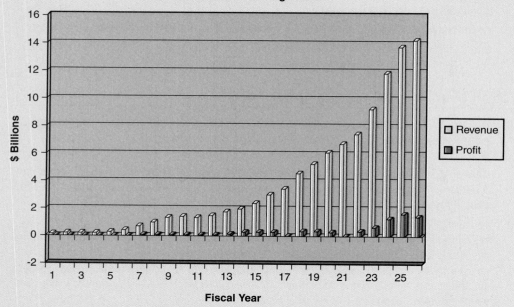

FIGURE 1-5 Twenty-five years of annual revenue and profit growth at Progressive Insurance

Joe Lewis and his friend Jack Green were two young lawyers trying to eke out a living in Cleveland, Ohio, during the Great Depression. In 1937, they started Progressive on their own. Times were tough for starting a new business—many people were out of work and typical factory workers earned just $36 per week, when they could find work at all. To keep expenses down, the friends used an old garage for the company headquarters. They charged $25 for an annual insurance policy, which was a lot of money given the state of the economy. In recognition of the hard times, they allowed their customers to pay their premiums in installments. In another insightful innovation, they offered clients a drive-in claims service.[30]

Peter Lewis, Joe's son, began his career in sales at Progressive shortly after graduating from Princeton in 1955. He was a bright, energetic young man who proposed many new ideas in an attempt to distinguish the company from its competitors.[31]

Peter Lewis often heard people complain about independent insurance agents who tried to persuade the company to cover high-risk drivers, including young drivers, people with points on their license, and people with a record of driving under the influence. At the time, other insurance companies wanted nothing to do with such drivers because they were considered unprofitable. Lewis, however, challenged the organization: "They're bringing us potential business. Can't we find a way to write these people?" It took a while for the idea to catch on—less than $100,000 worth of such policies were written in 1957. Over the next decade, however, this niche in the insurance market greatly expanded and became the foundation for future company growth.[32]

Early on, Progressive recognized the need to make significant investments in IT to achieve high-quality, efficient data processing. Such investments were planned well in advance as part of a five-year program, and each potential investment was subjected to vigorous cost/benefit analysis before proceeding. With each passing year, IT became a more important part of achieving operating efficiency and marketing success. Progressive's pricing strategies were increasingly dependent on using more data than its competitors. For example, Progressive segments its customer groups more precisely than competitors, giving it more price points for a given group. This precision requires superior customer and claims data and excellent data processing capabilities.

One of Progressive's first major IT initiatives was to develop an online computer system called PROTEUS (Progressive Online Transaction, Enquiry, and Update System). This system was developed to support the firm's policy processing;[33] it enabled workers to inquire into customers' policy files and to quote rates for customers online. The most significant productivity and service gain from PROTEUS was the ability to enter, endorse, and cancel business online.

In 1988, Progressive faced two major challenges to its continued growth and profitability. First, a consumer backlash in California led to the passage of a referendum called Proposition 103, which ordered all insurance companies to roll back rates by 20 percent and forced them to refund $1.2 billion in excessive premiums to Californians. Progressive's share of this total was $50 million.[34] Second, Progressive learned that Allstate had surpassed it in its specialty of high-risk auto insurance, making Allstate the firm's most threatening competitor.

Lewis sought out Ralph Nader, a longtime friend, Princeton classmate, and outspoken advocate of Proposition 103, to help him understand why consumers were so upset with insurers.[35] It became clear that the source of customer ire was uniformly poor claim service and the inability to do comparison shopping for the best rates. In what proved to be a strategic move, Progressive decided to streamline and improve its services in assessment, adjustment, repair management, and claims processing.[36]

In a second strategic move, the firm decided to move from being a niche provider of high-risk auto insurance to a broad-based provider of personal auto insurance. This move put Progressive in direct competition with such major carriers as State Farm and Allstate.

Progressive managers moved quickly to identify and implement several key IT innovations in support of these new business strategies. In 1989, Progressive introduced the next generation of its claims processing system, Progressive Automated Claims Management System (PACMAN). The goal was to expedite the claims process and squeeze out excess costs. PACMAN was a large, centralized, online database built on technology that was state of the art at the time. The system required many years of management time to define the system requirements, develop streamlined work processes, evaluate prototypes, and test all aspects of the system. Not only were claims representatives given extensive training for using the system, they were sent to an empathy training program to help them relate to the plight of their customers and focus on solving their problems. The new claims-handling process and automated system reduced the average time it took Progressive to process and pay a claim to six days. (The industry average was 42 days.) The improved process led to significant improvements in customer satisfaction and reduced customer defections by two-thirds.

To further speed up the claims process, Progressive managers implemented an Immediate Response® claims service in 1990. The service was available 24 hours a day, 7 days a week, and provided customers with personal service and support immediately after they reported a claim.

Customers could now call Progressive at any hour and talk with a trained claims representative, who could begin resolving the claim and offer immediate help. Progressive's claims representatives were on constant call to go to the scene of serious accidents, which helped to ease the trauma of injured people and their families during a difficult time. Claim reports were entered into the proprietary online computer system, allowing Progressive to verify coverage immediately and make the information available to a nearby claims representative. As a result of the new "24-7" reporting and response capability, Progressive was able to contact 90 percent of its customers and claimants within 24 hours and close one-third of its total claims within seven days.

These rapid settlements had several effects—the dramatic time reduction not only decreased the costs of claims processing and administration, it also cut the cost of litigation, shrank the amount of fraud, and slashed the number of customer defections to other insurance firms. Beginning in 1991, Progressive saw a major reduction in its general, sales, and administrative expenses as a percentage of total revenue, as shown in Figure 1-6. This led "CEO Peter Lewis to establish an outrageous objective: Because faster was better for the customer and cheaper for the company, Progressive would settle claims instantly!"[37]

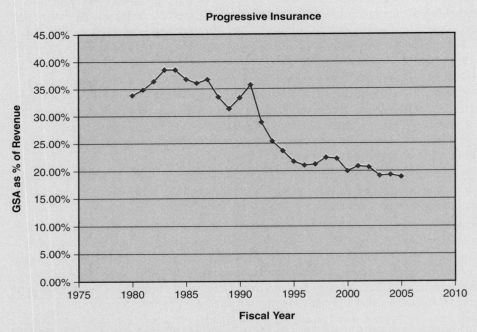

FIGURE 1-6 General, sales, and administrative (GSA) expenses at Progressive Insurance

To this end, Progressive introduced the Immediate Response Vehicle (IRV) in 1994. This specially marked and outfitted vehicle takes an experienced claims representative wherever the customer needs him—even to the scene of an accident. Upon arrival, the representative's first priority is to reassure the customer that Progressive understands his or her needs and is there to help. The rep then surveys the scene, takes digital photos of the damage, develops a damage estimate using data stored on a portable computer, and writes a check to cover the estimated costs—often within an hour of the accident.[38]

Also in 1994, Progressive introduced 1-800-AUTO-PRO, the first comparison rate service available for free by phone. With one call, consumers received a Progressive quote and comparison quotes for up to three competitors. Customers then were given the option of buying insurance directly over the phone or through a local agent. (Today this service is available by dialing 1-800-PROGRESSIVE.) Because of the technical difficulty of developing accurate rates for other companies, and concern over the potential loss of customers from a competitor's lower rates, Progressive first tested the program in California and then in Florida before rolling it out broadly.

In 1995, Progressive became the first major auto insurer in the world to launch a Web site. Policyholders can log in to update information, make payments, get vehicle recall information, and more. They can report accidents in minutes without picking up the phone. Customers also can schedule appointments with claims representatives and body shops while reporting claims online. Potential customers can view Progressive rates side-by-side with those of competitors.[39]

Never resting on its accomplishments, management identified further improvements for its claims service and rolled out the Claims Workbench in 1997. Progressive spent four years developing this software application, which enables claims representatives to move information between their wireless laptops and a mainframe to keep claims moving forward to resolution. The Claims Workbench allows claims representatives to have all pertinent information at their fingertips when they meet face-to-face with customers. Another application, Pathways, was developed to provide an encyclopedic listing of parts for nearly every car on the road. The claims representative uses Pathways to scroll through a database of parts, prices, and labor estimates to develop a settlement amount.

Lewis retired as CEO in 2000, after 45 years of outstanding service. The role of CEO was turned over to Glenn Renwick, who had served as the CIO. Even with the change in top management, Progressive continued its 25-year strategy of making ongoing investments in IT to improve operations.

Progressive worked closely with representatives of its independent agents to improve work processes, and worked with the IT organization to reduce costs and increase flexibility for product changes. The independent agents were provided electronic access to customer data, enabling them to provide more responsive service to their customers. These improvements increased the number of policy changes made directly by agents; enabled a substantial number of policies to be underwritten at the point of sale, including all underwriting data validation checks; increased the number of policyholders who chose installment plans and electronic funds transfer; and eliminated paper files and reports. Progressive's commitment to expanding technology and improving workflows has made its expense ratios comparable to or lower than any competitor who distributes through agents.

By 2005, Progressive had successfully introduced its next-generation Web-quoting platform, which provided much faster online quotes and led to an increase in Web application completion. Progressive also introduced "talk to me" functionality, allowing Internet customers instant telephone access to a licensed professional who can access their quote in real time and provide counsel.

Managers: Key to Information Technology Results

In 2006, Progressive announced two major company initiatives: to replace the customer and policy management system that served it well, because it is not consistent with Progressive's views of future needs; and to add a new data center that is not constrained by processing capacity and that assures a high level of system availability and disaster preparedness. The new policy management system will be implemented in 2008. Hundreds of workers are helping to define the new system through focus group sessions and prototype evaluations. Also in development is a clear implementation plan that includes extensive user acceptance testing and hands-on training.

"I have often described Progressive as a technology company in the auto insurance business. Much of what we have achieved has been made possible by our talented information technology staff, and we certainly would not have been able to support our growth without extraordinary commitment to our technology capacity and capabilities. We see the future and our agents being very dependent on technology and we are developing applications that allow us and our agents to do more, easier and at less cost. We are placing equal emphasis on infrastructure platforms to maintain strategic or cost advantages. Our continuous investment in technology over the past several years has positioned us well to remain a leader in technology solutions for service delivery to both our customers and agents, and we are committed to a level of technology investment that ensures we never constrain the business."

—Glen M. Renwick, Progressive President and CEO, in the 2003 annual report

Discussion Questions

1. Visit the Progressive Web site and review recent press releases and financial statements. Can you find evidence that the firm continues to implement effective IT solutions that increase revenue, decrease costs, or improve customer service? Write a paragraph that summarizes your findings.

2. Why does Progressive keep modifying its basic claims processing system? After three or four generations of this system, shouldn't it be right by now?

3. Progressive has a 25-year history of implementing change and new IT solutions. In a culture so attuned to change, is there less need for managers to help smooth the introduction and adoption of new IT systems? Defend your position.

4. Use the Change Management Continuum Model to plan a schedule and identify actions to support Progressive as it moves through the seven stages of commitment to implement its new customer and policy management system. Assume that the project started in January 2006 and will finish in June 2008.

Endnotes

[1] Anuradha Shukla, "Belarusbank Selects SAP for Banking Solution Portfolio," *TMCnet*, 20 September 2006.

[2] Marc L. Songini, "$20 M SAP System Remaking Belarus Bank's Operations," *Computerworld*, 9 October 2006.

[3] "Belarusbank Replaces Decentralized Banking System with SAP," *CRM Today*, 19 September 2006.

4 Andrew McAfee, "Mastering the Three Worlds of Information Technology," *Harvard Business Review*, November 2006.

5 Elena Malykhina, "One Midsize Company's Answer to Compliance Pain," *InformationWeek*, 1 May 2006.

6 IBM Case Studies, "Whirlpool Transforms Employees with E-Learning," *www-304.ibm.com* (accessed 30 December 2006).

7 "Cerner: Health IT Provider Uses Collaboration Tools to Improve Communication, Save Money," Microsoft Office Customer Solution Case Study, *http://whitepapers.techrepublic.com* (accessed 30 December 2006).

8 Darren Dial, "The End of E-Mail," *Inc.*, pages 41–42, February 2006.

9 "20 Great Ideas from InformationWeek 500 Companies," *InformationWeek*, 12 September 2006.

10 Marc Songini, "Company to Retire Legacy Systems for One Integrated One," *Computerworld*, 21 August 2006.

11 Bridget McCrea, "Accuracy Counts," *Logistics Management*, Volume 45, no. 10, pages 53–57, October 2005.

12 Richard Nolan and Warren F. McFarlan, "Information Technology and the Board of Directors," *Harvard Business Review*, October 2005.

13 "Information Technology Spending Scoreboard," itmWEB Media Corporation, *www.itmweb.com/blbenchspn.htm* (accessed 14 December 2006).

14 Tom Pisello, "IT Spending at Midsized Companies: How Much Does Size Matter," *CIO News*, 27 September 2005.

15 Keri E. Pearlson and Carol S. Sanders, *Managing & Using Information Systems: A Strategic Approach, 3rd edition,* John Wiley & Sons, Inc., page 4, 2006.

16 J.P. Kotter and L.A. Schlesinger, "Choosing Strategies for Change," *Harvard Business Review* 57, pages 106–114, March/April 1979.

17 D. R. Conner, *Managing at the speed of change: How resilient managers succeed and prosper where others fail*, Villard Books, pages 147–160, 1992.

18 CM Models: Commitment to Change Model at Enterprise Solutions Competency Center, U.S. Army PEO EIS Software Engineering Group, *www.army.mil/ESCC/cm/model2.htm*.

19 Viswanath Venkatesh, Michael G. Morris, Gordon B. Davis, and Fred D. Davis, "User Acceptance of Information Technology," *MIS Quarterly*, Volume 27, no. 3, pages 425–478, September 2003.

20 Press release, "Payless ShoeSource Selects Kronos for Retail to Standardize Workforce Optimization in More Than 4,000 Stores," *www.kronos.com*, 19 December 2005.

21 Marc L. Songini, "Scheduling App Leads to Rise in Retail Sales," *Computerworld*, page 17, 12 June 2006.

22 Andrew McAfee, "Mastering the Three Worlds of Information Technology," *Harvard Business Review*, pages 141–149, November 2006.

23 Bridget McCrea, "Accuracy Counts," *Logistics Management*, Volume 45, no. 10, pages 53–57, October 2005.

24 Toys "R" Us, About Us, *www4.toysrus.com/about/* (accessed 12 January 2007).

25 Heather Havenstein, "Federal ID Mandate Tests Health Insurers," *Computerworld*, 17 July 2006.

26 Meredith Levinson, "When Failure is Not an Option," *CIO Magazine*, 1 June 2006.

27 Allan Alter, "Putting the "I" Back in IT," *CIO Insight*, page 30, February 2006.

28 Laton McCartney, "How BearingPoint Lost Its Way," *Baseline*, no. 55, pages 36–52, February 2006.

29 Press release, "BearingPoint Granted Extended NYSE Trading Period," *www.bearingpoint. com*, 29 September 2005.

30 Chuck Salter, "Progressive Makes Big Claims," *Fastcompany.com*, no. 19, page 176, October 1998.

31 Suzanne Rivard, Benoit Aubert, Michel Patry, Guy Pare, and Heather Smith, *Information Technology and Organizational Transformation: Solving the Management Puzzle*, page 142, Elsevier Limited, 2004.

32 Chuck Salter, "Progressive Makes Big Claims," *Fastcompany.com*, no. 19, October 1998.

33 Wikipedia, the free online encyclopedia.

34 "Insurance Reform—Solving the Insurance Crises," The Foundation for The Taxpayer and Consumer Rights at *www.consumerwatchdog.org/insurance/prop103/* (accessed 10 December 2006).

35 Chuck Salter, "Progressive Makes Big Claims," *Fastcompany.com*, no. 19, October 1998.

36 Marcia Stepanek, "Q&A with Progressive's Peter Lewis," *BusinessWeek Online*, 12 September 2000.

37 Bill Davidson, *Breakthrough: How Great Companies Set Outrageous Objectives and Achieve Them*, page 16, Elsevier Limited, 2004.

38 Progressive Web site, About Progressive, Company Facts, History, *www.progressive.com* (accessed 7 December 2006).

39 Progressive Web site, About Progressive, Company Facts, History, *www.progressive.com* (accessed 7 December 2006).

STRATEGIC PLANNING

FDA ILLUSTRATES WHY MANAGERS MUST UNDERSTAND STRATEGIC PLANNING

The Food and Drug Administration (FDA) is responsible for extensive testing of new drugs before approving their use by the general public. This review process includes testing on animals, as well as clinical trials on healthy people and people who suffer from the disease a drug is intended to treat. Drug approval can take anywhere from 15 months to several years.

Fast-track designation, which was mandated by the FDA Modernization Act of 1997, was meant to accelerate approval of new drugs to treat serious or life-threatening diseases for which there was no existing medication. Experts reasoned that the potential loss of life from keeping a life-saving drug off the market outweighed the risks that the new drug might pose to consumers.

Many physicians, consumers, and advocacy groups claim that the FDA fast-track approval process has backfired, and fails to properly screen for drugs with dangerous side effects. They cite a Government Accounting Office report that more than 5 percent of FDA-approved drugs between 1997 and 2000 subsequently were withdrawn from the marketplace because dangerous side effects were discovered. For example, the arthritis pain reliever Reofecoxib (brand name Vioxx) was approved in 1999 and prescribed to 80 million people worldwide. Merck, the manufacturer, withdrew the product from the market in 2004 when it was linked to an alarming increase in heart attacks and strokes among patients who took it.

Throughout the 1990s, the FDA used decades-old legacy information systems and work processes that simply could not support fast-track drug approval. But in 2002, amid controversy over the fast-track process, the FDA and its new Chief Information Officer (CIO), Jim Rinaldi, established a multiyear strategic plan to upgrade its operations in support of fast tracking.

The first two years of the plan focused on streamlining the IT organization and its business processes, with the goals of simplification, standardization, and saving costs. A Central Office of Shared IT Services was formed to help consolidate more than a dozen service contracts, including e-mail services, help desk support, and desktop support. Two FDA-wide projects, an intranet and an automated workflow system, began in 2003 and were in the final stages of implementation in 2007. In 2004, the FDA began eliminating redundant systems and defining common data standards. The accumulated effect of all these projects enabled scientists to more easily access and share data, a prerequisite for an effective fast-track process. The FDA can now monitor a patient's reactions to a test drug during a clinical trial and detect any adverse effects that require additional follow-up. Such monitoring of reactions to test results were impossible in the past.

People close to the situation feel that the FDA's improved work processes and new IT systems will reduce the risk of approving drugs with dangerous side effects, and will shorten approval times. If the new strategy is successful, it could save countless lives.[1,2,3]

LEARNING OBJECTIVES

As you read this chapter, ask yourself:

- What is an effective strategic planning process, who needs to participate in it, and what are the deliverables of such a process?
- How is IT planning tied to overall strategic planning, so that business objectives and IT activities are well aligned?

This chapter defines strategic planning and outlines an effective process for good planning. It also provides an example of effective IT planning at United Parcel Service.

WHY MANAGERS MUST UNDERSTAND THE RELATIONSHIP BETWEEN STRATEGIC PLANNING AND IT

For more than 15 years, various surveys of business and IT executives have identified the need to improve alignment between business and IT as a top business priority. In this context, alignment means that the IT organization and resources are working on projects that support the key objectives of the business. This implies that IT and business managers have a shared vision of where the organization is headed and agree on its key strategies.

Unfortunately, according to the 2007 State of the CIO Survey conducted by *CIO Magazine*, four of five CIOs say they are not aligned with their organization's strategic goals.[4] This lack of alignment significantly increases the likelihood that IT is not being applied effectively and that valuable opportunities may go unexploited. If IT is not being used strategically, many managers consider it an overhead cost that should be minimized.

The odds of achieving good alignment are increased vastly if IT staff members have experience in the business and can talk to business managers in business terms rather than technology terms. IT people must be able to recognize and understand business needs and develop effective solutions. The CIO especially must be able to communicate well and should be accessible to other corporate executives. However, the entire burden of achieving alignment between the business and IT cannot be placed solely on the IT organization.

Business managers must communicate the vision, objectives, and strategies of the organization so that everyone can help define the actions required to meet organizational goals, including capitalizing on new opportunities, fixing organizational weaknesses, and minimizing the impact of potential threats. These activities focus the resources of the entire organization on the significant issues to be addressed. One approach that helps achieve this objective is effective strategic planning.

Strategic planning is a process that helps managers identify desired outcomes and formulate feasible plans to achieve their objectives by using available resources and capabilities. The strategic plan must take into account that the organization and everything around it is changing: consumers' likes and dislikes change; old competitors leave and new ones enter the marketplace; the costs and availability of raw materials and labor fluctuate, along with the fundamental economic situation (interest rates, growth in gross domestic product, inflation rates); and the degree of industry and government regulation changes. Strategic planning is typically an annual process timed to yield results that are used to prepare the annual expense budget and capital forecast. Organizations also may have semiannual or quarterly reviews that identify necessary adjustments.

The CEO must make long-term decisions about where the organization is headed and how it will operate, and has ultimate responsibility for strategic planning. Subordinates, lower-level managers, and consultants typically gather useful information and perform much of the underlying analysis. But the CEO must thoroughly understand the analysis and be heavily involved in setting high-level business objectives and defining strategies. The CEO also must be seen as a champion and supporter of the chosen strategies, or the rest of the organization is unlikely to "buy into it" and take the necessary actions to make it all happen.

There are a variety of strategic planning approaches, including issues-based, organic, and goals-based. An issues-based process begins by identifying and analyzing key issues that face the organization, setting strategies to address those issues, and identifying projects and initiatives that are consistent with the strategies. Organic strategic planning defines the organization's vision and values and then identifies projects and initiatives to achieve the vision while adhering to the values. Goals-based strategic planning involves defining the mission and vision of the organization, identifying objectives and goals that support the mission, setting strategies to achieve the goals, and identifying projects and initiatives (see Figure 2-1). Goals-based strategic planning is the most frequently used approach, and is discussed in detail in this chapter. This chapter also presents real data and an analysis based on applying a goals-based process to United Parcel Service.

Defining Vision and Mission

Senior management must create a **vision/mission statement** that communicates an organization's overarching aspirations and can guide it through changing objectives, goals, and strategies. The organization's vision/mission statement forms a foundation for making decisions and taking action. An effective statement consists of three components: a core ideology, a mission statement, and a vision of a desirable future. The core ideology identifies a few widely accepted principles that guide how everyone makes decisions in the organization. The mission statement concisely defines the organization's fundamental purpose for existing. It usually is stated in a challenging manner to inspire employees, customers, and shareholders. For example, Google's mission is "to organize the world's information and make it universally accessible and useful."[5] The organization's vision is a concise statement of what the organization intends to achieve in the future.

The most effective vision/mission statements inspire and require employees to stretch to reach its goals. These statements seldom change once they are formulated. For example,

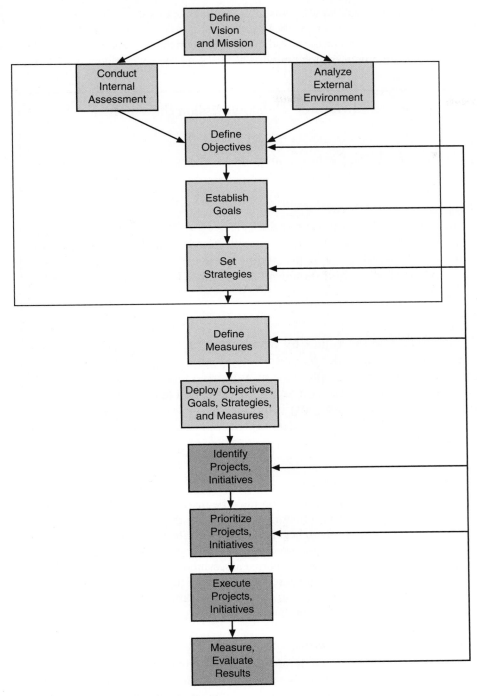

FIGURE 2-1 Strategic planning process

Procter & Gamble is the world's largest consumer products company with trusted, world-leading brands such as Tide, Pampers, Olay, Charmin, and Bounty. Every day, P&G products are used by more than half of the world's 6.5 billion consumers. Figure 2-2 illustrates P&G's vision/mission statement, which contains all three necessary components.[6]

Procter and Gamble's Vision/Mission Statement

Vision
Be, and be recognized as, the best consumer products and services company in the world.

Our Purpose
We will provide branded products and services of superior quality and value that improve the lives of the world's consumers. As a result, consumers will reward us with leadership sales, profit and value creation, allowing our people, our shareholders, and the communities in which we live and work to prosper.

Our Values
P&G is its people and the values by which we live. We attract and recruit the finest people in the world. We build our organization from within, promoting and rewarding people without regard to any difference unrelated to performance. We act on the conviction that the men and women of Procter & Gamble will always be our most important asset.

FIGURE 2-2   Procter & Gamble's vision/mission statement

Conducting Internal Assessment

All levels and business units of an organization must be involved in internally assessing its strengths and weaknesses. Preparing a historical perspective that summarizes the company's development is an excellent way to begin this step of strategic planning. Next, a multitude of data is gathered about internal processes and operations, including survey data from customers and suppliers and other objective assessments of the organization. The collected data is analyzed to identify and assess how well the firm is meeting current objectives and goals, and how well its current strategies are working. This process identifies many of the strengths and weaknesses of the firm.

Analyzing External Environment

Strategic planning requires careful study of the external environment surrounding the organization and assessing where the organization fits within it. This analysis begins with an examination of the industry in which the organization competes: what is the size of the market, how fast is it growing or shrinking, and what are the significant industry trends?

Next, the organization must collect and analyze facts about its key customers, competitors, and suppliers. The goal is two-fold: capture a clear picture of the strategically important issues that the organization must address in the future, and reveal the firm's competitive position against its rivals. During this step, the organization must get input from customers, suppliers, and industry experts, who can provide more objective viewpoints than employees. Members of the organization should be prepared to hear things they do not like, but that offer tremendous opportunities for improvement. It is critical that unmet customer needs are identified to form the basis for future growth. The most frequently used model for assessing the nature of industry competition is **Michael Porter's Five Forces Model** (see Figure 2-3).

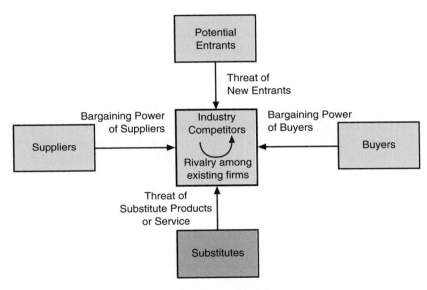

FIGURE 2-3   Michael Porter's Five Forces Model

Porter's Five Forces Model identifies fundamental factors that determine the level of competition and long-term profitability of an industry:

1. The threat of new competitors will raise the level of competition. Entry barriers determine the relative threat of new competitors. These barriers include the capital required to enter the industry or the cost to customers to switch to a competitor.
2. The threat of substitute products can lower the profitability of industry competitors. The willingness of buyers to switch and the relative cost and performance of substitutes are key factors in this threat.
3. The bargaining power of buyers determines prices and long-term profitability. This bargaining power is stronger when there are relatively few buyers but many sellers in the industry, or when the products offered are all essentially the same.
4. The bargaining power of suppliers can significantly affect the industry's profitability. Suppliers have strong bargaining power in an industry that has

many buyers and only a few dominant suppliers, or in an industry that does not represent a key customer group for suppliers.

5. The degree of rivalry between competitors is high in industries with many equally sized competitors or little differentiation between products.

Many organizations also perform a competitive financial analysis to determine how their revenue, costs, profits, cash flow, and other key financial parameters match up against those of their competitors. Most of the information needed to prepare such comparisons is available from corporate annual reports.

The analysis of the internal assessment and external environment frequently is summarized into a **Strengths, Weaknesses, Opportunities, Threats (SWOT) matrix**, as shown in Table 2-1. The SWOT matrix is a simple way to illustrate what the firm is doing well, where it can improve, what opportunities are available, and what environmental factors threaten the future of the organization. Typically, the internal assessment identifies most of the strengths and weaknesses, while the analysis of the external environment uncovers most of the opportunities and threats.

Innovation can create major opportunities for organizations to bring more highly valued products to the market, improve productivity, and raise profitability. For example, the Toyota Prius is a major product innovation. Toyota is collaborating more closely with suppliers not just to cut costs of individual parts, but also to capture savings across the entire vehicle manufacturing process by reaching back further into the design process. Innovation also can create sustainable competitive advantages by reinventing business processes and building new markets that meet untapped customer needs. For example, Agricultural Bank of China is now the country's second-biggest commercial bank, sustaining tremendous growth by making micro loans to peasant farmers. On the other hand, Wal-Mart's "every-day low prices" (enabled by low costs) defined its business model for nearly half a century. However, beginning around 2006, Wal-Mart's growth formula seemed to stop working, with less than 2 percent gains in same-store sales, its worst performance ever. While there were many possible reasons and contributing factors, clearly the firm is in need of innovation to re-energize its growth.[7]

TABLE 2-1 SWOT matrix analyzes the internal assessment and external environment

Strengths	Weaknesses
High profit margin	Has not introduced successful new product in
Strong sales growth	more than three years
Products available worldwide	High employee turnover
Opportunities	**Threats**
Further global expansion	Ripe for acquisition
Acquisition of competitors	Contract for more than half of workforce expires
Innovations	in six months

Defining Objectives

The terms *objective* and *goal* frequently are used interchangeably. For this discussion, we distinguish between the two by defining **objective** as a statement of a compelling business need that an organization must meet to achieve its vision and mission. For example, preserving consistency in revenue and earnings growth might be an objective for an established organization in a mature, slow-growing industry.

Establishing Goals

A **goal** is a specific result that must be achieved to reach an objective. In fact, several goals may be associated with a single objective. The objective states what must be accomplished and the associated goals specify how to determine whether the objective is being met.

Many organizations identify short-term, medium-term, and long-term goals. The organization starts by achieving relatively easy short-term goals, then tackles the more difficult medium-term goals, and finally addresses the long-term goals.

Other organizations encourage their managers to set Big Hairy Audacious Goals (BHAGs) that require a breakthrough in the organization's products or services to achieve. Such a goal "may be daunting and perhaps risky, but the challenge of it grabs people in the gut and gets their juices flowing and creates tremendous forward momentum."[8]

A key role of management is to recognize and drop goals that are no longer relevant. They also must recognize and resolve conflicting goals to avoid having the organization work at cross purposes.

As an example of setting objectives and goals, P&G has set a long-term objective of delivering a full decade of industry-leading revenue and profit growth.[9] To support that objective, it has defined specific goals to achieve industry-leading numbers in three areas: sales growth, earnings-per-share growth, and free cash-flow productivity.

Setting Strategies

A **strategy** describes specific actions an organization will take to achieve its vision/mission, objectives, and goals. Selecting a specific strategy focuses and coordinates an organization's resources and activities from the top down to accomplish its mission. Strategy development is a critical and highly creative step that is essential to the organization's future success. The purpose is to define a set of strategies that "gives the company an edge in the struggle with its rivals and a better chance at winning in the long-run competitive game."[10] In choosing from alternative strategies, managers should consider the long-term impact of each strategy on revenue and profit, the degree of risk involved, the amount and types of resources that will be required, and the potential competitive reaction.

In setting strategies, managers must draw on the results of the SWOT analysis and consider the following questions:

- How can we best capitalize on our strengths and use them to their full potential?
- How do we reduce or eliminate the negative impact of our weaknesses?
- Which opportunities represent the best opportunities for our organization?
- How can we exploit these opportunities?
- Will our strengths enable us to make the most of this opportunity?
- Will our weaknesses undermine our ability to capitalize on this opportunity?
- How can we defend against threats to achieve our vision/mission, objectives, and goals?
- Can we turn this threat into an opportunity?

The SWOT analysis, however, is only part of the input for setting strategies. The rest of this section summarizes a few different approaches that authors and researchers have developed for setting strategies.

For example, Michael Porter's three fundamental strategies define three basic approaches an organization can undertake in choosing how to compete in an industry:[11]

1. Become the cost leader with plant, equipment, labor costs, and business processes that help an organization become the low-cost provider of goods and services.
2. Provide goods and services that meet the needs of a set of customers in the marketplace better than other providers.
3. Focus on a specific niche in the marketplace and develop products or services for that niche while ignoring other customers.

Day[12] and Ansoff[13] defined **market options matrices** that identify an organization's product and market options. Ansoff's matrix suggests that an organization can grow by marketing new or existing products in new or existing markets, as shown in Figure 2-4. The matrix identifies interesting strategies, including attracting new customers for existing products, diversification into related and unrelated markets, and withdrawing from markets in which the organization's future is unlikely to turn around.

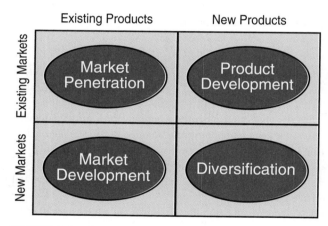

FIGURE 2-4   Ansoff market options matrix

The Boston Consulting Group developed the **growth-share matrix**[14] shown in Figure 2-5 to help organizations allocate their resources among various business units. This tool enables managers to divide their organization's collection of business units and products into four distinct groups, and offers advice for each group.

The four groups in Figure 2-5 are:

- *Stars*—High-growth business units or products that compete in markets where they have attained a large market share. They need further investment to sustain rapid growth or to ward off attacks by competitors. Stars can turn into cash cows.

Relative Market Share

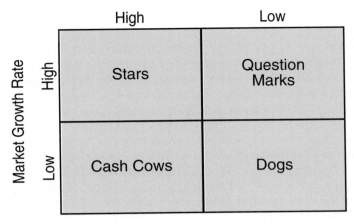

FIGURE 2-5 Boston Consulting Group growth-share matrix

- *Question marks*—Business units or products with a relatively low market share that operate in an arena with a high-growth rate. These products require extensive investment, but provide a relatively low return on investment because of the small market share. An organization should be prepared to either invest heavily in a question mark or invest nothing and let it die off. An organization cannot afford to invest in many question marks, so it must carefully choose the right business units and products to support.
- *Cash cows*—Low-growth business units or products that provide a solid foundation for the company by generating steady streams of cash and profits. They are mature and require relatively little investment.
- *Dogs*—Businesses or products that present little opportunity for growth and hold only a small market share. Management must avoid investing in dogs and must try to liquidate any existing dogs. They must be especially wary of expensive "turnaround" plans, in which overly optimistic ideas are advanced to make major investments in the hopes of salvaging products or organizational units that have been losing money for years.

P&G has identified its core strengths in the competencies that matter most in the consumer products industry: shopper and consumer understanding, branding, innovation, go-to-market capability, and global scale. P&G believes that its ability to combine these strengths creates significant and sustainable competitive advantage.[15] The company has defined the following strategies to focus on growth opportunities that build on P&G's strengths:[16]

- Continue to grow P&G core businesses by developing leading brands, focusing on big, growing markets and winning retail customers.
- Develop businesses with faster growth, higher margins, and better structures in which P&G has significant potential to achieve global leadership.

- Grow disproportionately in developing markets to serve more low-income consumers.

Defining Measures

Measures are metrics that track progress in executing chosen strategies to attain an organization's objectives and goals. They help managers determine if a strategy's ultimate purpose is being achieved. Results, determined by how well the measures are met, provide a feedback loop. Depending on the difference between the actual and desired measures, adjustments may be needed in the objectives, goals, strategies, measures, and the projects being worked.

Organizations must take care in setting these metrics because they run the risk of "getting what they measure" without accomplishing anything meaningful. For example, a consumer products firm may set a measure for each service representative to handle a certain high number of calls. Such a measure could result in extremely short interactions that may fail to satisfy the consumer and instead lead to consumer turnover—certainly not what was intended. If the objective is to improve consumer satisfaction, a more effective goal might be, "Answer 98 percent of all consumer phone calls within 90 seconds, and handle questions so effectively that we reduce the number of callbacks on the same issue from 27 percent of all calls to less than 8 percent." This example illustrates how setting a reasonable goal can require extensive data and a thorough analysis of the current operation.

P&G achieved the following measures for fiscal year 2006 against its long-term goals and objectives: sales growth of 6 percent, earnings-per-share growth of 12 percent, and free cash-flow productivity of 124 percent.

Deploying OGSM

Each year, the senior management team defines objectives (O) for the organization, establishes numerical goals (G) that index each objective, sets strategies (S) on how to reach the goals, and defines measures (M) to assess how well the strategies are being executed. This highest-level OGSM is then deployed to the organization's business units and functional units, where managers translate the information into their own unit's objectives and goals as input to their own OGSM processes (see Figure 2-6). In this manner, the organization's high-level objectives, goals, strategies, and measures are clearly communicated to all employees so that everyone is "on the same page." The various organizational units can then develop more detailed plans that align with the firm's objectives, goals, and strategies. Planning by individual business units must consider cross-functional needs and resources.

Corporate Level	O	G	S	M			
Business Unit Level			O	G	S	M	
Department Level				O	G	S	M

FIGURE 2-6   Deployment of the OGSM

After completing their current-year strategic planning, P&G executives were able to deploy the OGSM outlined in Table 2-2 to the entire organization.

TABLE 2-2   Procter & Gamble OGSM

P&G Objectives, Goals, Strategies, and Measures
Growth objective: Deliver a full decade of industry-leading revenue and profit growth
Growth goals: Industry-leading sales growth Industry-leading earnings-per-share growth Industry-leading free cash-flow productivity
Growth strategies: Continue to grow P&G core businesses by developing leading brands, focusing on big, growing markets, and winning retail customers Develop businesses with faster growth, higher margins, and better structures in which P&G has significant potential to achieve global leadership Grow disproportionately in developing markets to serve more low-income consumers
Growth measures for 2006: Sales growth of 6 percent Earnings-per-share growth of 12 percent Free cash-flow productivity of 124 percent

Identifying Projects and Initiatives

IT staff members constantly are picking up ideas for potential projects through their interactions with various business managers and from observing other IT organizations and competitors. They also keep abreast of new IT developments and consider how innovations and new technologies might be applied in their firm. As members of the IT organization review and consider the corporate OGSM; they can generate many ideas for IT projects that support corporate objectives and goals. They also recognize the need for IT projects that help other corporate units fulfill their business objectives.

Most organizations find it useful to classify various potential projects by type. One such classification is shown in Table 2-3.

TABLE 2-3   Project classification example

Project Type	Definition	Risks Associated with Project Type
Breakthrough	Creates a competitive advantage that enables the organization to earn a greater than normal return on investment than its competitors	High cost Very high risk of failure and potential business disruption
Growth	Generates substantial new revenue or profits for the firm	High cost High risk of failure and potential business disruption

TABLE 2-3 Project classification example (continued)

Project Type	Definition	Risks Associated with Project Type
Innovation	Explores the use of technology (or a new technology) in a new way	Risk can be managed by setting cost limits, establishing an end date, and defining criteria for success
Enhancement	Upgrades an existing system to provide new capabilities that meet new business needs	Risk that scope of upgrade may expand, making it difficult to control cost and schedule
Maintenance	Implements changes to an existing system to enable operation in a different technology environment (for example, underlying changes in hardware, operating systems, or database management systems)	Risk that major rework may be required to make system work in new technology environment; potential for system performance degradation
Mandatory	Needed to meet requirements of a legal entity or regulatory agency	Risk that mandated completion date is missed; may be difficult to define tangible benefits; costs can skyrocket

Prioritizing Projects and Initiatives

Typically, an organization identifies more IT-related projects and initiatives than it has the people and resources to staff. A combined process of setting and scheduling priorities is needed to define which projects will be staffed and when they will be executed. Many organizations create an IT investment board of business-unit executives to review potential projects and evaluate them from several different perspectives:

1. First and foremost, each viable project must relate to a specific organizational goal. These relationships make it clear that executing each project will help meet important organizational objectives (see Figure 2-7).

2. Can the organization measure the business value of the initiative? Will there be tangible benefits, or are the benefits intangible? **Tangible benefits** can be measured directly and assigned a monetary value. For example, the number of staff before and after the completion of an initiative can be measured, and the monetary value is the decrease in staff costs, such as salary, benefits, and overhead. **Intangible benefits** cannot directly be measured and cannot easily be quantified in monetary terms. For example, the increase in customer satisfaction due to an initiative is difficult to measure, and cannot easily be converted into a monetary value.

3. What kinds of costs (hardware, software, personnel, consultants, and so on) are associated with the project, and what is the likely total cost of the effort over multiple years? Consider not just the initial development cost but the total cost of ownership, including operating costs, support costs, and maintenance fees.

4. Preliminary costs and benefits are weighed to see if the project has an attractive rate of return. Unfortunately, costs and benefits may not be well understood at an early phase of the project, and many worthwhile projects do not have benefits that are easy to quantify.

5. Risk is another factor to consider. Managers must consider the likelihood that: the project will fail to deliver the expected benefits; the actual cost will be significantly more than expected; the technology will become obsolete before the project is completed; the technology is too "cutting edge" and will not deliver what is promised; or the business situation will change so that the proposed project is no longer necessary.

6. Some projects enable other projects. For example, a new customer database may be required before the order-processing application can be upgraded. Therefore, some sequencing of projects must be considered.

7. Is the organization ready and capable of taking on this project? Does the IT organization have the skills and expertise to execute the project successfully? Is the organization willing and able to make the required changes to receive their full value?

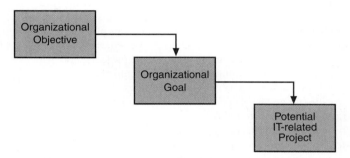

FIGURE 2-7   Ideal relationship between project, goals, and objectives

The following special feature explains how one organization weighs all these factors in determining which IT-related projects to resource and execute.

A MANAGER TAKES ACTION

Lowe's IT Portfolio Management Process

Lowe's is the second most popular home-improvement retailer behind Home Depot, with recent annual revenues exceeding $43 billion. Lowe's also receives excellent marks for customer service; it was ranked second among specialty retailers in the 2005 American Consumer Satisfaction Index. Its theme of "Let's Build Something Together" resonates with customers.

continued

Lowe's has a steady stream of IT-related efforts in its project pipeline, and spends about $400 million a year on IT. CIO Larry Stone and Stephen Boerst, Lowe's manager of IT strategy and planning, led efforts to upgrade the firm's process for prioritizing IT initiatives and allocating scarce resources among the many concurrent projects. They formed a Business Solutions Group of internal consultants, established an IT steering committee, and began using portfolio management software to improve decisions on projects.

The Business Solutions Group consists of experienced Lowe's business analysts who understand the firm's business processes and how IT might be used to make improvements. They consult with department heads to identify possible projects, assess the potential impact of each project on other business initiatives, and develop an estimate of costs and benefits. A business sponsor presents each project's business case to the IT steering committee for decisions on funding and staffing.

The IT steering committee consists of eight Lowe's executives, including the CEO and CIO. They meet twice a month to evaluate project proposals and decide which new projects will be funded. They also review active projects and decide if any should be accelerated, delayed, or terminated. Decisions about new projects are based on reviews of how well the project supports corporate objectives, estimated costs and benefits, and the level of risk. Decisions about active projects are made after considering projected costs and benefits based on results to date, any changes in the fundamental business justifications for executing the project, and any need to reallocate limited resources among competing projects.

The use of portfolio management software helps the group make decisions about projects. The software captures basic information about the schedules and resources required for every approved IT project, from routine maintenance efforts to major strategic projects. Each project at Lowe's is assigned to one of five categories: innovation, growth, enhancement, maintenance, and mandatory. Projects within each category are prioritized based on resources required, anticipated return on investment, and level of risk. Project milestones are then determined so that progress can be tracked from start to finish.

Using all this data, the software can generate several useful reports. A project status summary report uses color codes to highlight each project and show if it is within budget, on schedule, and within scope. A second report summarizes each project schedule, how the schedule affects other projects competing for the same resources, and how the schedule affects other projects that depend on the successful completion of the first one. A third report summarizes the use of human resources and funds of all IT-related projects on a weekly, monthly, or quarterly basis.[17,18,19]

Discussion Questions:

1. Dozens of potential projects are presented to the IT steering committee each quarter. Prepare a standard outline to follow for each project business case and make it easier to review projects consistently and effectively.

2. Identify and briefly discuss three scenarios that could lead to an active project being delayed or cancelled.

3. Is it possible that a project manager or project sponsor might attempt to conceal bad news about an active project to keep it from being cancelled? How could they hide the true status of their project? What should happen if deception is uncovered?

4. Prepare a list of skills and experiences that an ideal candidate should possess to be a member of the IT steering team.

Executing Projects and Initiatives

Once a set of projects is selected for execution, it must be staffed, provided the necessary resources, and properly initiated to ensure success. Throughout the life of a project, business managers have a key role in ensuring good results that meet business needs. They also must smooth the way for the successful adoption and use of the new systems and technology.

Unfortunately, the success rate for IT projects is not high. The Standish Group, a consulting firm that provides IT investment planning research and services, estimated in a 2004 report that the IT project success rate is only about 34 percent.[20] The next chapter provides advice for improving the success rate of your company's IT projects.

Measuring and Evaluating Results

The actual results of a project must be compared with the goals it expected to achieve. This comparison may indicate that a change is needed in the organization's measures, strategies, goals, or objectives, as depicted in Figure 2-1. Managers must be flexible and willing to reevaluate their positions based on the results and insights they gained from the project. They must reassess their course of action and decide whether to make adjustments in light of changing conditions.

For example, Wal-Mart is one of the world's leaders in implementing radio frequency identification (RFID) technology. In 2000, Wal-Mart mandated that its top 100 suppliers must start tagging all cases and pallets carrying merchandise by January 2005. Currently, 600 of its suppliers are RFID-enabled, and the firm expects that 1000 of its stores will soon use RFID. Wal-Mart originally had a dual goal to put the technology in 12 of its 137 distribution centers by the end of 2006, but changed its strategy to focus on enabling its retail stores with RFID. Wal-Mart believes the new strategy allows the retailer to better collaborate with suppliers and monitor the flow of products to avoid problems and react to spikes in demand. The new strategy will help Wal-Mart suppliers improve sales, reduce the number of out-of-stock items, and raise the efficiency of moving products from Wal-Mart backrooms to store shelves. Industry observers believe that the change in strategy could slow Wal-Mart's effort to improve the visibility of its supply chain; cases and pallets stored in distribution centers that do not use RFID are essentially "invisible."[21]

EFFECTIVE STRATEGIC PLANNING: UNITED PARCEL SERVICE (UPS)

The preceding section described an effective strategic planning process based on a defined mission/vision and OGSM. The following section uses the same process to outline a strategic plan developed for United Parcel Service.

Defining Vision and Mission

UPS senior management have created a well-defined vision/mission statement that essentially has remained the same for several years.

Vision: To bring the world's businesses together through synchronized commerce by coordinating their distribution systems, supply chains, and order management systems, helping them to compete better in an expanding global economy.[22]

Mission: To develop business solutions that create value and competitive advantages for customers of all sizes through product differentiation, market penetration, better customer service, and improved cash flow.[23]

Conducting Internal Assessment

The following basic background information helps you understand how UPS became the company it is today. Next, this section identifies UPS's current internal strategies, strengths, weaknesses, and threats.

Historical Perspective

In 1907, two teenagers working out of their basement in Seattle created what would become the world's largest package delivery service. Claude Ryan and Jim Casey started the American Messenger Company by performing a wide range of tasks, from carrying notes on foot or a bicycle to making home deliveries for drugstore customers. The company changed its name to United Parcel Service (UPS) in 1919.

UPS expanded its U.S. operations over the next 55 years until it served all 48 states of the continental United States. UPS went international in 1975 when it began offering services in Canada; it started operations the following year in Germany. Over the next decade, UPS expanded its service throughout the Americas and Europe. UPS extended service to the Middle East, Africa, and the Pacific Rim with the 1989 purchase of IML, a UK-based document and parcel delivery company.

Recognizing the growing need for overnight delivery, UPS expanded its Next Day Air service to all 50 states and Puerto Rico by 1985. That same year, UPS introduced international air package and document service between the United States and six European nations. In 1988, UPS won approval from the Federal Aviation Administration (FAA) to launch its own airline.

In 1992, UPS began tracking all ground shipments electronically using handheld Delivery Information Acquisition Devices (DIADs). All UPS drivers carry DIADs; they use them to scan bar-coded packages and upload delivery information to the UPS network. DIADs also can provide Internet access; they operate over multiple types of wireless networks and integrate with Global Positioning System networks to determine accurate locations.

The DIAD can capture a digital image of a recipient's signature to positively confirm their identity for the shipper. The DIAD also enables drivers to communicate with their operation centers and receive information about any changes in pickup schedules, traffic patterns, and other important messages.

Customers can visit the UPS Web site to track packages in transit; *www.UPS.com* services more than 10 million requests per day for online tracking information. Customers expect extremely high performance, predictable and reliable delivery times, and broad service offerings as they seek to minimize supply-chain costs and disruptions.

The Internet links sellers and buyers globally, and e-commerce radically changed how companies and consumers do business. UPS has been able to capture a large portion of the shipping associated with e-commerce. The company's online presence includes *www.UPS.com* and popular shipping and tracking systems such as WorldShip, Quantum View, and CampusShip. These systems give customers efficient tools to process, track, and manage shipments and supply chains.

During the late 1990s, UPS expanded its core business to offer services for customers that needed more efficient global supply chains. Specialty services were developed for companies in the automotive, consumer goods, and high-tech industries. Such companies were increasingly becoming international, and benefited greatly from UPS's expertise and experience in supply-chain management. Today, UPS Supply Chain Solutions provides logistical, global freight, financial, and mail services to enhance customers' business performance and improve their global supply chains.

UPS made several acquisitions to expand into new markets. The 1999 purchase of Challenge Air made UPS the largest express and air cargo carrier in Latin America. In 2001, UPS ventured into the retail business arena by acquiring Mail Boxes Etc. Within two years, more than 3000 Mail Boxes Etc. locations were rebranded as The UPS Store to offer UPS direct shipping services. The purchase of Menlo Worldwide Forwarding in 2004 enabled UPS to make heavy air freight shipments.

The internal assessment for UPS continues with an identification of current strategies.

Current Strategies

Global expansion—UPS continues to expand its service worldwide, providing customers in Asia, Europe, and South America with a range of time-definite and supply-chain services. UPS has expanded its Worldport air hub in Louisville, Kentucky and its European air hub in Cologne, Germany. Asia is a primary growth target for UPS; in 2005, it launched the first nonstop delivery service between the United States and Guangzhou, China. Also in 2005, UPS acquired the interest held by its joint-venture partner in China, giving it access to 23 cities that cover more than 80 percent of the country's international trade.

Provide all modes of service—UPS has established an integrated ground and air network that is the most extensive in the industry. It can handle all modes of service (express, ground, domestic, international, commercial, and residential) through one pickup and delivery system.

TABLE 2-4 Selected UPS financial data (in billions of dollars)

	2005	2004	2003
U.S. domestic revenue	$28.6	$27.0	$25.4
International package revenue	$8.0	$6.8	$5.6
Supply chain and freight revenue	$6.0	$2.8	$2.5
Total revenue	$42.6	$36.6	$33.5
Operating profit	$6.1	$5.0	$4.4
Operating margin (percent)	14.4	13.7	13.1
Net income	$3.8	$3.3	$2.9

TABLE 2-4 Selected UPS financial data (in billions of dollars) (continued)

	2005	2004	2003
Assets	$35.2	$33.1	$29.7
Long-term debt	$3.3	$3.3	$3.1

Identification of Strengths

The internal assessment defined the following strengths:

Financial strength—UPS has more than $3 billion in cash and cash equivalents, with a long-term debt of about $3 billion. Its operations generate a steady stream of cash and the operating margin is over 13 percent.

Makes effective use of technology—UPS has an IT organization of 4500 professionals and has invested an average of $1 billion per year on IT for the past 20 years. This investment has helped to develop a worldwide network supported by arguably the most sophisticated IT systems in the transportation industry. Customers can take advantage of the UPS IT platform by linking their systems directly into the UPS global network. This seamless integration allows UPS to provide useful solutions that help customers improve their profitability and access new markets.

Identification of Weaknesses

The internal assessment defined the following weaknesses:

Thin operating margin—The U.S. domestic operation is the cash cow for UPS, delivering two-thirds of the firm's revenue. This portion of the business had a 15.7 percent operating margin in 2005. Reinvestment of this money supports growth through acquisitions, new product development, and service enhancements. The 2005 operating margin for UPS's supply chain and freight business was just 2.6 percent, compared with an operating margin of 18.7 percent in the international package business. In the first quarter of 2006, the unit posted a $25 million operating loss.[24] In November 2006, UPS announced that it would cut 1200 positions in an attempt to increase the unit's profitability.[25]

Identification of Threats

The internal assessment defined the following threats:

Union contracts—About 241,000 UPS workers are represented by the Teamsters Union. The current Teamsters contract with UPS expires in 2008. Other workers are covered by the International Association of Machinist and Aerospace Workers. The Association of Parcel Workers of America is seeking to replace the Teamsters Union at UPS.[26] Also, about 3000 UPS pilots are represented by the Independent Pilots Association. The pilots approved a new five-year contract in 2006.[27]

Conducting External Assessment

The external assessment begins with an examination of the industry, then identifies UPS's strengths, opportunities, and threats. The section concludes with a competitive analysis.

Examining Industry

The external assessment begins by identifying that UPS competes in the package delivery industry. Customers increasingly are interested in the timing and predictability of deliveries; as a result, time-definite delivery has become essential for improved inventory management and overall distribution efficiency. Time-definite delivery has grown from 4 percent of the U.S. parcel delivery market in 1977 to more than 60 percent by 2005.[28]

Identification of Strengths

The external assessment identifies the following strengths:

Dominant market share—UPS is the world's largest package delivery company in both revenue and volume. It is also a global leader in providing supply-chain services.

Global reach—UPS's integrated global ground and air network is the most extensive in the industry. It is the only network that handles all levels of service (express, ground, domestic, international, commercial, and residential) through one integrated pickup and delivery system.

Strong brand image—UPS has strong brand recognition among customers and consumers. *Fortune* Magazine rated UPS as the "World's Most Admired Company" in 2006.

Identification of Opportunities

The external assessment identifies the following opportunities:

Acquisitions—Acquisitions offer UPS the opportunity to grow geographically, offer new services, and gain synergy. UPS invested $1.5 billion in acquisitions during 2005, including Overnite, a less-than-truckload (LTL) transportation service with more than 60,000 customers throughout the United States. This acquisition should provide UPS with synergistic opportunities within its small-package business. It also acquired Lynx Express, a leading independent parcel carrier in the United Kingdom; Messenger Service Stolica, a leading parcel and express delivery company in Poland; and the express operations of Sinotrans Air Transportation Development in China. UPS will continue to look for opportunities to expand its global presence, lower its costs, and enhance its service offerings and technology.

Growth in e-commerce—U.S. online retail sales are expected to grow from $34 billion in 2001 to $271 billion in 2008, a compounded average growth rate of 34.5 percent. This growth provides a major opportunity for UPS to provide effective solutions that meet customer needs and to provide shipment information that enhances retailers' business applications.

International growth—The international package delivery market continues to grow much faster than in the United States. UPS can continue to make significant investments in this arena. It can use its worldwide infrastructure and wide range of services to improve high-margin services. Both Europe and Asia offer significant opportunities for growth.

Identification of Threats

The external assessment identifies the following threats:

Rising oil prices—Global oil prices are highly volatile and impossible to predict over the short term. However, it seems likely that prices will climb over the long term. From

UPS's perspective, mandatory emissions systems in trucks would decrease fuel efficiency, compound the cost burden of high oil prices, and reduce income. Proposals for toll roads in several states, including California and Texas, also would increase costs.

Terrorism—The airplanes owned by UPS and other freight carriers are potential "weapons" for terrorists. UPS will be forced to spend more money to protect its equipment, both on the ground and in the air.

Competitive Analysis

FedEx, founded in 1973, is a strong competitor of UPS. It also offers overnight, ground, and heavy freight services as well as logistics services. FedEx won a $1.6 billion contract with the U.S. Postal Service in 2006 to carry Priority Mail and Express Mail.[29] FedEx has had consistent growth in net income over three decades. Its operating margin is under 9 percent, however, compared with 13 percent for UPS. UPS has greater customer coverage and offers its services at lower costs than FedEx. FedEx spends about 8 percent of its revenue on IT, compared with about 6 percent at UPS. Other financial comparisons between UPS and FedEx are shown in Table 2-5.

TABLE 2-5   Financial comparative analysis of UPS and FedEx (in billions of dollars)

	2005		2004		2003	
	Revenue	Operating Profit	Revenue	Operating Profit	Revenue	Operating Profit
UPS	$42.5	$6.1	$36.5	$5.0	$33.5	$4.4
FedEx	$29.4	$2.5	$24.7	$1.4	$22.7	$1.5

Based on the results of the internal and external assessments, UPS's major strengths, weaknesses, opportunities, and threats are summarized in Table 2-6.

TABLE 2-6   UPS SWOT analysis

Strengths	**Weaknesses**
Holds leading market share	Thin operating margin
Global reach	
Strong brand image	
Financially strong	
Makes effective use of technology	
Opportunities	**Threats**
Acquisitions	Union contracts
Growth in e-commerce services	Rising oil prices
International growth	Terrorism

Defining Objectives

Every for-profit organization considers future profitability a key objective. UPS had revenues of $43 billion and operating income of $6 billion in 2005. From 2002 to 2005, its revenues grew at a compounded annual growth rate of 10.8 percent, while operating profits grew at a rate of 14.5 percent. Management's objective is to preserve this consistency in revenue and earnings growth.

Establishing Goals

UPS's goal is to increase operating profits in all three business segments: domestic delivery, international delivery, and supply chain and freight.

Setting Strategies

The primary strategy to achieve these goals is to take advantage of UPS's competitive strengths while maintaining the firm's focus on meeting or exceeding customer requirements. The principal components of the UPS strategy are listed under "Growth strategy" in Table 2-7.

Defining Measures

Management defined several growth-related measures for 2006, as listed in the final section of Table 2-7.

TABLE 2-7   UPS vision, mission, objectives, goals, strategies, and measures

UPS Vision, Mission, Objectives, Goals, Strategies, and Measures
Vision: To bring the world's businesses together through synchronized commerce by coordinating their distribution systems, supply chains, and order management systems, helping them to compete better in an expanding global economy.[30]
Mission: To develop business solutions that create value and competitive advantages for all customers through product differentiation, market penetration, better customer service, and improved cash flow.
Growth objective: Preserve consistency in revenue and earnings growth.
Growth goal: Grow operating profits in all three operating divisions: domestic delivery, international delivery, and supply chain and freight.
Growth strategy: The primary strategy to achieve this goal is to take advantage of UPS's competitive strengths while maintaining the firm's focus on meeting or exceeding customer requirements. The principal components of this strategy are: • Build on its leadership position in the United States.   • Continue international expansion.   • Provide comprehensive supply-chain solutions.   • Leverage leading-edge technology and e-commerce advantages.   • Continue integration of UPS shipping technology into customers' operations.   • Pursue strategic acquisitions and global alliances.   • Focus supply-chain activities on the high-tech, healthcare, and retail sectors, where connectivity across all UPS businesses is strong.

TABLE 2-7 UPS vision, mission, objectives, goals, strategies, and measures (continued)

UPS Vision, Mission, Objectives, Goals, Strategies, and Measures
Growth measures for 2006:
• Increase earnings between 11 percent and 16 percent over 2005 results.
• Grow the small-package business by at least market rates in the United States.
• Grow the small-package business faster than market rates in major regions of the world.

Deploying OGSM to IT

Dave Barnes was promoted to senior vice president and CIO of UPS in January 2005. He began his career with UPS in 1977 as a part-time package handler. He has held multiple positions with UPS, both in the United States and abroad, including assignments in operations, finance, internal audits, and at UPS Airlines. Most recently, he was vice president of customer and operations applications portfolios. The position made him responsible for directing global application development and managing 3000 of UPS's 4700 IT workers. "The movement of goods and the movement of information are really tied together. You can't have one without the other in today's environment," said Barnes.[31] As head of the IT organization, Barnes must ensure that it understands the corporate OGSM. The IT organization works to define, execute, and measure projects that are consistent with the firm's mission, objectives, goals, and strategies.

Identifying and Prioritizing Projects and Initiatives

The UPS IT organization spends roughly $1 billion per year, and is always working on hundreds of IT-related projects. This section identifies and categorizes a few of these projects.

Breakthrough

The UPS Package Flow Technology is a multiyear breakthrough project to re-engineer the UPS package pickup and delivery process. The project is designed to support growth, improve productivity, reduce costs, and provide the platform for new services. Package flow technologies use forecasted and historical information to create a dispatch plan for every driver working in the package distribution center. The plan ensures that no driver is over-dispatched and minimizes any last-minute load changes to a driver's delivery truck.[32] This planning is key because UPS delivers multiple services using the same driver, unlike other carriers. A single UPS driver delivers overnight packages, collects COD payments, and delivers ground packages to commercial and residential customers on the route. The goal is to provide customers with a consistent, reliable service and enable UPS to better understand each customer's unique needs. A key component of the project is the printing of a pre-load assist label (PAL) at the package center. This label tells the package sorter which conveyor belt chute to use for transporting the package to the package car loading area.

According to UPS, "Once in the loading area, the PAL tells the loader the exact shelf location on the specific car where the package should be placed for optimum delivery. Before the creation of the PAL, a UPS employee loading a package car had to memorize thousands of addresses. To ensure that a loader correctly had loaded the package car, a driver often had to spend valuable time checking and rearranging packages in the back of the package car in the order he/she felt would be the most optimum for delivery. Now the driver knows exactly how many packages are in the back of the package car and in what order all those packages need to be delivered to ensure that all customer service levels are met."[33]

Savings from this project are estimated to be $750 million per year starting in 2008. The project also will increase revenue by several hundred million dollars per year by enabling further enhancements in service levels. Thus, the total tangible benefits could exceed $1 billion per year.[34]

Growth

UPS is dramatically expanding its Worldport, a sophisticated hub in Louisville, Kentucky, that is the heart of the company's global transportation network. The expansion is required to meet the demands of continued growth in UPS's air package volume around the world. This expansion will increase sorting capacity over the next five years by 60 percent to 487,000 packages per hour. The expansion plan calls for three new aircraft loading wings in the hub building, the installation of high-speed conveyors, and millions of dollars for new computer control systems.[35]

Innovation

UPS has extensively tested radio frequency identification (RFID) technology as a possible replacement for the bar coding it uses to sort packages. RFID would eliminate the need for line-of-sight reading, which is required for bar codes. Also, RFID tags can be scanned at greater distances than bar codes. The RFID chips, however, cannot accurately read tag information at high speeds beyond the range of RFID antennas. The first-pass accuracy rate is about 80 percent, which is much poorer than the 95 percent accuracy of bar-coded packages.[36] The firm continues to monitor RFID advances closely, and recently invested in G2 Microsystems, an RFID chip manufacturer.[37]

Enhancement

UPS drivers use handheld wireless computers called DIADs (Delivery Information Acquisitions Devices) to capture package identification data, location information, and customer signatures. The data eventually is relayed to the DB2 package tracking database. The EDD (Enhanced DIAD Download) is UPS-developed software that downloads an electronic manifest to the driver's DIAD at the start of each workday. This system enables drivers to view each scheduled package delivery displayed on the DIAD in the order of most efficient delivery. The EDD software can make the DIAD generate an alarm if the driver tries to deliver a package to the wrong customer or forgets to deliver packages to a customer.

Maintenance

UPS is continually evaluating and upgrading the DIAD device to obtain more memory for storing data and instructions, faster processors, and new network capabilities. UPS currently is evaluating and testing the fifth generation of the DIAD. Each new generation requires testing to ensure that existing software will operate on the new hardware. New software also is created strictly for the new DIAD device.

Mandatory

UPS developed software called Target Search to enable U.S. Customs and Border Protection agents to inspect packages that pass through the Worldport international hub in Louisville. The software captures and provides information about packages so agents can be more selective in choosing packages for inspection.

Table 2-8 summarizes the project categories of the UPS IT organization. These categories were introduced in Table 2-3.

TABLE 2-8 Classification of UPS IT-related projects

Project Type	Specific Projects
Breakthrough	Implement Package Flow Technology
Growth	Expand the Louisville Worldport facility
Innovation	Evaluate use of RFID chips to replace bar code technology
Enhancement	Upgrade DIADs with the new EDD software
Maintenance	Evaluate and upgrade DIADs, associated software, and networking capabilities
Mandatory	Deploy Target Search software

Executing Project, Then Measuring and Evaluating Results

The preceding set of projects and hundreds of other projects are being executed by UPS. During the execution of each project, actual results are compared to expected results. Necessary adjustments and changes are made to ensure that projects meet the desired business goals and are completed within a reasonable time frame and budget. Some approved projects even may be cancelled based on negative results and a realization that the project is no longer as attractive as it once seemed. For example, UPS carefully evaluated the use of RFID technology to replace its decades-old bar codes, but learned that the new technology could not yet meet its needs. Further efforts to implement RFID technology have been sidelined until the technology becomes cheaper and more effective.

The checklist in Table 2-9 provides a set of recommended actions for business managers to take to ensure that their organization follows an effective strategic planning process. The appropriate answer to each question is "yes."

TABLE 2-9   Manager's checklist

	Yes	No
Are the efforts of your IT group aligned with the organization's strategies and objectives?		
Do business managers clearly communicate your organization's vision/mission, objectives, goals, strategies, and measures? Does this communication help everyone define the actions required to meet organizational goals?		
Do you have an effective process to choose from alternative strategies that considers many factors, including the long-term impact of each strategy on revenue and profit, the degree of risk involved, the amount and types of required resources, and potential competitive reaction?		
Does your organization establish measures to track the progress of executing chosen strategies?		
Does your organization have an effective way to identify and prioritize potential projects?		
Does your organization measure and evaluate the results of projects as they progress, with an eye toward making necessary adjustments?		
Is your organization willing to cancel a project if results do not meet expectations?		

Chapter Summary

- Strategic planning is a process that helps managers identify desired outcomes and formulate feasible plans to achieve these objectives using available resources and capabilities. The strategic plan must take into account that the organization and everything around it is changing.

- Business managers have a key responsibility to communicate the organization's vision, objectives, and strategies so that everyone can help define the actions required to meet organizational goals.

- The vision/mission statement communicates an overarching aspiration that guides an organization and forms a basis for making decisions and taking action.

- Strategic planning requires careful study of the external environment surrounding an organization and assessing where the organization fits within it. Analysis of the internal assessment and external environment are frequently summarized into a SWOT matrix.

- Management must define objectives for what their organization must accomplish and associated goals that specify how to determine if the objectives are being met.

- Management also must define the strategies or specific actions the organization will take to achieve its vision/mission, objectives, and goals. There are several approaches for identifying strategies, including Porter's three fundamental strategies, the market options matrix, and the growth-share matrix.

- Measures are metrics that track an organization's progress in executing chosen strategies to attain its objectives and goals.

- An organization's highest-level objectives, goals, strategies, and measures (OGSM) must be deployed to its various business units and functional units so that everyone knows what is expected and how to achieve it. Members of an IT organization support the corporate OGSM by generating ideas for IT projects that are consistent with the identified strategies.

- Many organizations create an IT investment board of business-unit executives to review potential projects and evaluate them from different perspectives. Once a set of projects is selected for execution, they must be staffed, provided with necessary resources, and properly initiated to ensure success.

- An analysis of actual project results may indicate that a change is needed in the organization's measures, strategies, goals, or objectives. Managers must be flexible and willing to reassess their course of action when necessary. In light of changing conditions, they can decide whether to make adjustments to a project or perhaps even terminate it.

Discussion Questions

1. What role should the CEO take during the strategic planning process? Identify other key players and their roles.

2. How often should the vision/mission statement be updated or revised?

3. Discuss how you would gather data about an organization's internal processes and operations. Identify specific tools or techniques you might apply to gain useful insights.

4. Identify the five fundamental forces that determine an industry's level of competition and long-term profitability. Give an example of how each factor can affect profitability.

5. How would you distinguish between an organizational weakness and a threat to the organization? How would you distinguish between a strength and an opportunity?

6. What would it mean if an organization could not identify any opportunities? What if it could not identify any threats?

7. What is the difference between an objective and a goal? Give an example of each.

8. Would you recommend that an organization set BHAGs? Why or why not?

9. Define the term *strategy* and discuss how a strategy is related to objectives and goals. Provide an example of an organizational strategy.

10. Briefly describe the market options matrix and identify the four basic options it suggests.

11. Briefly describe the growth-share matrix and discuss how an organization might use it to allocate resources.

12. Define the term *measure* and discuss how a measure is related to an objective and a goal.

13. Discuss what it means to deploy an organization's OGSM. What should the deployment accomplish?

14. Give an example of an IT-related project that an IT organization might use in an attempt to achieve its own OGSM. Give an example of an IT project that might help another department achieve its OGSM.

15. In comparing two potential growth projects, should the project with the better rate of return always be chosen? Discuss your answer fully.

16. Give an example of project results that might prompt an organization to reconsider one or more of its goals, strategies, or measures. Give an example of project results that might prompt an organization to cancel an active project.

Action Memos

1. In the past three years, your firm has not done well. Its market share and profitability have dropped and the stock price has fallen more than 25 percent. A new CEO has been hired to lead a turnaround process for the firm. She has called a meeting of her direct reports to discuss the need for a new direction at the firm. She also wants to create a strategic plan that captures this new vision and reinvigorates the employees, executives, shareholders, and the financial community. She has asked each executive to briefly discuss their ideas for the plan. It is your turn to speak. What do you say?

2. You are the CFO of a mid-sized manufacturer. The IT organization has just completed its annual strategic planning and budgeting process. Their plans and $30 million budget were forwarded to you for review by the recently hired CIO. Frankly, you cannot understand the plan, nor do you see a close connection between the proposed projects and the strategic goals of the organization. You are especially concerned because little of the budget is allocated for growth and breakthrough projects. The CIO is on the phone, asking to meet with you to discuss his plans and budget. How do you respond?

Web-Based Case

Go to the Johnson & Johnson corporate Web site at *www.jnj.com*. You may be surprised to find that they do not have a corporate mission statement. Instead, for more than 60 years, a simple, one-page document—Our Credo—has guided Johnson & Johnson in fulfilling its responsibilities to its customers, employees, stockholders, and the community. Compare and contrast the Johnson & Johnson credo with the typical vision/mission statement. Do you think senior management should spend the effort and time to develop a vision/mission statement? Why or why not?

Case Study

Strategic Plan: Company of Your Choice

Choose a company that interests you and document their strategic plan, including:

- Vision, mission, objectives, goals, strategies, and measures
- An industry analysis
- A SWOT analysis

Identify two IT-related projects that would be consistent with this plan. Recommend one of the two projects for implementation. (*Hint*: Choose a publicly traded company to ensure that you can find sufficient information to develop the strategic plan.)

Endnotes

[1] Allan Holmes, "Rx for Risk," *CIO Magazine*, 15 March 2006.

[2] Rick Whiting, "FDA to Require Drug Makers to Adopt Supply-Chain Tracking System," *InformationWeek*, 12 June 2006.

[3] "FDA" in Wikipedia; "Vioxx" in Wikipedia.

[4] Allan Holmes, "The ROI of Alignment," *CIO Magazine*, 1 January 2007.

[5] Corporate Information, Company Overview at Google Web site, *www.google.com/corporate/* (accessed 24 February 2007).

[6] "Purpose, Value, and Principles," Procter & Gamble Web site, *www.pg.com/company/who_we_are/ppv.jhtml* (accessed 28 February 2007).

[7] Anthony Bianco, "Wal-Mart's Midlife Crises," *BusinessWeek*, pages 46–56, 30 April 2007.

[8] James Collins and Jerry Porras, *Built to Last, Successful Habits of Visionary Companies*, Harper Collins Publishers, page 9, 1994, 1997.

[9] Procter & Gamble 2006 Annual Report.

[10] William Lasher, *Strategic Planning for a Growing Business*, Texere, an imprint of Thomson/South-Western, page 99, 2005.

[11] Michael E. Porter, *Competitive Strategy*, The Free Press, New York, 1980.

[12] G. S. Day, *Strategic Market Planning*, West Publishing, St. Paul, MN, 1980.

[13] I. Ansoff, *Corporate Strategy*, Penguin, Harmondsworth, 1989.

[14] "The Growth-Share Matrix," The Boston Consulting Group Web site, *www.bcg.com/this_is_bcg/mission/growth_share_matrix.html* (accessed 28 February 2007).

[15] Procter & Gamble 2006 Annual Report.

[16] Procter & Gamble 2006 Annual Report.

[17] "Portfolio Management," Serena Web site at *www.pacificedge.com* (accessed 8 February 2006).

[18] Fortune 2005 Global 500, Lowe's, *CNNMoney.com* (accessed 8 February 2006).

[19] Cindy Waxer, "How Lowe's Grows," *CIO Magazine*, 1 December 2005.

[20] "Standish: Project Success Rates Improved over 10 Years," *Software Magazine*, 15 January 2004.

[21] Marc L. Songini, "Wal-Mart Shifts RFID Plans," *Computerworld*, 26 February 2007.

[22] United Parcel Service 2005 Annual Report – Letter to the Shareholders.

[23] United Parcel Service 2005 Annual Report – Letter to the Shareholders.

[24] John D. Boyd, "UPS Seeing Red," *Traffic World*, page 43, 1 May 2006.

[25] "Laying Off," *Air Cargo World*, Volume 96, no. 11, page 8, November 2006.

[26] Barry B. Burr, "Labor Dissidents Try to Oust Teamsters at UPS," *Pensions & Investments*, Volume 34, no. 15, pages 6 and 34, 24 July 2006.

[27] John D. Boyd, "UPS Pilots OK Deal," *Traffic World*, page 30, 11 September 2006.

[28] Nabil Alghalith, "Competing with IT: The UPS Case," *The Journal of American Academy of Business,* Cambridge, Volume 7, no. 2, September 2005.

[29] Ian Putzger, "Hardly Making it In," *Traffic World*, pages 23–24, 21 November 2006.

[30] United Parcel Service 2005 Annual Report – Letter to the Shareholders.

[31] Ann Bednarz, "UPS Veteran Takes Over as CIO," *Network World*, 10 January 2005.

[32] Clyde E. Witt, "Package Flow Technology at UPS," *Material Handling Management*, pages 18–20, April 2005.

[33] UPS Pressroom – Fact Sheets – "Package Flow Technologies: Innovation at Work," UPS Web site, *www.pressroom.ups.com/mediakits/factsheet/0,2305,1134,00.html* (accessed on 8 March 2007).

[34] Satish Jindel, "PFT Provides UPS a $1 Billion Payback," *Traffic World*, page 6, 8 May 2006.

[35] UPS Pressroom, "UPS Sets $1 billion + Expansion of Global Air Hub," *www.pressroom.ups.com*, 17 May 2006.

[36] William Hoffman, "Parcel Tagging Out," *Traffic World*, page 18, 10 April 2006.

[37] Christian Annesley, "UPS Rules Out RFID Tagging for Small Packages," *Computer Weekly*, page 10, 20 June 2006.

CHAPTER **3**

PROJECT MANAGEMENT

ON THE IMPORTANCE OF PROJECT MANAGEMENT

"According to polling by Forrester Research Inc., the pressure to deliver projects on time and on budget puts project management skills atop the list of hiring priorities for a majority of CIOs in 2007."

—Linda Tucci, senior news writer, "CIOs Scramble for Project Managers," *SearchCIO.com*, 19 March 2007.

BROWN-FORMAN: GOOD PROJECT MANAGEMENT PROCESS DELIVERS OUTSTANDING RESULTS

Brown-Forman is an American-owned spirits and wine company with annual sales of $3 billion in 135 countries. The firm employs 4,000 people worldwide and 1,300 in its Louisville, Kentucky headquarters.

A key objective of the firm is to improve the global reach of its brands. One strategy identified to achieve this objective is to create highly efficient standard business processes that can be reapplied across the firm's subsidiaries.

Brown-Forman cash management processes enable the firm to receive and pay for goods and services, and to support the daily management of working capital, investments, and transactions with domestic and foreign subsidiaries. Previously, Brown-Forman relied on a "home-grown" Excel spread-sheet system with many inefficient manual interfaces. For example, workers had to re-enter bank information from the company's subsidiaries into spreadsheets to determine daily cash positions. The system clearly was not a good solution to reapply across its subsidiaries.

Executive vice president and CFO Phoebe A. Wood identified replacing the antiquated cash management system as a high priority. The project was initiated and executed following a standard project management approach that Brown-Forman had developed over the years. This approach clearly defined a time-proven multiphase process for employees that greatly improved the probability of project success.

The project's brief but critical initiation phase involved several business managers, including project leader Bob Walker from the Treasury Department, an IT manager, an assistant treasurer, and a treasury analyst. They focused on defining the scope of the project and key business objectives. This phase clearly identified numerous opportunities for improvement and indicated that the project was reasonable to pursue.

During the planning phase of the project, members of the accounting, finance, and IT areas and the system's ultimate end users further defined requirements in each area and established specific and measurable goals. They also evaluated several options for meeting business needs. They learned that the software firm SAP had developed state-of-the-art cash management systems for other consumer products firms. Walker and key members of his team visited Colgate-Palmolive, were impressed with what they saw, and decided to create the new cash management system based on the SAP software solution. The team then developed a preliminary budget and schedule, complete with milestones. They also identified potential change in management issues that might arise.

Next, the team explored the capabilities and limitations of the standard, uncustomized SAP software. They developed basic prototypes to demonstrate how the new work processes would operate and how the software would work. Necessary areas of customization were identified. Once this blueprint was approved, the implementation team built the system, including necessary customizations.

When the cash management system was coded and fully tested, a series of training sessions were held for key users. The key users defined and executed a systems acceptance test plan to ensure that the system met their expectations. The key users then prepared and delivered end-user training that was tailored to each user's role so that everyone was ready to "go live."

The final phase involved changing from the old Excel-based system to the new processes associated with the SAP software. Brown-Forman used a "flash cut-over," in which the old system was used on a Friday and the new system was up and running on the next Monday.

With the new SAP-based system and work processes, Brown-Forman cut the total effort associated with its cash management processes by over 60 percent. They also reduced the hours required for month-end closing from 48 hours to just 1 hour. The project came in on time and on budget. Thus, following a well-defined project management process enabled Brown-Forman to achieve outstanding results.[1,2]

LEARNING OBJECTIVES

As you read this chapter, ask yourself:
- What is project management, and what are the key elements of an effective project management process?
- How can an effective project management process improve the likelihood of project success?

This chapter clarifies the importance of project management and outlines a tried and proven process for successful project management.

WHY MANAGERS MUST UNDERSTAND PROJECT MANAGEMENT

Projects are the way that much of an organization's work gets done. On the basis of data released by the Bureau of Economic Analysis, part of the U.S. Department of Commerce, the Project Management Institute estimates that project spending in the U.S. public and private sectors combined exceeds $2 trillion per year, an amount equivalent to a quarter of America's GDP. Worldwide, project-related spending is estimated to be almost $10 trillion per year.[3]

Unfortunately, the success rate for IT-related projects is not high. The Standish Group, a consulting firm that provides IT investment planning research and services, estimated in its 2004 report that the IT project success rate is only about 34 percent (see Figure 3-1). "Success" in this context means that the project met schedule and budget constraints and delivered all critical features and requirements. About 51 percent of all projects are "challenged," meaning that they are over time, over budget, and/or lacking critical features and requirements. Meanwhile, about 15 percent of all IT projects are estimated to be clear failures. These findings are based on a study of 40,000 projects over the past 10 years.[4] This chapter provides advice for improving the success rate of your IT projects.

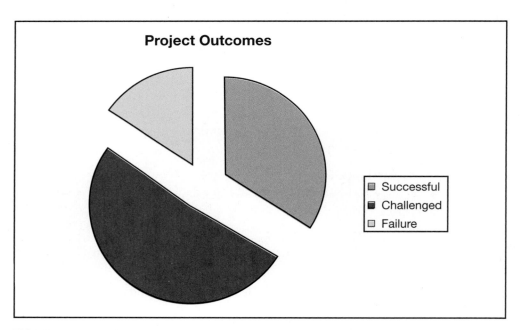

FIGURE 3-1   The Standish Group's 2004 findings for IT-related projects

Researchers Hamel and Prahalad defined the term **core competency** to mean something that a firm can do well and that provides customer benefits, is hard for competitors to imitate, and that can be leveraged widely to many products and markets.[5] Today, many organizations recognize project management as one of their core competencies, and see their ability to better manage projects as a way to achieve an edge over competitors and deliver greater value to shareholders and customers. As a result, they spend considerable effort identifying potential project managers and then training and developing them. For many managers, their ability to manage projects effectively is now a key to their success within an organization.

WHAT IS A PROJECT?

A **project** is a temporary endeavor undertaken to create a unique product, service, or result.[6] Each project attempts to achieve specific business objectives and is subject to certain constraints such as total cost and completion date. As previously discussed, there must be a clear connection between business objectives, goals, and projects; also, projects must be consistent with business strategies. For example, an organization may have a business objective to improve customer service by offering a consistently high level of service that exceeds customers' expectations. Initiating a project to reduce costs in the customer service area by eliminating all but essential services would be inconsistent with this business objective.

At any point in time, an organization may have dozens or even hundreds of active projects aimed at accomplishing a wide range of results. Projects are different from operational activities, which are repetitive activities performed over and over again. Projects are not repetitive; they come to a definite end once the project objectives are met or the project is cancelled. Projects come in all sizes and levels of complexity, as you can see from the following examples:

- A senior executive led a project to integrate two organizations following a corporate merger.
- A consumer goods company executed a project to launch a new product.
- An operations manager led a project to outsource part of a firm's operations to a contract manufacturer.
- A hospital executed a project to load software on physicians' PDAs, which enabled them to access patient data anywhere in the hospital.
- A computer software manufacturer completed a project to improve the scheduling of help desk technicians and reduce the time on hold for callers to its 1-800 phone support services.
- A staff assistant led a project to plan the annual sales meeting.
- A manager completed a project to enter her departmental budget into a preformatted spreadsheet template.

Project Variables

Five highly interrelated parameters define a project—scope, cost, time, quality, and user expectations. If any one of these parameters changes for a project, there must be a corresponding change in one or more of the other parameters. A brief discussion of these parameters follows.

Scope

Project scope is a definition of which work is and is not included in a project. Project scope is a key determinant of the other project factors, and must carefully be defined to ensure that the project meets its essential objectives. In general, the larger the scope of the project, the more difficult it is to meet cost, schedule, quality, and stakeholder expectations.

For example, the Big Dig is the unofficial name of Boston's Central Artery/Tunnel Project, whose scope included rerouting Interstate 93, the primary controlled-access highway through the center of Boston, into a 3.5-mile tunnel under the city. The scope of the project also included the extension of Interstate 90 to Logan International Airport and the construction of the Rose Kennedy Greenway in the space vacated by the previous I-93 elevated roadway. Another major component of the project scope included construction of the Zakim Bunker Hill Bridge over the Charles River. This 10-lane, cable-stayed hybrid bridge is the widest of its type ever built and the first to use an asymmetrical design.

When initially conceived, the project was estimated to cost $2.8 billion. By the time construction began, the estimate had ballooned to $5.8 billion despite eliminating a rail connection between Boston's North Station and South Station from the scope of the project. As of the end of 2006, the actual cost of the project had exceeded $14 billion. Not only were cost and schedule goals not met, the project failed to meet quality expectations. The tunnels have been plagued with hundreds of water leaks that damaged the steel supports and forced temporary closures for repairs. In the summer of 2006, the tunnel connecting I-90 to the Ted Williams Tunnel had to be closed after a partial collapse of the ceiling caused a fatal accident.[7] Many observers believe that the broad and changing scope of the project (poor scope management) contributed to its many problems.

Cost

The cost of a project includes all the capital, expenses, and internal cross-charges associated with the project's buildings, operation, maintenance, and support. Capital is money spent to purchase assets that appear on the organization's balance sheet and that are depreciated over the life of the asset. Capital items typically have a useful life of several years. A building, office equipment, computer hardware, and network equipment are examples of capital assets. Computer software also can be classified as a capital item if it costs more than $1000 per unit, has a useful life exceeding one year, and is not used for research and development.

Expense items are nondepreciable items that are consumed shortly after they are purchased. Typical expenses associated with an IT-related project include the use of outside labor or consultants, travel, and training. Software that does not meet the criteria to be classified as a capital item is classified as an expense item.

Many organizations use a system of internal cross-charges to account for the cost of employees assigned to a project. For example, the fully loaded cost (salary, benefits, and overhead) of a manager might be set at $120,000 per year. The sponsoring organization's budget is cross-charged this amount for each manager who works full-time on the project. So, if a manager works at a 75 percent level of effort on a project for 5 months, the cross-charge is $120,000 × 0.75 × 5/12 = $37,500. The rationale behind cross-charging is to enable sound economic decisions for whether employees are assigned to project work or operational activities. If employees are assigned to a project, cross-charging helps organizations determine which one makes the most economic sense.

Organizations have different processes and mechanisms for budgeting and controlling each of the three types of costs: capital, expense, and internal cross-charge. Money from the budget for one type of cost cannot be used to pay for an item associated with another type of cost. Thus, a project with lots of capital remaining in its budget cannot use the available dollars to pay for an expense item even if the expense budget is overspent.

Table 3-1 summarizes and classifies various types of common costs associated with an IT-related project.

TABLE 3-1 Typical IT-related project costs

	Development Costs		
	Capital	Internal cross-charge	Expense
Employee-related expenses			
Employees' effort		X	
Travel-related expenses			X
Training-related expenses			X
Contractor and consultant charges			X
IT-related capital and expenses			
Software licenses (greater than minimum to be classified as expense)	X		
Software licenses (less than minimum to be classified as expense)			X
Computing hardware devices	X		
Network hardware devices	X		
Data entry equipment	X		
Total development costs	X	X	X

Time

The timing of a project is frequently a critical constraint. For example, in most organizations, projects that involve finance and accounting must be scheduled to avoid any conflict with operations associated with the closing of end-of-quarter books. Projects often must be complete by a certain date to meet an important business goal or a government mandate.

A full decade after the Health Insurance Portability and Accountability Act (HIPAA) became federal law, health insurers were still struggling to meet all the requirements of this far-reaching legislation. One requirement of HIPAA mandates that all healthcare providers obtain a 10-digit National Provider Identifier (NPI) number to serve as each provider's unique, standard identifier. This system is designed to improve the efficiency and effectiveness of electronically transmitting health information. Large insurers had to begin processing documents with the NPIs by May 2007. Failure to use the NPI identifier means that patient and provider requests cannot be processed, including requests for payment of services. Conforming to this mandate required extensive changes in many systems used both by healthcare providers and insurers. For example, Blue Cross and Blue Shield of North Carolina spent more than 10,000 hours making necessary changes to its 11 internal systems that must process NPIs.[8]

Quality

The **quality** of a project can be defined as the degree to which the project meets the needs of its users. The quality of a project that delivers an IT-related system may be defined in terms of the system's functionality, features, system outputs, performance, reliability, and maintainability. For example, Avaya Inc. provides communication solutions that enable organizations to communicate effectively with their customers, suppliers, business partners, and employees. These solutions are comprised of hardware, software, and other services. The company recently launched Customer Interaction Express (CIE), new full-featured software for contact centers to serve its customers in the Middle East and North Africa. In creating this software, Avaya met numerous quality targets that were important to its potential customers. The software includes features that enable it to handle customer interactions via phone, e-mail, fax, and even text messaging. CIE shows both real-time and historical data about how customers contact the firm across all means of interaction. The software is designed to perform well for firms with 100 to 1000 employees that operate contact centers with 10 to 150 customer service agents. It also is designed to be highly reliable in support of around-the-clock customer service operations.[9] Failure of the software to meet any of these quality requirements would lessen the likelihood of its adoption by potential customers.

User Expectations

As a project begins, the stakeholders will form expectations or already have expectations about how the project will be conducted and how it will affect them. For example, based on previous project experience, the end users of a new IT system may expect that they will have no involvement with the system until it is time for them to be trained. However, the project manager may need the end users to help define system requirements, evaluate system options, develop user documentation, and define and conduct the user acceptance test.

As another example, end users may expect to be present at weekly project status meetings and expect to hear progress reports firsthand. Perhaps the project manager did not consider involving them in the status meetings or even plan to have weekly meetings, however. Both examples illustrate the huge differences in expectations that can exist between stakeholders and project members. It is critical to a project's success to identify expectations of key stakeholders and team members; if there are differences, they must be resolved to avoid future problems and misunderstandings.

The five project parameters—scope, cost, time, quality, and user expectations—are all closely interrelated, as shown in Figure 3-2. For example, if the time allowed to complete the project is decreased, it may require an increase in project costs, a reduction in project quality and scope, and a change of expectations among the project stakeholders, as shown in Figure 3-3.

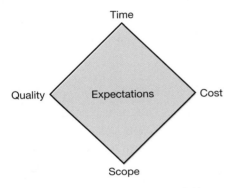

FIGURE 3-2   Original project definition

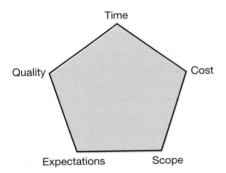

FIGURE 3-3   Revised project definition

WHAT IS PROJECT MANAGEMENT?

Project management is the application of knowledge, skills, and techniques to project activities to meet project requirements.[10] Project managers must attempt to deliver a solution that meets specific scope, cost, time, and quality goals while managing the expectations of the **project stakeholders**—the people involved in the project or those affected by its outcome.

The essence of artistic activity is that it involves high levels of creativity and freedom to do whatever the artist feels. Scientific activity, on the other hand, involves following defined routines and exacting adherence to laws. Under these definitions, part of project management can be considered an art, because project managers must apply intuitive skills that vary from project to project and even from team member to team member. "There's a certain salesmanship and a lot of psychology involved with performing the project management job well," said project manager consultant Mark Reilley. "You're selling change to end users and to bosses. You also have to be a cheerleader for the team and get to know what motivates people to perform."[11]

Project management is also part science because it uses time-proven, repeatable processes and techniques to achieve project goals. Thus, one challenge to successful project management is recognizing when to act as an artist and rely on one's own instinct, versus when to act as a scientist and apply fundamental project management principles and practices. The following section covers the nine areas associated with the science of project management.

PROJECT MANAGEMENT KNOWLEDGE AREAS

According to the Project Management Institute, project managers must coordinate nine areas of expertise, including scope, time, cost, quality, human resources, communications, risk, procurement, and integration.

Scope Management

Scope management includes defining the work that must be done as part of the project and then controlling the work to stay within the scope to which the team has agreed. Key activities include initiation, scope planning, scope definition, scope verification, and scope change control.

To avoid problems associated with a change in project scope, a formal scope change process should be defined before the project begins. The project manager and key sponsors should decide whether they will allow scope changes at any time during the project, only in the early stages of the project, or not at all. The trade-off is that the more flexibility you allow for scope changes, the more likely the project will meet user needs for features and performance. The project will be more difficult to complete within changing time and budget constraints, however. After all, it is harder to hit a moving target.

The change process should capture a clear definition of the change that is being requested, who is requesting it, and why. If the project team has decided not to allow any scope changes during the project, then each new requested scope change is filed with other requested changes. Once the original project is complete, the entire set of requested scope changes can be reviewed and the project team can decide which, if any, of the changes will be implemented and when. (Often, it makes sense to group related changes.) A follow-on project can then be considered to implement the recommended changes. The scope, cost, schedule, and benefits of the project must be determined to ensure that it is well defined and worth doing.

If the project team has decided to allow scope changes during the project, then time and effort must be allowed to assess how the scope change will affect the interrelated project variables, such as cost, schedule, quality, and expectations. This impact on the project must be weighed against the benefits of implementing the scope change, and the team must make a yes/no decision whether to implement the scope change. Of course, there may be alternatives for implementing the scope change, and the pros and cons must be weighed for each. The time required just to research scope changes can add considerable cost and additional time to the original project. Each scope change should be approved formally or rejected by the project manager and key stakeholders.

Time Management

Time management includes estimating a reasonable completion date, developing a workable project schedule, and ensuring the timely completion of the project. It requires identifying specific tasks that project team members and/or other resources must complete; sequencing these tasks, taking into account any task dependencies or hard deadlines; estimating the amount of resources required to complete each task, including people, material, and equipment; estimating the elapsed time to complete each task; analyzing all this data to create a project schedule; and controlling and managing changes to the project schedule. A project schedule is needed to complete a project by a defined deadline, avoid rework, and avoid having people who do not know what to do or when to do it.

The development of a work breakdown structure is a critical activity needed for effective time management. A **work breakdown structure (WBS)** is an outline of the work to be done to complete the project. You start by breaking the project into various stages or groups of activities that need to be performed. Then, you identify the tasks associated with each project stage. A task typically requires a week or less to complete and produces a specific deliverable—tangible output like a flowchart or end-user training plan. Then the tasks within each stage are sequenced. Finally, any predecessor tasks are identified—for example, tasks that must be completed before a later task can begin.

Next you must determine how long each task will take.

Thus, building a WBS allows you to look at the project at a high level or at a great level of detail to get a complete picture of all the work that must be performed. Development of a WBS is another approach to defining the scope of a project—work not included in the WBS is outside the scope of the project.

Table 3-2 shows a sample WBS for a project whose goal is to establish a wireless network in a warehouse and install RFID scanning equipment on forklift trucks for the tracking of inventory. The three phases of the project in Table 3-2 are "Define Warehouse Network," "Configure Forklift Trucks," and "Test Warehouse Network." Figure 3-4 shows the associated schedule in the form of a Gantt chart, with each bar in the chart indicating the start and end dates of each major activity (heavy black lines) and task (lighter lines).

TABLE 3-2 Work breakdown structure

Task		Duration	Start	End	Predecessor Tasks
1	Implement Warehouse Network	28d	5/08/09	6/14/09	
2	Define Warehouse Network	25d	5/08/09	6/09/09	
3	Conduct Survey	3d	5/08/09	5/10/09	
4	Order RF Equipment	14d	5/11/09	5/30/09	3

TABLE 3-2   Work breakdown structure (continued)

Task		Duration	Start	End	Predecessor Tasks
5	Install RF Equipment	6d	6/01/09	6/08/09	4
6	Test RF Equipment	2d	6/09/09	6/10/09	5
7	Configure Forklift Trucks	19d	5/08/09	6/01/09	
8	Order RFID Scanners for Trucks	12d	5/08/09	5/25/09	
9	Install RFID Scanners on Trucks	5d	5/26/09	6/01/09	8
10	Test RFID Scanners	2d	6/02/09	6/03/09	9
11	Test Warehouse Network	28d	5/08/09	6/14/09	
12	Develop Test Plan	2d	5/08/09	5/09/09	
13	Conduct Test	3d	6/11/09	6/15/09	6,10,12

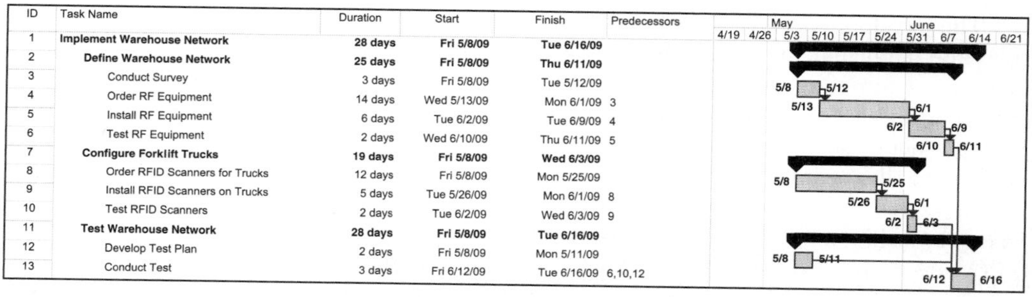

ID	Task Name	Duration	Start	Finish	Predecessors
1	**Implement Warehouse Network**	**28 days**	**Fri 5/8/09**	**Tue 6/16/09**	
2	**Define Warehouse Network**	**25 days**	**Fri 5/8/09**	**Thu 6/11/09**	
3	Conduct Survey	3 days	Fri 5/8/09	Tue 5/12/09	
4	Order RF Equipment	14 days	Wed 5/13/09	Mon 6/1/09	3
5	Install RF Equipment	6 days	Tue 6/2/09	Tue 6/9/09	4
6	Test RF Equipment	2 days	Wed 6/10/09	Thu 6/11/09	5
7	**Configure Forklift Trucks**	**19 days**	**Fri 5/8/09**	**Wed 6/3/09**	
8	Order RFID Scanners for Trucks	12 days	Fri 5/8/09	Mon 5/25/09	
9	Install RFID Scanners on Trucks	5 days	Tue 5/26/09	Mon 6/1/09	8
10	Test RFID Scanners	2 days	Tue 6/2/09	Wed 6/3/09	9
11	**Test Warehouse Network**	**28 days**	**Fri 5/8/09**	**Tue 6/16/09**	
12	Develop Test Plan	2 days	Fri 5/8/09	Mon 5/11/09	
13	Conduct Test	3 days	Fri 6/12/09	Tue 6/16/09	6,10,12

FIGURE 3-4   Gantt chart depicting the beginning and end of project tasks

Cost Management

Cost management includes developing and managing the project budget. This area involves resource planning, cost estimating, cost budgeting, and cost control. As previously discussed, a separate budget must be established for each of the three types of costs—capital, expense, and internal cross-charge—and money in one budget cannot be spent to pay for another type of cost.

One approach to cost estimating uses the WBS to estimate all costs (capital, expense, and cross-charge) associated with the completion of each task. This approach can require a fair amount of detail work, such as determining the hourly rate of each resource assigned to the task and multiplying by the hours the resource will work on the task, estimating the cost per unit for supplies and multiplying by the number of units required, and so on. If possible, the people who will complete the tasks should be allowed to estimate the time duration. This approach helps them to better understand the tasks they are expected to complete, gives them some degree of power in defining how the work will be done, and obtains their "buy-in" to the project schedule and budget. You can develop a project duration based on the sequence in which the tasks must be performed and the duration of each task. You can also sum the cost of each task to develop an estimate of the total project budget. This entire process is outlined in Figure 3-5, and the resulting budget is depicted in Table 3-3.

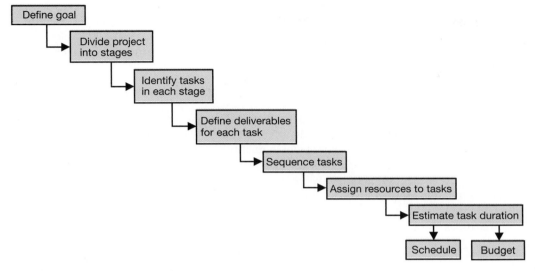

FIGURE 3-5   Development of a WBS leads to creation of a schedule and budget

TABLE 3-3   Project budget

Task		Capital	Expense	Cross-charges
1	Implement Warehouse Network			
2	Define Warehouse Network			
3	Conduct Survey		$2,400	
4	Order RF Equipment	$9,000		
5	Install RF Equipment		$7,800	
6	Test RF Equipment			$ 960
7	Configure Forklift Trucks			
8	Order RFID Scanners for Trucks	$12,500		
9	Install RFID Scanners on Trucks			$2,400
10	Test RFID Scanners			$1,200
11	Test Warehouse Network			$ 960
12	Develop Test Plan			
13	Conduct Test			$1,440
	TOTAL Costs	$21,500	$10,200	$6,960

As an example, suppose that a company plans to implement a new software package for its accounts payable process. The company must spend $15,000 on computer hardware (capital) and pay the software vendor $20,000 for its time and effort to implement the software (expense). The vendor also must be paid $25,000 for the software package license (capital). In addition, one business manager will spend six months full time leading the effort. Six months worth of the fully loaded cost of the manager (say, $120,000 per year) must be charged to the cross-charge budget of the accounting organization. The cross-charge is a total of $60,000.

Quality Management

Quality management ensures that the project will meet the needs for which it was undertaken. This process involves quality planning, quality assurance, and quality control. **Quality planning** involves determining which quality standards are relevant to the project and determining how they will be met. **Quality assurance** involves evaluating the progress of the project on an ongoing basis to ensure that it meets the identified quality standards. **Quality control** involves checking project results to ensure that they meet identified quality standards.

When it comes to developing IT-related systems, "defects in requirements are the source of the majority of defects that are identified during testing, and problems with requirements are among the top causes of project failure,"[12] according to a recent report by Forrester Research, a technology and market research company that provides advice to global leaders in business and technology. Thus, most organizations put a heavy emphasis on accurately capturing and documenting what the business needs from the system and managing changes in user requirements during the project. Software is often employed, including everything from general-purpose word-processing software to specially designed requirements management software from vendors such as Borland Software, Compuware, IBM Relational, Serena Software, and Teleogic AB.[13]

Human Resource Management

Human resource management is about making the most effective use of the people involved with the project. It includes organizational planning, staff acquisition, and team development. The project manager must be able to build a project team staffed with people with the right mix of skills and experience, and then train, develop, coach, and motivate them to perform effectively on the project.

The project manager may be assigned all members of the team, or may have the luxury of selecting all or some team members. Team members should be selected based on their skills in the technology needed for the project, their understanding of the business area affected by the project, their expertise in a specific area of the project, and their ability to work well on a team. Often, compromises must be made. For example, the best available subject-matter expert may not work well with others, which becomes an additional challenge for the project manager.

Experienced project managers have learned that forming an effective team to accomplish a difficult goal is quite a challenge in itself. For the team to reach high levels of performance, it takes considerable effort and a willingness to change on the part of the entire team.

A useful model to describe how teams develop and evolve is the **forming-storming-norming-performing model**, which was first proposed by Bruce Tuckman (see Figure 3-6).[14]

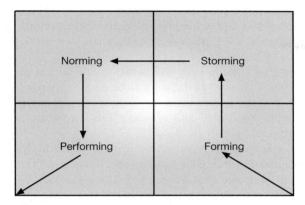

FIGURE 3-6   Tuckman's forming-storming-norming-performing model

During the forming stage, the team meets to learn about the project, agrees on basic goals, and begins to work on project tasks. Team members are on their best behavior and try to be pleasant to one another while avoiding any conflict or disagreement. Team members work independently of one another and focus on their role or tasks without understanding what others are attempting to do. The team's project manager in the formation stage tends to be highly directive and tells members what needs to be done. If the team remains in this stage, it is unlikely to perform well and it will never develop breakthrough solutions to problems or effectively solve a conflicting set of priorities and constraints.

The team has moved into the storming stage when it recognizes that differences of opinion exist between team members and allows these ideas to compete for consideration. Team members will raise such important questions as: What problems are we *really* supposed to solve? How can we work well together? What sort of project leadership will we accept? The team might argue and struggle, so it can be an unpleasant time for everyone. An inexperienced project manager, not recognizing what is happening, may give up, feeling that the team will never work together effectively. The project manager and team members must be tolerant of one another as they explore their differences. The project manager may need to continue to be highly directive.

If the team survives the storming stage, it may enter the norming stage. During this stage, individual team members give up their preconceived judgments and opinions. Members who felt a need to take control of the team give up this impulse. Team members adjust their behavior toward one another and begin to trust one another. The team may decide to document a set of team rules or norms to guide how they will work together. Teamwork actually begins. The project manager can be less directive and can expect team members to take more responsibility for decision making.

Some teams advance beyond the norming stage into the performing stage. At this point, the team is performing at a high level. Team members are competent, highly motivated, and knowledgeable about all aspects of the project. They have become interdependent on one another and have developed an effective decision-making process that does not require the project manager. Dissent is expected, and the team has developed an effective process to ensure that everyone's ideas and opinions are heard. Work is done quickly and with high quality. Problems that once seemed unsolvable now have "obvious" solutions. The team's effectiveness is much more than the sum of the individual members' contributions. The project manager encourages participative decision making, with the team members making most of the decisions.

No matter what stage a team is operating in, it commonly will revert to less advanced stages in the model when confronted with major changes in the work to be done, a change in project leadership, or substantial changes in the team's make-up. Senior management, the project champion, and the sponsor must recognize and consider this important dynamic when contemplating project changes.

Another key aspect of human resource management is getting the project team and the sponsoring organization to take *equal* responsibility for making the project a success. The project team members must realize that on their own they cannot possibly make the project a success. They must ensure that the sponsoring organization becomes deeply involved in the project and takes an active role. The project team must actively involve the end users, provide information for them to make wise choices, and insist on their participation in major decisions. The sponsoring organization must remain engaged in the project, challenge recommendations, ask questions, and weigh options. It cannot simply sit back and "let the project happen to them." Key users need to be identified as part of the project team with responsibility for developing and reviewing deliverables. Indeed, some organizations require that the project manager come from the sponsoring organization. Other organizations assign co-project managers to IT-related projects—one from the IT organization and one from the sponsoring organization.

Communications Management

Communications management involves the generation, collection, dissemination, and storage of project information in a timely and effective manner. It includes communications planning, information distribution, performance reporting, and managing communications to meet the needs of project shareholders. The key stakeholders include the following people:

- A project **champion** is a senior-level executive who is a strong advocate for the project. A champion can convince other senior managers of the project's merits to gain their approval to fund and staff it. Every major project needs a champion to advocate its benefits, assist and mentor the team, and remove any roadblocks to keep the project on track.
- The project **sponsor** is a senior manager in the organization that will be most affected by the project's implementation. The budget for the project will be deducted from the sponsor's organization. The project sponsor will "own" the project and must live with its consequences. The project sponsor must feel strongly that the project is right to do and that his/her organization is capable of using the results. The project champion and sponsor might be the same person.

- The **end users** are the people most directly affected by the project. To complete their work, they probably will have to learn new work processes and tools created by the project.
- The project team members devote their time and effort to complete the project successfully.

In preparing a communications plan, the project manager should recognize that the various stakeholders have different information needs in the project. A useful tool for identifying and documenting these needs is the stakeholder analysis matrix shown in Figure 3-7. This matrix identifies the interests of the stakeholders, their information needs, and important facts for managing communications with the champion, sponsor, project team members, and key end users who are associated with the project. The project manager should include his or her manager in this analysis. Based on analysis of this data, the preferred form and frequency of communication is identified for each stakeholder.

Stakeholder Analysis

Key Stakeholders	Ray Boaz	Klem Kiddlehopper	John Smith	Forklift Drivers
Organization	Project champion and VP of Supply Chain	Project sponsor and warehouse manager	Experienced forklift driver	15, all different
Useful Facts	Very persuasive Trusted by CEO	Risk taker, very aggressive, will push this through, no matter what	Has driven forklift truck for 5 years Well respected by peers	Not the sharpest knives in the drawer Not highly motivated to make project a success
Level of interest	High	High	Medium	Low
Level of influence	High	Medium	High	Low
Suggestions on managing relationship	Demands respect, somewhat formal Speak in business terms, never get technical; no surprises!	Poor listener, forgets details, put it in writing	Must keep John enthusiastic about project	Don't ignore Attend occasional shift change-over meeting
Information Needs	ROI Budget Schedule	Schedule and potential operational conflicts	Schedule, especially timing of training, safety, and productivity issues	Schedule, especially timing of training, safety issues
Information medium, format and timing	Bi-weekly face-to-face meeting	Weekly e-mail Newsletter Bi-weekly face-to-face	Newsletter Catch-as-catch can	Won't read newsletter, provide brief update at weekly safety meeting

FIGURE 3-7 Stakeholder analysis matrix

If the project team is unable to recruit both a project champion and sponsor, the problem may be that management does not see clearly that the benefits of the project outweigh its costs, or that the project appears to run counter to organizational goals and strategies. A potential project without both a champion and a sponsor is highly unlikely to get the needed resources, and for good reason. No project should be started without both a champion and a sponsor.

Risk Management

"Things will go wrong, and at the worst possible time," according to a variation of Murphy's Law, a popular adage. "**Project risk** is an uncertain event or condition that, if it occurs, has a positive or a negative effect on a project objective."[15] Known risks are risks that can be identified and analyzed. For example, in creating a new IT-related system that includes the acquisition of new computing and/or networking hardware, a known risk might be that the hardware will take longer than expected to arrive at the installation site. If the hardware is delayed by several weeks, it could have a negative effect on the project completion date. Countermeasures can be defined to avoid some known risks entirely, and contingency plans can be developed to address unavoidable known risks if they occur. Of course, there are also unknown risks that nobody can anticipate.

A hallmark of experienced project managers is that they follow a deliberate and systematic process of **risk management** to identify, analyze, and manage project risks. Having identified potential risks, they can make plans to avoid them entirely. When an unavoidable risk occurs, they already have defined an alternative course of action to minimize the impact on the project. They waste no time executing the backup plan. Unknown risks cannot be managed directly, however; an experienced project manager will build some contingency into the project cost and schedule estimate to allow for their occurrence.

While inexperienced project managers realize that things may go wrong, they fail to identify and address known risks and do not build in contingencies for unknown risks. Thus, they are often unsure of what to do, at least temporarily, when there is a project setback. In their haste to react to a risk, they may not implement the best course of action.

The project manager needs to lead a rigorous effort to identify all risks associated with the project. The project team, stakeholders, and other experienced risk experts should participate in the effort. These risk experts can include seasoned project managers and members of the organization's risk management department. After each risk is identified and defined, as shown in Table 3-4, the group should attempt to classify the risk by the probability that it will occur and the impact on the project if the risk does occur. Both the probability and the impact can be classified as high, medium, or low, as shown in Table 3-5.

TABLE 3-4 Identification of project risks

Risk	Example
R1	The required new servers arrive at the installation site more than two weeks late.
R2	Business pressures make key end users unavailable to develop the user acceptance test by the date it is needed.
R3	Business pressures make end users unavailable during time scheduled for training.
R4	One or more end-user computers have insufficient memory or CPU capacity to run the new software efficiently (or at all).
Rn	....

TABLE 3-5 Assessment of project risks

		Impact on Project		
		Low	Medium	High
Probability Risk Occurs	High	R10		R2, R3
	Medium	R5, R6	R*n*	R1
	Low	R8, R11	R7, R9	R4

Dark = High risk/high impact; risk management plan is needed
Lightest = Medium or high risk and impact; risk management plan recommended
Lighter = Low or medium risk and impact; risk management plan not needed

The project team then needs to consider which risks need to be addressed with some sort of risk management plan. Generally, the team can ignore risks with a low probability of occurrence and low potential impact. Risks with a high probability of occurrence and a high potential impact need to have a risk owner assigned. The **risk owner** is responsible for developing a risk management strategy and monitoring the project to determine if the risk is about to occur or has occurred. One strategy is to avoid the risk altogether, while another is to develop a backup plan. The risk management plan can be documented as shown in Table 3-6.

TABLE 3-6 Risk management plan

Risk	Description	Risk Owner	Risk Strategy	Current Status
R2	Business pressures make key end users unavailable to develop the user acceptance test by the deadline.	Jon Andersen, manager of end users in the business area	Try to avoid this problem by starting development of the user acceptance test three weeks earlier than originally planned. Monitor progress carefully.	Key users have been identified and have started developing the test.
R3	Business pressures make end users unavailable during the time scheduled for training.	Jon Andersen, manager of end users in the business area	Try to avoid this problem by hiring and training four temporary workers to fill in for end users as they participate in training.	Three of four temporary workers have been hired. Their training is scheduled to begin next week.
R1	The required new servers arrive at the installation site more than two weeks late.	Alice Fields, team member responsible for hardware acquisition	Set a firm delivery deadline with the vendor, with a substantial dollar penalty for each day that the equipment is late.	The contract with the penalty clause has been signed by the vendor, who agrees to provide a shipment status update each Tuesday and Friday.

One of the biggest risks associated with a project is that considerable time, energy, and resources might be consumed with little value to show in return. To avoid this potential risk, an organization must ensure that a strong rationale exists for doing the project. The project must have a direct link to an organizational strategy and goal, as shown in Figure 3-8. For example, assume that an organization has been losing sales because of customer dissatisfaction. As a result, it has set a goal of increasing the retention rate of existing customers and defined one of its key strategies as improving customer service to world-class levels. Furthermore, the organization has set specific related measures that it wants to achieve, as shown next. A project that can deliver results to achieve these measures is then directly linked to the organizational strategy and goal.

- *Goal*—Increase the retention rate of existing customers.
- *Strategy*—Improve customer service to world-class levels.
- *Measure*—Reduce customer turnover from 25 percent per year to 10 percent by June 2010 by responding to 95 percent of customers' inquiries within 90 seconds, with less than 5 percent callbacks about the same problem.
- *Project*—Implement a state-of-the-art customer call center with "24 × 7" availability and a well-trained staff.

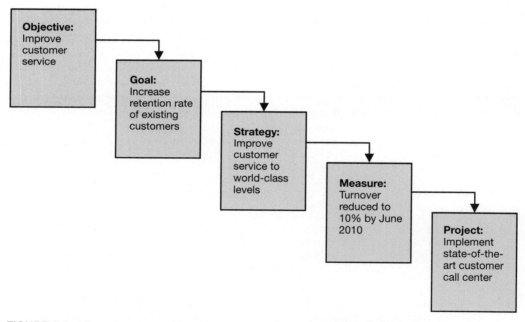

FIGURE 3-8   Example of a project that is well linked to an organizational strategy and goal

Procurement Management

Procurement management involves acquiring goods and/or services for the project from sources outside the performing organization. This activity is divided into the following processes:

- *Plan purchase and acquisition*—This process determines what is needed and when.
- *Plan contracting*—This process documents requirements for products and services and identifies potential providers.
- *Request seller responses*—This process obtains bids, information, proposals, or quotations from potential providers.
- *Select seller*—During this process, offers are reviewed, the preferred provider is identified, and negotiations are started.
- *Contract administration*—This process manages all aspects of the contract and the relationship between the buyer and the provider. The process includes tracking and documenting the provider's performance, managing contract changes, and taking any necessary corrective actions.
- *Contract closure*—This process completes and settles terms of any contracts, including resolving any open items.

The make-or-buy decision is a key decision made during the plan purchase and acquisition process. The **make-or-buy decision** involves comparing the pros and cons of in-house production versus outsourcing of a given product or service. In addition to cost, two key factors to consider in this decision are: 1) Do we have a sufficient number of employees with the skills and experience required to deliver the product or service at an acceptable level of quality and within the required deadlines? 2) Are we willing to invest the management time, energy, and money required to identify, recruit, train, develop, and manage people with the skills to do this kind of work? Outsourcing is discussed further in Chapter 4.

A contract is a legally binding agreement that defines the terms and conditions of the buyer-provider relationship, including who is authorized to do what, who holds what responsibilities, costs and terms of payment, remedies in case of breach of contract, and the process for revising the contract. Three basic types of contracts exist:

- **Fixed-price** or **lump-sum contracts**—With this type of contract, the buyer and provider agree to a total fixed price for a well-defined product or service. For example, the purchase of a large number of laptop computers with specified capabilities and features frequently involves a fixed-price contract.
- **Cost-reimbursable contracts**—This type of contract requires paying the provider an amount that covers the provider's actual costs plus an additional amount or percentage for profit. Three common types of cost-reimbursable contracts exist. In a cost-plus-fee or cost-plus-percentage of cost contract, the provider is reimbursed for all allowable costs and receives a percentage of the costs as a fee. In a cost-plus-fixed-fee contract, the provider is reimbursed for all allowable costs and receives a fixed fee. In a cost-plus-incentive-fee contract, the provider is reimbursed for all allowable costs. In addition, a predetermined fee is paid if the provider achieves specified performance objectives—for example, the provider's hardware must be received, installed,

and operational by a specific date. In such contracts, buyers run the risk of paying more for the work but are rewarded by having their objectives met or exceeded. Providers run the risk of reduced profits if they fail to deliver, but can be rewarded for superior performance.

- **Time and material contracts**—Under this type of contract, the buyer pays the provider for both the time and materials required to complete the contract. The contract includes an agreed-upon hourly rate and unit price for the various materials to be used. The exact number of hours and precise quantity of each material are not known, however. Thus, the true value of the contract is not defined when the contract is approved. If not managed carefully, time and material contracts actually can motivate suppliers to extend projects to maximize their fees.

Poor procurement management can result in serious project problems and even a project's outright cancellation, as discussed in the following special feature.

A MANAGER (FAILS TO) TAKE ACTION

The Department of Veterans Affairs

The Department of Veterans Affairs (VA) was established in 1989. It succeeded the Veterans Administration, and has responsibility for providing federal benefits to veterans and their dependents. It is the second-largest of 14 U.S. Cabinet departments, with more than 25 million active veterans and roughly 70 million people eligible for VA benefits.

In 2002, the VA awarded an 11-year, $103 million contract for the creation and operation of a security incident response center that would become "the nucleus of all VA information and Internet security operations nationwide, providing continuous protection, detection and response capabilities against threats, remotely exploitable vulnerabilities and real-time incidents on all VA-affiliated networks."[16]

As a result of mismanagement and errors in the planning, award, and administration of the contract, the VA wasted $92 million in just three years before the project was killed for lack of available funding. An audit by the VA Inspector General reported the following key findings:

- The VA selection process to identify a qualified provider for the project was faulty, with little due diligence put into assessing provider qualifications or ensuring that the quoted prices were reasonable. Just one week before the contract was awarded to a firm called the Veterans Affairs Security Team, the company incorporated as an LLC (limited liability corporation), which means that the owners were protected from liability for acts and debts of the firm.

continued

- A major scope change was made to add a Managed Security Services (MSS) component to address the "acquisition, installation, integration, configuration, and monitoring of VA's enterprise infrastructure; vulnerability assessment and penetration testing; cyber security intelligence gathering and support of network operations; and support of the Enterprise Cyber Security Infrastructure Project."[17] This scope change was managed poorly and led to significant problems. There is no indication that the VA even considered separating the two efforts to allow for improved project management and control.
- The work to add the MSS component was not done under the terms of the original fixed-price contract. Instead, the new work was governed by an "indefinite delivery, indefinite quantity" contract that essentially opened the floodgates to uncontrolled spending and implemented features and capabilities with little regard to cost and budget.
- Vagueness in the modifications to the original contract contributed to the VA making overpayments and duplicate payments of more than $8 million. "The VA also spent nearly $35 million on equipment and supplies, but has no record of what the equipment is or where it may be."[18]

Discussion Questions:

1. What recommendations should the VA Inspector General make to recover the losses from this project and to avoid repeating these mistakes during future VA projects?
2. How might the need to create an MSS have been handled differently to avoid the problems that arose here?
3. What processes and tools might the auditor have used to gather the data in the audit report?
4. What knowledge did you gain from this example that will help you if you take responsibility for procurement management of a major project?

Project Integration Management

Project integration management is perhaps the most important knowledge area because it requires the assimilation of all eight other project management knowledge areas. Project integration management requires the coordination of all appropriate people, resources, plans, and efforts to complete a project successfully. Project integration management comprises seven project management processes:

1. Developing the project charter that formally recognizes the existence of the project, outlines the project objectives and how they will be met, lists key assumptions, and identifies major roles and responsibilities.
2. Developing a preliminary project scope statement to define and gain consensus about the work to be done. Over the life of the project, the scope statement will become fuller and more detailed.

3. Developing the project management plan that describes the overall scope, schedule, and budget for the project. This plan coordinates all subsequent project planning efforts and is used in the execution and control of the project.

4. Directing and managing project execution by following the project management plan.

5. Monitoring and controlling the project work to meet the project's performance objectives. This process requires regularly measuring effort and expenditures against the project tasks, recognizing when significant deviations occur from the schedule or budget, and taking corrective action to regain alignment with the plan.

6. Performing integrated change control by managing changes over the course of the project that can affect its scope, schedule, and/or cost.

7. Closing the project successfully by gaining stakeholder and customer acceptance of the final product, closing all budgets and purchase orders after confirming that final disbursements have been made, and capturing knowledge from the project that may prove useful for future projects.

As an example of a firm that excels in project integration management, consider Atos Origin, an international IT services company that employs 47,000 workers in more than 50 countries and has annual revenues over $6.5 billion.[19] The firm was selected by the International Olympic Committee as the worldwide IT partner of the 2008 Beijing Olympics. It has the primary responsibility of project integration, consulting, systems integration, operations management, information security, and software applications development for the games. Based on its experience with previous Olympics (Salt Lake City in 2002, Athens in 2004, and Turin in 2006), Atos Origin has developed an effective project management process broken into five stages: planning, designing, building, testing, and operating. Work started in October 2004 with the planning and designing phases of the comprehensive sports IT project. The system building stage of the project began in early 2006.[20]

The biggest challenge related to the Olympics is its inflexible deadline—absolutely everything must be ready and working by 8 p.m. on August 8, 2008, when the Beijing Summer Olympics begins. (Eight is a lucky number in China.) Other challenges include the sheer size and complexity of the effort. Atos Origin must coordinate the work of hundreds of subcontractors to deliver a highly reliable IT infrastructure and IT services in support of the world's most widely viewed sporting event. Custom software, thousands of workstations and laptops, tens of thousands of phones, hundreds of servers, and multiple operations centers and data centers must all operate together flawlessly.[21]

The manager's checklist in Table 3-7 provides a set of recommended actions for business managers to improve the success rate of their organization's projects. The appropriate answer to each question is yes.

TABLE 3-7 Manager's checklist

Actions	Yes	No
Are project scope, cost, time, quality, and user expectations treated as highly interrelated variables—in other words, changing one affects the others?		
Is an internal cross-charge system used to account for the cost of employees assigned to a project?		
Is the project scope well defined and managed?		
Is a detailed work breakdown structure prepared to define the project schedule and cost?		
Is the project's estimated cost well defined and controlled?		
Is the project team performing quality planning, quality assurance, and quality control?		
Is there a heavy emphasis on defining user requirements?		
Does the project manager take action to form and maintain an effective working team?		
Do the project team and sponsoring organization take equal responsibility for the success of the project?		
Have a project champion and project sponsor been identified for the project?		
Has a communications plan for all key stakeholders been defined for the project?		
Has the project manager followed a deliberate and systematic process of risk management to identify, analyze, and manage project risks?		
Is it clear that a strong rationale exists for doing the project? Does the project have a direct link to an organizational strategy and goal?		
Is a process in place to manage project procurement?		
Has responsibility for the seven project integration management processes been defined?		
Are the seven project integration management processes being performed?		

Chapter Summary

- The success rate of all IT projects is only about 34 percent. In addition, 51 percent of all IT projects are "challenged" and about 15 percent are failures.
- Today, many organizations have recognized project management as one of their core competencies.
- Five highly interrelated parameters define a project—scope, cost, time, quality, and user expectations. If any one of these project parameters is changed, there must be a corresponding change in one or more of the other parameters.
- Project management is the application of knowledge, skills, and techniques to project activities to meet project requirements. Project managers must attempt to deliver a solution that meets specific scope, cost, time, and quality goals while managing the expectations of the project stakeholders—the people involved in the project or those affected by its outcome.
- According to the Project Management Institute, project managers must coordinate nine areas of expertise, including scope, time, cost, quality, human resources, communications, risk, procurement, and integration.
- The forming-storming-norming-performing model describes how teams form and evolve.
- Experienced project managers follow a deliberate and systematic process of risk management to avoid risks or minimize their negative impact on a project.
- Three main types of contracts define the terms and conditions of the buyer-provider relationship: fixed-prices, cost-reimbursable, and time and materials.
- Project integration management is a critical knowledge area of project management that involves chartering, scoping, planning, executing, monitoring and controlling, change control, and project closing.

Discussion Questions

1. Why do you think that the success rate of IT projects is so low?
2. What is meant by the scope of a project? How can the scope of a project be defined?
3. Can the scope of a project be changed without affecting cost or schedule? Explain fully.
4. Do you think that organizations should try to develop and use a system of internal cross-charging? Why or why not?
5. What is the difference between quality assurance and quality control?
6. Describe the ideal project champion and project sponsor.
7. Is there a difference between project time management and personal time management? Can someone be "good" at one but not the other? Explain your answer.
8. Discuss the "team dynamics" for a highly effective (or ineffective) team of which you were a member. Can you explain why the team performed so well (or poorly) using the forming-storming-norming-performing model?
9. What sort of behaviors would indicate that the sponsoring team is not fully engaged in a

project and instead is looking to the project team to make the project a success? What is the danger with this attitude?

10. Identify some of the challenges of performing project integration management on a project in which team members are distributed globally and cannot physically meet in one location. How might these challenges be overcome?

11. Imagine that you are hiring a firm to complete a large but undetermined amount of project work for your firm. Which form of contract would you prefer and why?

12. How would you respond to a project team member who feels that risk management is a waste of time because the future cannot be predicted? Instead, this person prefers to react to problems as they occur.

Action Memos

1. You are a senior manager of your firm and have served as a project champion on many projects in the past. You have just received a voice mail from a project manager of a current project you are championing: "I thought about developing a communications plan for the project as you suggested. It seems the easiest way to keep everyone informed is through the use of a monthly newsletter. I will publish the first one in about four weeks and distribute it via e-mail to you and other senior managers within the firm." How would you respond to this new project manager?

2. You and a small group of managers from the sponsoring organization have just completed defining the scope, schedule, and cost for an important project in your firm. You estimate that the project will take 12 people about 10 months and cost just over $2 million. You just received an e-mail from your manager insisting that the project schedule be shortened by three months because senior management is impatient with the improvements this project will deliver. He promises to "free up" four additional resources to be assigned to your project in the next month or so. How do you respond?

Web-Based Case

Of the nine project management knowledge areas, choose the one in which you are most interested. Do research on the Web to find one technique or tool related to the chosen knowledge area. Prepare a brief summary report that describes how you would use the technique or tool to help manage a project.

Case Study

The FBI Stumbles Developing a Virtual Case File System

In March 2005, the FBI shelved a $170 million software development project designed to improve the efficiency and effectiveness of its investigations. The software, called Virtual Case File (VCF), was to serve three functions: manage investigative records, share and analyze information, and provide electronic approval for the flow of paperwork. While the FBI already had taken action to replace its cumbersome and obsolete Automated Case Support (ACS) system with VCF by 2000, the September 11 terrorist attacks exposed the FBI's inability to "connect the dots" and focused public attention on the urgent need to overhaul its antiquated information sharing technology.

Four years in the making, VCF was one-third of a larger IT modernization initiative known as Trilogy. Its other two components consisted of a hardware and software upgrade and the construction of secure LAN and WAN networks. In May 2001, the FBI outsourced these two components of Trilogy to DynCorp, which finally completed the project in April 2004, 22 months late and $138 million over budget. As a result, the FBI did not have the technological infrastructure in place to deploy VCF until 2004.

The Department of Justice had deemed the project too large to be handled by one contractor, so in June 2001, the FBI outsourced the development of VCF to Science Applications International Corporation (SAIC), a San Diego-based science and technology firm that primarily services U.S. government agencies. In December 2003, after scrambling to meet its deadline for the project, SAIC delivered 700,000 lines of unworkable code. The FBI rejected it upon delivery.

Over the next few months, the FBI assembled a list of nearly 400 problems. When SAIC announced that it would fix the problems for an additional cost of $56 million, the FBI refused the offer. Instead, in June 2004, the agency began testing a scaled-down version of the product in its New Orleans field office. The testing checked the VCF's ability to provide electronic approval for documents uploaded into the old ACS system. The FBI also hired the Aerospace Corporation for an additional $2 million to review the project. By early 2005, the agency realized that nothing was salvageable and accepted Aerospace's recommendations to scrap the whole project.

"There is a long history of failures in large software projects, especially when you're converting an existing system," said Steve Bellovin, Professor of Computer Science at Columbia University. "This is almost a textbook example of how to not do it."[22]

In answering the question "Where do we go from here?" in his testimony before Congress, FBI Director Robert Mueller III stated that "we will take with us a number of valuable 'lessons learned.'"[23] The Aerospace report, as well as several earlier audits and studies, pointed out exactly what these lessons should be and where the FBI and SAIC went wrong.

Ever-Expanding Scope, Time Crunch, and Spiraling Costs

In June 2001, when the FBI awarded the contract to SAIC, the purpose of the project was to update and consolidate the ACS and four other important FBI investigative applications. These updates were intended to allow agents to access these five integrated applications through the Internet. Following September 11, Mueller recognized that the original scope of the project would not allow the FBI to fulfill new objectives under its counterterrorism mandate. The FBI, an organization that collected information after a crime, now had to catch terrorists before the crime. In December 2001, Mueller asked SAIC to abandon the development of the Web interface and begin work on a new case management system from scratch. The FBI decided not to incorporate off-the-shelf products, which not only lengthened production time but magnified the number of bugs in the final product.

In addition, following September 11, the FBI was under the gun to produce the new case management system as quickly as possible. The FBI abandoned the project's three-year time frame and moved up final deadlines. The deployment of VCF was divided into two phases, with deadlines set for December 2003 and June 2004. With six months of development time already lost to the Web interface, SAIC had only 22 months to develop a much more extensive product.

The target dates of the Trilogy project's other two components were pushed up as well. The overall costs of the project spiraled from $379.8 million to $581.1 million.

91

Trilogy Costs Based on Glen A. Fine's statement before the Senate Committee on Appropriations	
2000	$379.8 million
Early 2002	$458 million
Late 2002	$568.7 million
2003	$581.1 million

In addition, SAIC estimates that it put $3.9 million of its own money into the project.

No Blueprint, Changing Specifications, and Micromanagement

In March 2002, Sherry Higgins, FBI's new director of the newly created Office of Program Management, asked Special Agent Larry Depew to become the VCF project manager. Depew was a technophile who had programmed his own case management database to handle an investigation in the early 1990s. Yet, he had no experience in IT project management. He had been one of a team of seven agents to evaluate SAIC's original Web interface, and had recommended sacking it.

Depew organized a team of FBI agents to work with SAIC engineers and specify requirements for the VCF. For six months, Depew's team of FBI subject experts met with SAIC engineers to define and redefine user needs through a software development process called Joint Application Development (JAD). But the JAD approach went horribly awry as FBI agents overstepped their boundaries, dictating details that should have been decided by experienced engineers. For example, agents went as far as proposing a design for a portion of the interface. Depew, who led the sessions, would decide what was inside or outside the scope of the project. The result was an unwieldy, 800-page system requirement document.

Unfortunately, the FBI's attempts to define and redefine system requirements did not stop there. As SAIC delivered parts of the product, FBI's team of agents demanded more design changes in a "we-will-know-it-when-we-see-it" approach. SAIC would then retroactively make these changes to related parts of the software. Inevitably, this process led to inconsistent implementation of the altered system requirements.

"This cycle was repeated over and over again and prevented SAIC from defining system acceptance criteria and suitable test standards," said SAIC Executive Vice President and General Manager Arnold Punaro.[24]

Due to the time pressure created by the September 11 attacks, SAIC agreed to go forward without an enterprise architecture—a comprehensive blueprint that describes how the current and future structure of an organization, its IT systems, and its processes align with strategic goals. The decision turned the project into a high-risk venture.

"Here is where SAIC made honest mistakes," admitted Punaro. "We should have made known that this approach was too ambitious."[25]

Contractor Responsibilities

The defining and changing of systems requirements was not the only area in which SAIC failed to insist that the FBI follow sound software engineering practices. SAIC let the FBI's inexperienced IT team make several risky decisions, the worst of which was the intended "flash cutover." Rather than implement a phased migration in which parts of the new system could be tested and repaired until the whole system was deemed fully functional, the FBI planned to switch from their old ACS system to VCF overnight, with no overlapping period.

University of Pennsylvania Computer Science Professor Matt Blaze was on the National Research Council (NRC) committee that reviewed the VCF project in 2004. He said the NRC committee was horrified when it learned of the planned flash cutover. "I remember thinking," said Blaze, "that that would be a very good day for a crime spree."[26]

Former SAIC vice president David Kay said, "SAIC was at fault because of the usual contractor reluctance to tell the customer, 'You're screwed up. You don't know what you're doing.'"[27]

Yet, SAIC's mistakes may have gone beyond this problem. In response to the tighter deadlines and the demanded changes, SAIC brought more employees into the project, including new hires, and adopted a risky parallel development approach. SAIC personnel working on VCF increased to 250, a possible violation of what is known as Brook's Law. This popular axiom states that "Adding manpower to a late software project makes it later." The reasoning behind the law is that the time spent communicating to new, inexperienced team members outweighs the time it would have taken to complete the tasks at hand. In addition, to decrease development time, SAIC divided its staff into eight development teams working in parallel on different parts of the software. These parts would be integrated in the final phase of development. As later project reviews noted, this approach resulted in a failure to apply coding standards uniformly across the product.

When the original contract was signed, the Trilogy project was a relatively minor endeavor for the SAIC. Established in 1969, SAIC today is one of the largest science and technology government contractors. For the FBI alone, SAIC had developed CODIS, a national DNA database used by law enforcement agencies around the world, and NICS, the national criminal background check system. The IT revolution of the 1990s and the September 11 attacks created a boom in the industry. SAIC's business, income, and staff expanded rapidly just before and after SAIC was awarded the contract, as shown in the following table.

Year	SAIC Revenue	SAIC Employees
1970	$243,000	20
1975	$31,862,000	955
1980	$150,894,000	3,300
1985	$420,297,000	5,709
1990	$1 billion	11,449
1995	$1.9 billion	17,853
2000	$5.5 billion	39,000
2005	$7.2 billion	42,000

FBI's CIO Change-Over and Decentralized IT Structure

In reviewing the FBI's failure with VCF, FBI Director Robert Mueller admitted, "We lacked skill sets in our personnel such as qualified software engineering, program management, and contract management. We also experienced a high turnover in Trilogy program managers and chief information officers."[28] During the three years of Trilogy's development, the FBI had five different CIOs, as shown in the following table. The high turnover rate created an additional obstacle to defining system requirements and setting goals.

CIO	Term
Bob Dies	November 2001–May 2002
Mark Tanner	May 2002–July 2002
Darwin John	July 2002–May 2003
Wilson Lowery Jr.	May 2003–December 2003
Zalmai Azmi	December 2003–present

In addition, the FBI's IT management was so decentralized that when Zalmai Azmi arrived, the FBI did not have a centralized budget for IT management. As CIO, Azmi managed a whopping budget of $5,800. Azmi instituted measures to centralize IT decision making while finally getting the organization to produce an enterprise architecture. Azmi has also focused on boosting the skill level of IT personnel.

In March 2006, after wasting over $100 million and five years, the FBI awarded Lockheed Martin a $305 million contract to create a new case management system. The project is under intense scrutiny as Congress, IT professionals, and the media wonder whether Azmi's changes will be enough to ensure success. Some lessons have been learned. The project is to be deployed in four phases over six years, and the system will rely in part on commercial, off-the-shelf software. Yet the question remains: what other lesson does the FBI need to learn before it can obtain the IT system needed to "connect the dots" and possibly prevent another September 11?

Discussion Questions

1. Which of the actions recommended in Table 3-7 were not performed on this project?

2. What decisions did the FBI and SAIC make during development that increased risk? How can these be avoided in the next project that the FBI takes on?

3. What changes should the FBI make to its IT organization to improve its ability to handle large IT projects?

4. FBI CIO Zalmai Azmi has focused on building IT project management skills within the agency. Should the FBI consider outsourcing IT project management instead?

Endnotes

1 Russ Banham, "Monumental Challenge," *CFO Magazine*, Volume 22, no. 2, pages 58–62, February 2006.

2 "mySAP ERP Financials at Brown-Forman," SAP Solutions at *www.sap.com* (accessed 25 March 2007).

3 Dr. Malcolm Wheatley, "The Importance of Project Management," for Project Smart, *http://projectmanagement.ittoolbox.com/documents/*, 4 April 2005.

4 "Standish: Project Success Rates Improved over 10 Years," *Software Magazine*, 15 January 2004.

5 Gary Hamel and C.K. Prahalad, "The Core Competence of the Corporation," *Harvard Business Review*, Volume 68, no. 3, pages 79–93, May–June 1990.

6 Project Management Institute, Inc., *A Guide to the Project Management Body of Knowledge (PMBOK Guide)*, page 5, 2004.

7 "Big Dig," Wikipedia (accessed 27 March 2007).

8 Heather Haverstein, "Federal ID Mandate Tests Health Insurers," *Computerworld*, 17 July 2006.

9 "Avaya Unveils Call Center Software for Customers in the Middle East," *CRM News*, 29 March 2007.

10 Project Management Institute, Inc., *A Guide to the Project Management Body of Knowledge (PMBOK Guide)*, page 6, 2004.

11 Mark Reilley quoted by Julia King, "Project Management in the Trenches," *Computerworld*, 14 February 2004.

12 Kym Gilhooly, "Lost in Translation: How Requirements Management Tools Can Bridge the Communications Gap," *Computerworld*, 6 November 2006.

13 Ibid.

14 Bruce Tuckman, "Developmental Sequence in Small Groups," *Psychological Bulletin*, Volume 63, pages 384–389, 1965.

15 Project Management Institute, Inc., *A Guide to the Project Management Body of Knowledge (PMBOK Guide)*, page 5, 2004.

16 "Department of Veterans Affairs Awards Landmark Cyber Security Contract to Small Business Venture Led By SecureInfo Corporation," *BusinessWire*, 9 August 2002.

17 "Review of VA Central Incident Response Capability Contract Planning, Award, and Administration," Virginia Office of Inspector General, *www.va.gov/oig/52/reports/2007/VAOIG-04-03100-90.pdf* (accessed 26 February 2007).

18 Jaikumar Vijayan, "Failed VA Security Contract Was An 'Open Checkbook', Report Says," *Computerworld*, 29 March 2007.

19 Elena Malykhina, "IT Team Seeks Olympic Gold," *InformationWeek*, 7 June 2004.

20 "1000-Day Countdown to Beijing 2008 Olympic Games: Atos Origin on Schedule to Deliver the Games," 18 November 2005, press release from *www.atosorigin.com* (accessed 12 May 2007).

21 "Beijing 2008 Olympic Games: Beijing Organizing Committee for the Games of the XXIX Olympiad and Atos Origin Jointly Inaugurate the Integration Lab," 30 January 2007, press release from *www.atosorigin.com* (accessed 12 May 2007).

22 Peter Neumann, Steve Bellovin, Matt Blaze, Robert Charette, "IT Conversations," moderated by Harry Goldstein, IEEE Spectrum Radio, *www.itconversations.com/shows/detail1688.html*, 6 December 2006.

23 "Statement of Robert S. Mueller, III, Director, Federal Bureau of Investigation Before the United States Senate Committee on Appropriations Subcommittee on Commerce, Justice, State and the Judiciary," *www.fbi.gov/congress/congress05/mueller020305.htm*, 3 February 2005.

24 "Arnold Punaro's Record Testimony Prepared for the Subcommittee on Commerce, Justice, State and the Judiciary, U.S. Senate Committee on Appropriations," *www.saic.com/cover-archive/law/trilogy.html*, 3 February 2005.

25 Ibid.

26 Peter Neumann, Steve Bellovin, Matt Blaze, Robert Charette, "IT Conversations," moderated by Harry Goldstein, IEEE Spectrum Radio, *www.itconversations.com/shows/detail1688.html*, 6 December 2006.

27 Dan Eggen and Griff Witte, "The FBI's Upgrade That Wasn't," *Washington Post*, *www.washingtonpost.com/wp-dyn/content/article/2006/08/17/AR2006081701485.html*, 18 August 2006.

28 "Statement of Robert S. Mueller, III, Director, Federal Bureau of Investigation Before the United States Senate Committee on Appropriations Subcommittee on Commerce, Justice, State and the Judiciary," *www.fbi.gov/congress/congress05/mueller020305.htm*, 3 February 2005.

CHAPTER 4

BUSINESS PROCESS AND IT OUTSOURCING

WHAT ROLE DOES OUTSOURCING PLAY IN BUILDING A SUCCESSFUL ORGANIZATION?

"You've got to realize that information technology is a catalyst for change at General Motors. The whole end goal...is to build cars and trucks, not to build great IT—it's just an enabler. Whoever can help me do that, do that together...that's going to be the company that wins at General Motors."

—Ray Szygenda, Group Vice President and CIO of General Motors[1]

ELI LILLY: WHY MANAGERS GET INVOLVED IN OUTSOURCING

Eli Lilly is a multinational pharmaceutical company that operates mainly in the United States and Europe. Its headquarters is in Indianapolis, Indiana, and it employs roughly 44,500 people. Recent annual revenues were nearly $16 billion.

In drug research and development, the cost and time required to bring a new product to market have been increasing for decades. Drug companies are struggling to reverse these trends while simultaneously creating new drugs that are more effective for specific patient groups. Eli Lilly believes that effective solutions cannot be found simply by "tinkering at the margins of our current business model but rather [by] finding wholly new ways to realize and deliver the full value of our products."[2] Eli Lilly is taking the "approach of mapping the critical path for all phases of drug development and implementing more efficient alternatives. In many cases, the result will be a more global approach."[3]

Researchers at U.S. pharmaceutical firms use clinical trials to evaluate new drugs on patients in strictly controlled settings and are required to submit the findings of these trials to obtain approval from the U.S. Food and Drug Administration. They use a clinical data management system to capture, record, and analyze the large volume of data associated with a clinical trial. At the end of the clinical trial, the data in the system is sent to regulatory authorities for independent analysis and approval of the trial's findings.[4]

In November 2006, Eli Lilly awarded a multiyear contract for business process outsourcing to Tata Consulting Services (TCS), Asia's largest IT services firm. TCS has annual revenues of more than $4 billion and nearly 100,000 employees.[5] Under the contract, Tata will establish a new medical information sciences center in northern India to work with Lilly's Medical Information Sciences disciplines and provide a wide range of services in clinical data management, statistical analysis, and medical writing.[6] The new center will open with a staff of more than 100 professionals, including doctors, biochemists, software engineers, and microbiologists. The staff is expected to grow over time.[7]

"The goal of our relationship with TCS has several dimensions beyond reducing cost and risk, including gaining access to a global talent pool, increasing scalability and flexibility of our resources, and maintaining a global workflow that is operational 24 hours a day," according to Dr. Steven Ruberg, Group Director of Global Medical Information Sciences at Lilly.[8]

"TCS was selected because of its rich and varied experience as well as for its strong management capabilities, and ability to provide customers with talented persons and innovative approaches to address their needs," said Jay Turpen, Director of Sourcing Capabilities at Lilly.[9] TCS has worked with a number of leading pharmaceutical and healthcare companies, including the Danish pharmaceutical firm Novo Nordisk for the data management of its clinical trials in India.[10]

TCS's deal with Lilly is yet another indication that India's largest outsourcing firms are ready to expand the scope of their operations beyond their well-established basic IT and business services.[11]

LEARNING OBJECTIVES

As you read this chapter, ask yourself:

• How do managers determine which business processes are good candidates for outsourcing?

• How can I ensure the success of an outsourcing project?

Outsourcing has become a frequent management strategy to achieve lower costs, improve organizational focus, and upgrade capability. Many outsourcing contracts are multiyear, multimillion-dollar deals that require approval by an organization's board of directors. Unfortunately, many problems are associated with outsourcing, including quality problems, legal issues, negative impact on customer relationships, and data and security leaks. The potential for problems is so great that only about half of all outsourcing efforts are considered successful. Thus, the stakes are extremely high and the potential for a major business setback is great.

The probability of having a successful outsourcing project can be increased greatly if the business managers who lead the effort are forewarned about potential problems. These managers must be able to choose projects and activities that are appropriate for outsourcing and avoid those that are not. They also must follow an effective outsourcing process to minimize risks and ensure success.

Institutions that have failed to apply best practices to their outsourcing operations have, in some instances, experienced a significant decline in performance, putting the future value from their outsourcing venture at risk.

WHAT ARE OUTSOURCING AND OFFSHORE OUTSOURCING?

Outsourcing is an arrangement in which one company contracts with another organization to provide services that could be provided by company employees. When the people doing the work are located in another country, the arrangement is called **offshore outsourcing**. Either way, the responsibility for control of the outsourced business function or process is shared between the firm contracting for services and the outsourcing service provider. Frequently, the employees who were performing the work internally are transferred and become employees of the service provider. Outsourcing can cover large and small projects alike. For example, a large organization can approve a billion-dollar,

10-year contract for a company like IBM to manage its entire IT services, or a single contractor might be hired to cover the responsibilities of a worker on maternity leave.

In 1989, Kodak outsourced its data center operations to IBM in a 10-year, $250 million deal. While the Kodak IT contract was certainly not the first outsourcing deal, the contract was big enough and broad enough in scope to draw worldwide attention.[12] Soon many large companies with large and experienced IT organizations, such as DuPont, J.P. Morgan, and Xerox, were employing outsourcing. In some cases, the entire IT operation of a company—including planning, business analysis, and the installation, management, and servicing of the network and workstations—is outsourced to a single firm. For example, Sears, Roebuck and Computer Sciences Corporation (CSC) entered into a 10-year, $1.6 billion outsourcing contract in June 2004, prior to Sears' merger with Kmart. The pact called for CSC to provide support for the retail giant's desktop and server computers, voice and data networks, and systems that support Sears' Web sites.[13]

Today outsourcing takes many forms and is by no means limited to information technology outsourcing (ITO). Nor is outsourcing used only by large corporations; small and medium-sized organizations have turned to outsourcing to meet their needs. Many organizations contract with service providers to handle complete business processes such as accounting and finance, customer services, human resources, and even research and development, in what is called business process outsourcing (BPO). They also outsource selective components of business processes such as benefits management, claims processing, customer call center services, and payroll processing. Contracts often include an IT component to provide BPO.

Outsourcing can involve the sale of hardware, software, facilities, and equipment used in current operations to the outsourcing service provider. The outsourcing provider then uses these assets to impart services back to the client. Depending on the value of the assets involved, this sale may result in a significant cash payment from the service provider to the customer or vice versa. For example, Galaxy Nutritional Foods, Inc. is a producer and marketer of plant-based dairy alternatives for the retail and food service markets. Galaxy decided to outsource the manufacture of its food products to Schreiber Foods, and expects to achieve substantial and ongoing cost reductions as a result. However, prior to the completion of the outsourcing, Galaxy had to report a nonrecurring charge of $7.9 million. This charge reflected the difference between the carrying cost of production equipment on Galaxy's books and the amount to be received from Schreiber Foods upon sale of the equipment.[14]

A more advanced stage of outsourcing involves evaluating all aspects of an organization's business activities to take advantage of an outsourcer's best practices, business contacts, capabilities, experience, intellectual property, global infrastructure, or geographic presence by tapping resources and providing capabilities anywhere around the globe. Outsourcing firms that can provide these services are referred to as **global service providers (GSP)**. They fill a higher-level need than outsourcing firms that simply provide low-cost staff augmentation services.[15] GSPs provide high-value services such as performing certain core business processes and enabling new revenue opportunities around the world. A **core business process** is one that provides valuable customer benefits, is hard for competitors to imitate, and can be leveraged widely across many products and markets. It takes the unique knowledge and skills of the organization's workers to operate these processes effectively. Core processes typically have a direct impact on the organization's customers, are major cost drivers, or are essential for providing services.[16] For example, a core business process for Honda is the design of engines. Honda was able to leverage this process to develop a wide range of quality products, including ATVs, automobiles, lawn mowers, marine motors, motorcycles, personal watercraft, scooters, snow blowers, and trucks.

As an example of a firm that has chosen outsourcing, consider Banco Pichincha. Ecuador's largest private bank has more than 1.5 million clients, a loan portfolio of more than $1.5 billion, and more than 230 branches in Ecuador, Peru, Colombia, Panama, Spain, and the United States. The bank signed a 5-year, $140 million outsourcing contract with Tata Consultancy Services to redesign and develop the bank's core banking solutions and provide BPO services for the bank's operations. Antonio Acosta, joint president of the bank, said, "We chose TCS as our strategic partner on the strengths of its end-to-end technology capabilities, reputation in providing certainty of results, deep domain expertise in banking, and committed scale of operations in this region."[17] To provide the necessary business process services, TCS will set up a new company in Ecuador with a staff of 500 people supported by TCS's offshore BPO center in Chile and global delivery centers across the world. TCS will retain all Banco Pichincha's staff that currently are working in these processes and bring in their best human resources and practices worldwide to train TCS employees in Ecuador.[18]

There are more than 1000 outsourcing firms. Table 4-1 presents the 2006 Top 10 best-managed global outsourcing vendors, as compiled by the Brown-Wilson Group.[19] A total of 12,755 qualified outsourcing purchasers and users and 3340 outsourcing corporate employees answered the survey. The survey rated companies on their consistent strength in four areas: human capital performance, CEO commitment, corporate direction, and leadership impact.

TABLE 4-1 Brown-Wilson's 10 best-managed global outsourcing firms

Rank	Company	2005 Estimated Revenue($ billion)	Employees	Services
1	Affiliated Computer Services	$4.4	52,000	BPO, ITO, HRO*
2	Satyam	$1.1	29,000	BPO, ITO
3	Cognizant	$0.9	25,000	ITO, BPO
4	Perot Systems	$2.0	18,000	ITO
5	Infosys	$2.2	58,000	ITO, BPO
6	Patni	$0.5	12,000	ITO
7	Tata Consultancy Services	$2.9	54,000	ITO
8	HCL	$0.8	13,000	ITO
9	GenPac	$1.0	30,000	BPO
10	Mellon	$4.3	17,000	BPO, HRO*

* HRO is short for Human Resources Outsourcing.

Why Do Organizations Outsource?

Organizations decide to outsource for many reasons. Three of the most frequently cited reasons are to cut or stabilize costs, improve the firm's focus on core operations, and upgrade the firm's capabilities and services.

To Cut or Stabilize Costs

The top reason to outsource is to cut or stabilize costs. Outsourcing service providers typically have a lower cost structure due to greater economy of scale, specialization, or expertise, which means they can perform the work at a much lower cost than their clients. In addition, the fundamental costs of doing business in a developing country—employee health care, retirement, and unemployment; taxes; and environmental and regulatory compliance—are much lower than those in a developed country.[20] Such cost advantages tip the scales in favor of outsourcing and offshore outsourcing. Thus, organizations that do not outsource probably have greater recruiting, training, research, development, marketing, and deployment expenses. These costs must be passed along in the form of higher prices to the customer, placing the firms at a competitive cost disadvantage.

One firm that exploits outsourcing to its advantage is Pizza Inn, headquartered near Dallas, Texas. Pizza Inn operates more than 360 restaurants domestically and internationally, and had recent annual sales of about $150 million. The firm outsourced its warehouse management and delivery services and realized a significant reduction in its operating costs. As a result, Pizza Inn was able to reduce the prices of products distributed to its franchisees, thus improving their profitability and, in turn, boosting the firm's operating income. In the fiscal quarter following completion of the outsourcing, general and administrative expenses were slashed 30 percent ($391,000) due primarily to lower payroll costs, plus a reduction in property taxes and insurance expenses.[21]

To Improve Focus

Another reason for outsourcing is to enable an organization to focus on its most important priorities. It is highly ineffective to divert the time and energy of key company resources to do routine work that does not require their unique skills and intimate knowledge of the firm, its products, its services, and its customers. Outsourcing "frees up" a large amount of resources and management effort that can be redirected to other more strategic issues within the company.

For example, many of the services required to operate an insurance firm, such as billing, human resources, and transaction processing, are important but not essential to future growth. So, insurance firms increasingly are turning to outsourcing to enable them to reduce costs and grow, as they can now focus on their core business. For example, AIG Entrepreneur specializes in property and casualty insurance for small and medium-sized enterprises. It signed a 10-year, $100 million agreement with the outsourcing firm Accenture to provide IT hardware and software plus insurance support services. The goal is to simplify, automate, and optimize AIG business processes to increase profitability, improve operational performance, and enhance services. Annuity, life insurance, pension, property, and casualty services will be provided through Accenture's multilingual service center in Bucharest, Hungary. Underwriting, policy, and claims services will be supported by Accenture computer hardware and software.[22]

To Upgrade Capabilities and Services

Often, an outsourcing service provider can perform a business process better than its clients ever could. The outsourcing provider might be highly efficient, with world-class capabilities and access to new technology, methods, and expertise that would not be cost effective for its clients to acquire and maintain. Thus, outsourcing a function can provide a considerable upgrade in capabilities and service.

For example, Mumbai International Airport Limited contracted with Tata Consultancy Services to implement and manage the IT infrastructure at Chhatrapati Shivaji International Airport (CSIA). The primary goal was not savings, but to introduce the best technology solutions to build the country's busiest airport into a model, world-class experience that rivals any airport in the world.[23]

As another example, logistics service providers have developed a wide array of services that enable their clients to improve operating efficiency and effectiveness, reduce inventory, and increase customer service by reducing delivery times and providing delivery status at any point in the pipeline. The providers usually can deliver all these services at a lower cost. As a result, many organizations have outsourced their logistics operations to third-party logistics providers to manage complex global supply chains.

Issues Associated with Outsourcing

While companies can gain many potential benefits from outsourcing, these gains do not come without potential problems. Four areas of risk include quality problems, exposure to legal liabilities, negative impact on business partner and customer relationships and satisfaction, and potential data and security breaches.[24]

In a recent *InformationWeek* research survey of 420 IT professionals, half of them rated their companies' outsourcing efforts a success (see Table 4-2).[25]

TABLE 4-2 Results of outsourcing projects

Evaluation	Percent of Outsourcing Efforts
Success	50 percent
Neutral	33 percent
Disaster	17 percent

According to the Diamond Management and Technology Consultants 2006 outsourcing study, 47 percent of outsourcing "buyers experienced an abnormal contract termination in the past year, while only 2 percent stated that their outsourcing expectations were exceeded."[26]

While outsourcing may prove beneficial for many companies, several potential issues must be addressed. Any organization that considers outsourcing must be aware of these issues and develop solutions for them.

Quality Problems

Outsourcing part or all of a business process introduces significant risks that the service provider will create quality problems. For example, the toymaker RC2 is not well-known, although it has major licensing deals with Sesame Street, Winnie the Pooh, Disney, Nickelodeon, and Thomas & Friends. The firm works with third-party suppliers in China and Hong Kong to manufacture its products. RC2 and industry observers were shocked in June 2007 when the firm issued a recall for 1.5 million Thomas the Tank wooden trains and related components that had been contaminated with lead paint. The manufacture of the toys had been outsourced to a factory in Dongguan, China.[27]

Legal Issues

The details of the outsourcing arrangement are documented in a formal contract. The contract describes how responsibilities are divided between the client and the outsourcing firm, what services are to be provided, what service levels must be met, and how problems between the two firms will be resolved. Many outsourcing contracts are multiyear, multimillion-dollar deals that require approval by a board of directors. The average length of an outsourcing contract is five years,[28] so the life of the contract can extend well beyond the reign of the executives who crafted it. As might be expected, ending such megadeals prematurely can generate expensive legal fees.

For example, Sears Holdings, the corporate parent of Sears and Kmart, ended its $1.6 billion, 10-year outsourcing contract with Computer Sciences Corporation after less than one year. Sears Holdings stated that it terminated the contract for cause, while CSC attempted to hold Sears liable for up to $100 million in contract termination fees.[29] It took years to settle the dispute. J.P. Morgan Chase & Co. scrapped a $5 billion service agreement with IBM following its merger with Bank One Corp. Suncorp-Metway Ltd., an Australian banking, insurance, and investment firm that focuses on retail consumers and small and medium-sized enterprises, terminated its outsourcing contract with CSC following its acquisition of AMP General Insurance.[30] These examples illustrate the need to include terms of disengagement in the original contract to avoid spending excessive time and money in court.[31]

Negative Impact on Customer Relationships and Satisfaction

Outsourcing can greatly reduce the amount of direct communication between a company and its customers. This prevents a company from building solid relationships with its customers, and often leads to dissatisfaction on one or both sides. For example, based on an unusually heavy volume of customer complaints, Dell decided to stop routing U.S. technical support calls for its OptPlex and Latitude notebook computers to a call center in Bangalore, India. Dell customers complained of language difficulties and delays in reaching senior technicians when speaking to support personnel. The drop in customer satisfaction was noticeable enough to be measured and reported by both Consumer Reports and Technology Business Research.[32]

Data Security and Integrity Issues

Another key outsourcing issue is concern over maintaining data security and integrity to safeguard against data security lapses. For instance, the state of Florida contracted work on its payroll and human resources system to Convergys, a U.S.-based outsourcing service company. Convergys, in turn, subcontracted work to index state personnel files to GDXdata, another U.S.-based firm. GDXdata allegedly outsourced the indexing work to a firm in India, a violation of the GDXdata contract with Convergys. Florida state employees had to be warned that their personal data might have been compromised, including sensitive information about the state's law enforcement agents.[33] Besides the security issues, this example illustrates the need for outsourcing firms to put limits on additional outsourcing and subcontracting.

Special Issues Associated with Offshore Outsourcing

Firms that consider establishing offshore outsourcing agreements must be aware that major differences between outsourcing and offshore outsourcing must be taken into account. The most obvious issues are how to control and manage the work being performed when your outsourcing partner may not speak your language and is guided by different cultural values and industry standards. This issue is only intensified by thousands of miles of separation across multiple time zones and the extreme difficulty of meeting face to face. Such separation creates a high potential for lost productivity due to communication problems and increased opportunity for misunderstandings.

Other issues also are associated with offshore outsourcing:

- *Cost advantage*—Salaries in developing countries such as China, India, Latin America, and the Philippines are increasing at more than 15 percent per year. At these rates, the cost advantage to outsource to such countries is being reduced.
- *Turnover*—The rate of employee turnover is as high as 50 percent at outsourcing firms in some countries.[34] Thus, there is a high potential that key employees at the service provider for your account or project might leave, causing significant project disruptions or delays.
- *Intellectual property rights*—Various countries have widely divergent stances on the protection of corporate data, copyrights, patents, and trade secrets. Not only must you consider whether the country has laws to protect your firm's intellectual property, you must ask whether the laws are enforced. For example, one U.S. software manufacturer outsourced the code development for a new release of its software to an Indian-based firm. To protect itself, the U.S. firm required the employees of the Indian firm to sign a nondisclosure agreement—a contract in which parties agree not to disclose important corporate information. However, an employee of the Indian firm stole a copy of the code and tried to sell it to a competitor. Despite solid evidence gathered by the FBI, prosecutors in India have failed to convict the man, who continues to work.[35]
- *Important technology issues*—The outsourcing firm must be able to provide a high level of system availability and network uptime and guarantee that all processing applications operate efficiently and reliably. High IT reliability, availability, and efficiency are essential so that business processes can be executed on a timely basis without significant service interruptions. The potential for problems is exacerbated by offshore outsourcing with service providers in developing parts of the world.

Now that we have identified many of the issues associated with outsourcing and off-shore outsourcing, we will outline an effective outsourcing process that manages these issues.

Planning an Effective Outsourcing Process

Outsourcing is like any other business initiative: it takes planning, knowledge, and skill to execute well. As already discussed, roughly 50 percent of outsourcing efforts are considered successful, while the other half are evaluated as so-so or outright disasters. Many of the organizations that were successful carefully planned and executed their outsourcing efforts following a multistep process. This process is shown in Figure 4-1 and discussed in the following sections.

Establishing a "Smart" Outsourcing Strategy

The critical component to obtaining successful results from any outsourcing activity is executive-level understanding and support for a smart sourcing strategy. **Smart sourcing** is based on analyzing the work to be done, its associated current processes, and level of effectiveness and resources required, and then determining the best way to do the work in the future—whether with internal employees, on-shore or offshore outsourcing firms, or some combination. Organizations that move to smart sourcing recognize that outsourcing is not just about lowering labor costs. Outsourcing can achieve strategic competitive advantages by reducing time to market for new products, cutting the time required for problem resolution, and freeing up resources to enable greater innovation. Armed with this more complete understanding of the potential of outsourcing, the organization can make better strategic decisions about appropriate activities and projects for outsourcing, as well as which outsourcing firms they will hire. As this chapter discusses later, smart sourcing requires an organization to work in a true partnership with the outsourcing provider. This partnership must be built on a high level of collaboration, mutual trust and respect, and a sharing of common goals.

Evaluating and Selecting Appropriate Activities and Projects for Outsourcing

Many outsourcing projects have failed to meet expectations, especially when work was relocated simply to cut labor costs or to clean up a poorly performing operation. Generally, shifting seriously flawed operations to a less expensive organization does not solve fundamental problems. Thus, an organization must carefully consider which process and projects it should assign for outsourcing.

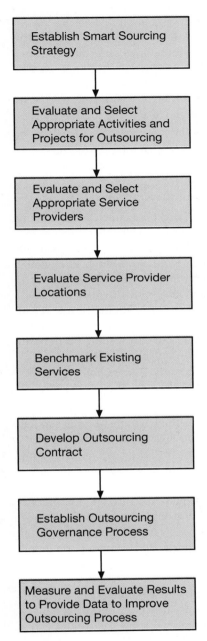

FIGURE 4-1 Multistep process for successful outsourcing

The most significant outsourcing risk is dealing with increased management complexity. This level of risk is heightened as the organization increases the scope of processes being outsourced. Many organizations hesitate to outsource processes that are considered mission critical, that are tightly linked to other key processes, that clearly differentiate them from the competition, or that strongly influence sales. Thus, an organization's initial experience with outsourcing probably should not involve a critical, core business process. Organizations can ask the following key questions to separate core business processes from their less critical processes:[36]

- How critical is the project or process to unique strategic differentiation?
- How competitive and innovative is the organization in this business area?
- How cost effective are activities in this business area?
- How much customer value does the project or process provide?

Many companies start with a short-term, low-risk outsourcing pilot effort, perhaps moving responsibility for a small business process to an outsourcing provider that appears to be an attractive, long-term outsourcing partner. They may employ an experienced outsourcing consultant to help get started, provide ongoing feedback, and help evaluate the pilot results. At least six months are required to gain experience with the service provider and work through various start-up issues so that a fair assessment can be made. After this initial experience, the company may want to expand the scope of its outsourcing efforts. It can do so with the experience gained from the initial pilot and try not to repeat the same mistakes. It also will have substantial experience with at least one outsourcing vendor and be in a better position to know what the company needs in an outsourcing partner.

Evaluating and Selecting Appropriate Service Providers

When outsourcing a major business process or project, an organization should think in terms of hiring a *partner*, not just a provider. Thus, choosing the best outsourcing service provider is not based solely on the lowest price quoted or the highest savings promised. Ideally, the organization can choose an outsourcing firm with which they can build a strong strategic partnership based on a mutually sustained commitment to achieve specific business goals. The customer must use due diligence in carefully researching the potential partner's capabilities and reputation. This research can be conducted through discussions with current and former customers of the firm, seeking input from industry trade groups and consultants, on-site visits to the vendor's facilities, and review of public records related to the firm. These records include Dun & Bradstreet credit reports, filings and reports from the Securities and Exchange Commission (SEC), and articles in trade magazines and the press.

Companies also can research outsourcers through documents generated as a result of the Sarbanes-Oxley Act. This legislation was enacted in response to several major accounting scandals at Enron, WorldCom, Tyco, and other companies in the late 1990s and early 2000s. Under Sarbanes-Oxley, a report filed with the SEC by a publicly held firm must contain a statement signed by the CEO and CFO attesting that the report's information is accurate. Penalties for false attestation include up to 20 years in jail and significant monetary fines. As a result, firms spend considerable time and energy to document and test internal control processes. But what if a fundamental business process is outsourced to a third party? An SAS 70 audit (Statement of Auditing Standards No. 70, Service Organizations) is a tool that can help evaluate an outsourcing firm's internal controls. Under such an audit, the service firm prepares a written document describing its control goals and objectives. An outside service auditor then examines the document and the service firm's operations to render an opinion on several issues:

- Are the control goals clearly stated?
- Are the controls suitably designed to achieve the service organization's stated control objectives?
- Are the controls actually being used?
- Are the controls operating effectively (Type 2 SAS 70 audit)?

Firms considering outsourcing need to spend considerable time and effort to thoroughly review the outsourcing firm's SAS 70 audit and ensure that they understand the firm's control goals and implementations. They must be comfortable that the internal controls implemented by their potential partner are adequate. Failure to share the results of an SAS 70 audit should be a warning signal in dealing with an outsourcing vendor.

SAS 70 can help evaluate a firm's internal controls, but it does not fully address information security control. ISO (International Standards Organization) 17799 identifies "best practice" information security controls and their objectives. An organization considering outsourcing can use this standard to evaluate the service provider's security policy and measures more fully.

In summary, organizations should choose outsourcing firms based on several factors, as listed in Table 4-3.

TABLE 4-3   Factors for evaluating outsourcing partners

Factors
Proven experience in business process outsourcing
Reputation
Knowledge of the industry
Expertise in the organization's processes
Price
Freedom from major lawsuits and customer complaints
Financial viability
Trustworthiness
Proven high level of innovative and continuous improvement
Proven ability to deliver services effectively to the countries in a company's base of operations
Use of best-in-class processes and technology
Thorough review of the outsourcing firm's SAS 70 audit reveals no problem
Review of the outsourcing firm's security versus ISO 17799 best practices reveals no major outages

Evaluating Service Provider Locations

Any outsourcing service provider, no matter what its base of operations, can be affected by economic turmoil, natural disasters, and political disturbances. The potential for these risks is greater in some places than others. Be sure that you understand the base of operations that will service your needs. Ideally, your outsourcing partner can provide services from several geographic locations if necessary. Your company should investigate the capability for avoiding business interruption whether the outsourcing firm is "on shore" (in your own country) or offshore.[37]

For example, Bangalore, with a population of about 6.5 million, is India's fifth-largest metropolitan area. It often is compared with the Silicon Valley in the United States because many outsourcing service providers have offices there, including Infosys, IBM, Tata Consultancy Services, and Wipro. India's outsourcing industry is an important source of national income, and so it is a prime target of terrorist groups. Indian authorities arrested a suspected member of a terrorist group for plotting attacks on several Bangalore outsourcing firms in 2006. In October 2006, these outsourcing firms closed their Bangalore facilities due to a public sector strike related to a border dispute. In January 2007, they shut down operations because of riots between Muslims and Hindus. In February 2007, they did not open their offices due to a labor strike over a court decision on water distribution from a nearby river.[38]

Other factors when considering location include the availability and reliability of high-speed communications networks and power grids, the availability of sufficiently trained workers, and the effectiveness of the outsourcing firm's national legal system in protecting intellectual property, including copyrights, trade secrets, and patents. Of course, the challenges of outsourcing become even more difficult when the work is being done in a country that has significant language, cultural, and time zone differences. Such considerations may force a firm to change its initial choice of outsourcing service partner.

Table 4-4 summarizes the factors to consider when evaluating service provider locations.

TABLE 4-4   Factors for evaluating the location of an outsourcing provider

Factors
Potential for business disruption has been addressed adequately by the provider through use of effective backup and alternate business recovery sites
Access to high-speed, reliable communications networks is readily available
Access to reliable power grids is readily available
Provider has access to an adequate supply of sufficiently trained workers
National legal system supports and enforces the protection of intellectual property

Benchmarking Existing Service Levels

Before signing an outsourcing contract, an organization should benchmark its existing service levels so that it knows how well the services are being delivered and it knows the associated costs. This benchmark can then be used to establish a reasonable baseline for negotiating target results and costs from the outsourcing service provider. The agreed-to targets are then used to define the service-level agreement (SLA) of the contract.[39]

A key to effective benchmarking is choosing the right measures to evaluate the performance of the process. Remember, you get what you measure. For example, reasonable metrics for a call center might be to measure average hold time for customers or the number of abandoned calls. Measuring these parameters and trying to improve performance would lead to better results for the firm and the customer. On the other hand, setting a measure for the average number of calls handled per customer service agent may lead to counterproductive behavior. The agent might not fully listen to the customer and cut the call short to get to the next call.

The time and cost required to perform a benchmark depends on the size, scope, and complexity of the process being measured and the number of metrics used. It can cost more than $100,000 to hire an outside consulting firm to perform a benchmark. Doing the benchmarking with employees can be much less expensive and take less time because they already are familiar with the people and the process. However, employees need to be trained to perform the benchmark process, and internal benchmarks can be tainted by bias, especially if the people doing the measuring are part of the in-house process.[40]

Increasingly, experienced organizations include broad measures of desired business outcomes into the performance measurements they expect outsourcing partners to deliver. These measures define valuable business benefits that the organization wants from the outsourcing initiative, including increased speed to market, reduced product or service defects and rework, and lower working capital requirements made possible by higher efficiencies.[41]

Developing an Outsourcing Contract

The development of an outsourcing contract is a job for experienced procurement and legal professionals. Although numerous issues should be addressed, only a few are covered in this section.

The ownership of assets and facilities is an important factor in determining the cost of the outsourcing contract. There are three basic alternatives:

- The firm can transfer ownership of the assets along with operational responsibility to the outsourcing service provider. The provider typically offers a financial incentive to do this, such as a reduction in charges or a cash transfer to cover the value of the assets.
- The firm can transfer the assets to a third party (financial services firm) under some sort of leaseback arrangement.
- The firm can retain ownership of the assets while the provider takes on the operational responsibility. Experienced members of the client's finance and accounting organization must become involved in analyzing the various options.

For example, Electronic Data Systems (EDS) was awarded a $1 billion, 8-year contract to provide a variety of IT services to KarstadtQuelle AG, which is headquartered in Essen, Germany. The firm's core activities include Karstadt department stores throughout Germany, domestic and international mail-order companies, and its tourism business (Thomas Cook). As part of the arrangement, EDS will gain a 75 percent stake in Itellium, the firm's in-house IT subsidiary. EDS will update Itellium's IT infrastructure to form a new outsourcing center that will serve KarstadtQuelle and other European retailers. As another example, IBM won a contract to manage the IT resources of Switzerland's Banque Cantonale Vaudoise, and is following a similar strategy to create a Swiss-based outsourcing center for European banks.[42]

The current trend is to reduce the size and complexity of outsourcing contracts. Instead of entering into all-encompassing outsourcing contracts with a single firm, organizations are opting for simpler, more business-specific arrangements that employ multiple service providers. For example, one service provider might handle network operations, another might manage servers in data center environments, and a third could handle the help desk.[43] The goal is not only to cut costs by fostering competition among the vendors, but to take advantage of each vendor's areas of specialization and technical expertise. This approach requires that the vendors work well together, cooperating to solve problems and not pointing fingers. It also requires additional overhead in the form of specialists who help manage the relationship with each provider.

Unfortunately, allocating pieces of a major outsourcing contract among several firms is not a foolproof approach to containing project costs. For example, the UK National Programme for IT in the National Health Service (NHS) has the ambitious goal of electronically linking 50 million patients to 30,000 doctors and 270 healthcare providers by 2010. Healthcare records, appointment details, prescription information, and up-to-date research into illnesses and treatment will be made available both to patients and health professionals. Information will be available over a secure link whenever and wherever health-related decisions must be made. In 2003, the UK government awarded contracts to several IT outsourcing service companies to provide the regional IT systems and services needed to support the project. The UK government allocated $4 billion for the program in 2002, but it is estimated that the program will cost more than $55 billion![44]

When entering into an offshore outsourcing agreement, it is critical to determine what legal system and which country will have jurisdiction over any contract disputes. Each party in the contract, of course, prefers to have its own country rule.

Establishing an Outsourcing Governance Process

Governance of an outsourcing contract involves formal and informal processes and rules to manage the relationship between the two organizations. Governance defines procedures such as periodic formal reviews between the outsourcing company and its service provider, and explicit escalation procedures in the event of a disagreement. The goal of such procedures is to ensure that the outsourcing initiative succeeds, even as personnel, business needs, and operating conditions change.

Governance requires dedicated, trained vendor relationship professionals to manage the working relationship between the organization and outsourcing service provider. These relationship managers engage the service provider and work collaboratively to find problems and fix them. Good relationship managers should have excellent communication, problem solving, and negotiation skills. They also need a thorough knowledge of the business processes and technologies involved. Similarly, the outsourcing service provider has invested many years in the recruitment, development, training, and retention of relationship managers. They expect to encounter client relationship managers of similar status, experience, and knowledge that they can work with as equals.

Measuring and Evaluating Results

A key component of governance is to implement ongoing monitoring and analysis of outsourced business processes using an appropriate set of metrics. Such a program will determine if an organization is realizing the full benefits of outsourcing and reduce the degree of operational risk.[45] It also will enable the firm to hold its outsourcing provider accountable for implementing corrective action as needed. In a typical contract, if the service provider's performance and costs don't meet the SLA standards, financial penalties can be assessed and the contract can even be terminated.[46] How frequently measures are taken and how quickly changes are implemented depends on the importance of the business process in meeting true business goals.

In 2003, Procter & Gamble signed four major outsourcing contracts, each with a different provider: a $3 billion, 10-year deal with Hewlett-Packard to take over much of the firm's IT services; a 5-year deal with Jones Lang LaSalle for facilities management; a 5-year deal with Sykes Enterprises to outsource customer care, customer relationship management (CRM) applications, and global fulfillment services; and a $400 million, 10-year deal with IBM for human resource services. An organization of 100 people was formed to focus solely on managing all outsourcing relationships and to avoid losing control of service levels or the scope of the outsourced project. The employees were trained by experts from P&G's Purchasing Department in the best practices for dealing with suppliers. The group includes experts in disciplines from IT to CRM and facilities management. The P&G team uses software to track hundreds of service levels and enter data into a performance scorecard to see how well each vendor is performing. Each member of the team is assigned to "work on just one of the outsourcing contracts, but they interact and keep tabs on how their colleagues manage the other vendors."[47]

The ongoing tracking and measurement of important metrics enables the organization to use the data as feedback, so that each step in the outsourcing process can be improved based on the result of the project.

Read the following special feature to see what can happen when an organization does not take an effective approach to identifying and meeting its outsourcing needs.

A MANAGER TAKES (THE WRONG) ACTION

Swansea City Council Outsources IT

Swansea is a historic city of 226,000 people on the coast of Wales, United Kingdom. Its beginnings trace back to the Vikings in 1013. The city's location makes it a popular tourist resort, and it also is home to some high-tech industries.

In late 2004, the Cabinet of the Swansea City Council developed ambitious plans to launch an e-government initiative to enable residents to conduct business with the council face to face, by phone, or over the Internet. The initiative would extend hours of operation, create contact centers to make it easier for residents to interact with the council,

continued

establish a call center that would enable 200 phone numbers to be replaced with a one-stop center and a single phone number, and provide a set of improved Internet services known as service@swansea. The initiative also would generate about £72 million in savings by making the council more efficient and by cutting costs through reduced paperwork and duplication. The savings then could be invested to further improve council services such as education and improving the environment.

The city council decided to outsource not only the work of implementing the e-government initiative, but to outsource the operation and support of the council's IT services. The value of the outsourcing contract was estimated at £119 million; it would be funded by transferring the money the council already spends on IT and adding a new investment of £2 million per year over the next decade.

The IT staff went on strike in August 2004 during outsourcing negotiations, complaining that the council had refused to fully consult the staff and that they had been insufficiently consulted over being transferred as part of the outsourcing. Instead of outright transfer to the outsourcing firm, the staff wanted the council to second them to the outsourcing firm. Secondment, or the detachment of employees from their regular organization for temporary reassignment, would protect pension arrangements and other staff benefits that may not be available from an outsourcer. There was talk of escalating the strike to all 5000 union staff. This potential escalation was condemned by deputy council leader Gerald Clement, who accused the union of "holding the council and the people of Swansea to ransom."[48]

In September 2004, it was learned that the IT outsourcing project would cost £150 million over 10 years, an increase of nearly 50 percent over the original estimate. The new estimate included the addition of £5 million per year to cover the cost of acquiring a call center, a city center drop-in shop for residents, and transfer costs in switching to a private company to run the project.[49]

After six weeks of the strike, and as the 5000 members of the union were set to vote on joining the strike, the Swansea City Council called in a third party, Acas, to assist in settling the strike. The IT workers finally suspended their strike after eight weeks and returned to work. Even though 63 percent of the 5000 union members voted to strike against outsourcing, the council still planned to award the outsourcing contract for the service@swansea project in December.

The council let it be known that employment arrangements for the IT staff would not be finalized and communicated to the staff until February 2005. A director of an outsourcing consultancy who was not involved in the Swansea situation stated, "Looking at signing a deal in December but only sorting out the staff issues in February 2005 is courting disaster. The affected IT people deliver support for critical services. Without clarity on their future, how can Swansea expect anything but further issues?"[50]

Final selection of the outsourcing firm was delayed until January 2005 and the contract was expected to be signed in April 2005. Capgemini, a large IT, consulting, outsourcing, and professional services company with more than 75,000 employees in 30 countries, was the winner, although the contract represented its first major outsourcing contract in local government.[51]

continued

In April 2005, a decision was made to transfer 70 percent of the IT staff to Capgemini. A council spokesman said that the IT department's proposals were considered but the plan to transfer IT staff to the supplier was preferred because it offered better value and faster return on investment. In addition, secondment created greater legal risks.[52]

Council official Mary Jones said, "Our [IT] staff will join one of the world's leading IT companies, where they will be able to develop their skills and careers. We have heard their concerns and have agreed with Capgemini to a range of measures to ensure protection of their terms and conditions for the period of the 10-year contract."[53] These concerns included the right not to have to work outside the Swansea local area and the ability to remain on the local government pension plan. The contract was expected to be signed in June, with the staff transfers completed by July.

In September 2005, the Swansea City Council announced that it would give its IT staff just two weeks notice of their new terms and conditions of employment prior to their transfer to Capgemini on October 1. However, there were further delays in the signing of the contract and the number of staff to be transferred was reduced to 66 from 73.[54]

The contract with Capgemini finally was signed in January 2006 after a delay of 18 months. In addition, the council decided to implement a less ambitious e-government project and cut the value of the contract to £83 million. In January 2007, the Swansea City Council cancelled the e-government component of the outsourcing contract, slashing the contract value to £40 million.

After one year of operating under the outsourcing contract, frustrated council members complained that computer systems and support services cost more and delivered less under the new outsourcing arrangements. Council members have given up using the help desk, which was transferred to Aberdeen, Scotland. Councillors complained that help desk workers did not even know where Swansea was or what a councillor was. In addition, the e-mail system was slow and difficult to use, and often was unavailable for use altogether.

Discussion Questions:

1. Businesses and public organizations in the UK have long been successful adopters of outsourcing. In fact, the UK is recognized as being the most advanced country in Europe in terms of outsourcing. So what went wrong with the Swansea City Council's efforts, which is described as a "PR disaster" for outsourcing? Identify three serious mistakes you think the city council made.

2. Should the city council have handled the outsourcing of ongoing operations and outsourcing the building of the new e-government initiative as two separate issues? Discuss fully.

3. How should the city council proceed from here?

The manager's checklist in Table 4-5 provides a set of recommended actions for business managers to consider when contemplating outsourcing. The appropriate answer to each question is yes.

TABLE 4-5 Manager's checklist

Actions	Yes	No
Do you know what your organization's core business processes are?		
Is there a clear goal to be achieved through outsourcing, such as cost stabilization, improved focus, or upgraded capabilities?		
Have you fully considered the four key areas of risk associated with outsourcing?		
If you are considering offshore outsourcing, do you understand the special issues involved?		
Has your organization established a smart sourcing strategy?		
Does your organization follow a logical, well-defined outsourcing decision-making process similar to the approach outlined in this chapter?		

Chapter Summary

- Outsourcing is an arrangement in which one company contracts with another organization to provide services that could be provided by company employees. When the people doing the work are located in another country, the arrangement is called offshore outsourcing.

- Outsourcing takes many forms, including IT outsourcing, business process outsourcing, and human resources outsourcing.

- Global service providers offer high-value services, including handling certain core business processes and creating revenue opportunities around the world.

- Organizations turn to outsourcing to cut or stabilize costs, improve the firm's focus on its core operations, and upgrade the firm's capabilities and services. In a recent survey, roughly half of the companies contacted rated their outsourcing effort a success.

- Outsourcing is a high-risk activity that raises the potential for quality problems, exposure to legal liabilities, negative impact on business partner and customer relationships and satisfaction, and potential data and security breaches.

- Other issues are associated with offshore outsourcing. For example, it can be difficult to manage and control work done by people who do not speak your language, who are guided by different cultural values and standards, and who reside several time zones and thousands of miles away. Also, the cost advantage for offshore outsourcing is eroding; wages in many developing countries are increasing at more than 15 percent a year. In addition, countries have varying ideas about the protection of intellectual property, and there are potential problems with the availability and reliability of power grids, IT networks, and their applications.

- An eight-step process to successful outsourcing includes: 1) establishing a smart outsourcing strategy, 2) evaluating and selecting appropriate activities and projects for outsourcing, 3) evaluating and selecting appropriate service providers, 4) evaluating service provider locations, 5) benchmarking existing service levels, 6) developing the outsourcing contract, 7) establishing an outsourcing governance process, and 8) evaluating and measuring results.

Discussion Questions

1. What are some differences and key issues that distinguish outsourcing from offshore outsourcing?

2. How would you distinguish a global service outsourcing provider from a staff augmentation outsourcing firm?

3. Give three primary reasons why organizations turn to outsourcing.

4. Can a firm be successful without outsourcing? Discuss the question fully and identify an example to support your position.

5. What key issues must be addressed when considering whether to outsource?

6. What are the key areas of risk when a firm enters into an outsourcing effort? How can these risks be reduced by following an effective outsourcing process?

7. What additional risks arise when a firm is considering offshore outsourcing? Is there any way to minimize these risks?

8. What is the difference between outsourcing and smart sourcing?

9. Give an example of a business process that would be appropriate to consider for a firm's first venture into business process outsourcing. Give an example of an inappropriate business process.

10. What is the difference between an outsourcing service *provider* and a *partner*? Does it matter if your firm treats your outsourcing firm as a provider or a partner? Why or why not?

11. Discuss the impact of the Sarbanes-Oxley Act on choosing and managing an outsourcing firm.

12. Can you choose an outsourcing service provider without knowing the locations from which your operations will be serviced? Why or why not?

Action Memos

1. General Motors group vice president and CIO Ralph Szygenda is a strong advocate of the multisourcing approach. His groups employ more than a dozen outsourcing firms to provide IT-related services and equipment. Such multisourcing requires that vendors work well together. To that end, GM works with its vendors to improve integration and adopt standards.[55] GM employs about 2000 IT staff around the world to focus on managing outsourcing vendor relationships. Senior GM IT managers meet with each outsourcing provider on a monthly basis to identify projects and issues that need attention. This approach to outsourcing has been credited with reducing GM's IT expenses by $1 billion per year over the past six years.[56]

   One of your classmates has just finished an on-campus job interview with a GM representative. During the interview, the recruiter discussed the need for qualified people to serve on a team that manages an outsourcing vendor relationship. Your classmate has sent you a text message that says you should cancel your interview with GM because the role discussed seemed meaningless. How would you react? Would you cancel the interview? What would you say to your classmate?

2. You are an experienced and respected claims manager in a midsized financial services firm. Your firm is about to close a five-year deal to outsource the servicing of its claims payment process to a respected firm in India. You have mixed feelings as you stare in amazement at the text message you have just received from your manager: "We need an experienced manager to relocate to India for 6 to 9 months to make sure that we get off to a good start with this outsourcing project. I'd like to put in your name for this opportunity. I'm on my way into a meeting to discuss potential candidates. What do you think?"

   How would you respond? How might you want to follow up with your manager later?

Web-Based Case

Make a list of three publicly held outsourcing firms: one located in your country and two located outside your country. Attempt to perform steps 3 and 4 of the outsourcing process discussed earlier in this chapter, using data you can find on the Web. As a reminder, steps 3 and 4 are evaluating and selecting appropriate service providers, and evaluating service provider locations.

Discussion Questions

1. Based on the data you can find on the Web, which firm and which location would you choose, and why?

2. What additional data would you need to make a truly well-informed decision? Where and how could you get this data?

Case Study

Accenture: A Model of Western Outsourcing Success

During the past five years, India-based outsourcers such as Tata Consultancy Services, Wipro, and Infosys Technologies have been pulling the rug from under the feet of the high-tech outsourcing giants. Between 2002 and 2006, the market share of these Indian companies rose from 0.5 to 7 percent. With significantly cheaper labor and overhead costs, these companies have been able to underbid IT leaders like IBM and Hewlett-Packard. Yet one Western outsourcing company continues to grow and thrive: Accenture.

Accenture was formed in 1989 when a group of partners broke away from nationally and internationally based Arthur Anderson firms to form a new organization specializing in consulting and technology. Over the next decade, Accenture expanded from systems integration to consulting, outsourcing, and related technology services.

Accenture's Recent Projects

Powered by SAP software, Accenture will implement the largest total cost management system to date for the U.S. Army. The system will serve more than 79,000 users and supplant more than 80 Army legacy systems.

Best Buy partnered with Accenture to build new supply-chain and analytical capabilities while reducing its server requirements by 39 percent.

BT, one of the largest global communication providers, outsourced much of its UK human resources to Accenture. Accenture reduced absenteeism by 33 percent, increased productivity, and reduced human resource costs by 20 percent.

By 2000, Accenture was netting more than $9.5 billion annually and had more than 75,000 employees in 48 countries. The company outgrew its locally owned independent partnership structure and went public in 2001. At the same time, when most IT profits were plunging during the dot com bust, Accenture scored double-digit growth.

The key to their success? "They never took their eyes off business development," said Rauline Ochs, a former vice president at BEA Systems, an enterprise software developer that partners with Accenture to deploy BEA platforms for their clients.[57]

Accenture carefully researches market developments and implements strategy based on this research. The company's business strategy is highly focused on technological innovation, maintaining its global presence, and expanding its vertical assets.

Expanding Vertical Assets

A horizontal market consists of businesses that produce the same or interchangeable goods or services. By contrast, companies that are based on vertical structure may own everything from the mine or factory where the raw material is produced to the plant where the final product is shipped. Whether serving the automotive, retail, or pharmaceutical industries, Accenture focuses on business process outsourcing (BPO), supplying services to handle complete business processes. For example, Accenture's IT staff may create databases and algorithms for a drug trial. Accenture doctors can then conduct the tests and review the clinical data. The result of this vertical approach is that companies do not have to oversee and coordinate the work of a series of contractors, but simply can turn to one consulting firm to manage the entire project.

Customers increasingly are looking for contractors with expertise in their specialized areas. It reduces their time to market and gives them access to new talent pools that they could not build in-house. By outsourcing a project, companies also can concentrate on their core business and more effectively pursue their own strategic goals.

DuPont had an IT organization that successfully supported the company's strategic goals, but IT was costing the company $1.2 billion annually. By 1996, the organization managed to reduce its IT-related expenses by 45 percent through infrastructure consolidation and layoffs. However, DuPont wanted to be able to focus its energy on developing science innovation in food, health care, home products, apparel, transportation, and electronics.

"We had eliminated work and consolidated but didn't invest in people or new technology or, especially, process management," said Maryann Holloway, DuPont's director of global alliance management and IT operations.[58]

In a 10-year, $4 billion deal that has been called the "big bang" of the conservative chemical industry, DuPont contracted with Accenture and Computer Sciences Corporation (CSC) to develop and maintain IT systems in more than 40 countries. Accenture took over the management of DuPont's manufacturing and engineering systems; order processing; materials and resource planning; and safety, health, and environmental analysis systems. The deal set a precedent in outsourcing IT, both in the scope and duration of the contract.

"This was a real precursor to multisourcing," said Holloway, who explained that a great deal of hard work is still involved in integrating the service providers.[59]

By stabilizing its fixed IT costs, DuPont was able to enjoy greater financial flexibility and concentrate on its core business: the pursuit of scientific innovations in its own industry. In addition, DuPont was able to reduce IT costs by another 5 percent.

Accenture has developed its position as a business process outsourcer. However, some business analysts have warned that Accenture is spreading itself too thin. Accenture maintains five operating groups: Financial Services, Communications and High Tech, Products, Resources, and Government. The company must build specialization across a broad range of areas, as shown in the following table.

Accenture's Areas of Expertise by Industry	
Aerospace and defense	Government
Airlines	Health and life sciences
Automotive	Industrial equipment
Chemical	Media and entertainment
Communication	Metals
Consumer goods and services	Mining
Electronics and high tech	Public transportation
Energy	Retail
Financial services	Travel
Forest products	Utilities
Freight and logistics	

The risk is that an investment in one area may not pay off if the company cannot offer that service to its other clients. "Even IBM focuses on a core of four processes, and EDS recently whittled down its offerings after discovering that its $3 billion BPO business was too scattered across multiple processes to give it any significant momentum," cautioned Forrester analysts.[60] Yet, Accenture has continued to aggressively pursue vertical expansion.

Technological Innovation

Given its vertical strategy, Accenture not only must research and develop emerging technologies, but anticipate how new technologies can benefit its clients and improve business processes. In addition, Accenture researches business processes in and of themselves: its well-regarded research program has reviewed more than 6000 companies to determine what factors provide business with strategic advantage over time. Accenture's staff track the latest trends, such as the recent growth of Indian companies as they move into the global market by acquiring or merging with companies in Europe, North America, and elsewhere. Accenture's staff produce reports on the Japanese insurance market, supply chain management for the aerospace industry, and trends in retail outsourcing.

At the same time, R&D's staff develop new advances in fields such as human-computer interaction, business process analytics, and intelligent device integration. During fiscal year 2006, the company had 1368 patents pending. To fuel this innovation, Accenture opened its fourth technology lab in 2006, in Bangalore, India.

Global Positioning

Accenture operates a Global Delivery Network with more than 40 centers in strategic locations. Beyond traditional Western hubs like London, Madrid, and Chicago, Accenture has facilities in countries such as the Czech Republic, Latvia, Slovakia, and Brazil, which allows it to provide on-shore, nearshore, or off-shore options to Latin American, Northern European, and Eastern European clients. Just under half of Accenture's 2006 revenue came from the European Middle East Africa (EMEA), with most of the rest coming from the Americas. Accenture also is building inroads into Asia and the Pacific, with centers in Shanghai and Dalian, China to gain access to the world's fastest-growing economy. By setting up shop in China, Accenture not only has access to cheap labor but to the country's rapidly expanding business markets. Accenture recently picked up contracts with China National Offshore Oil Corp., TCL, Lenovo, and Diageo Shanghai.[61]

Revenues Before Reimbursements by Geography (in billions of US$)	
Americas	$7.7
EMEA	$7.6
Asia Pacific	$1.3
Total	$16.6

When the Indian consulting companies began to pull profits from the "Big Six" Western companies, Accenture responded by accelerating its move into India. By August 2006, 35,000 of its 160,000 employees were located in India.

Accenture's Advantage

As Accenture competes with Tata, Infosys, and Wipro in India, China, and elsewhere, it has one enormous advantage over these companies: its vertical structure. Indian companies have not built broad expertise in specific industries yet. They cannot offer the same range of specialized outsourcing and consulting services. Therefore, the Indian firms cannot compete with Accenture for the increasing number of customers who are looking for consultants or outsourcers with expertise in their own industry.

Accenture is in a position to do what Indian companies are only now positioning themselves to do: hire professionals in India so that they can offer vertical solutions at lower prices. Originally, most employees in India staffed low-level IT posts and positions that did not require customer contact, but this has changed. In March 2007, Accenture announced plans to hire 2000 more Indian professionals. By the year's end, Accenture will have more employees in India than in the United States.

"We are actively recruiting management consultants, professionals with deep industry and functional expertise, and graduates of India's top business schools," announced Mark Foster, Accenture's group chief executive for Management Consulting and Integrated Markets.[62]

Other Western companies are adopting similar approaches. Cisco plans to have 20 percent of its "top talent" in India within five years. IBM is expanding its Bangalore research group. Once criticized for its overambitious investment in specialized industries, Accenture recently has been hailed as a model of success. No business analyst can argue with Accenture's high annual growth rate and expanding market share.

Accenture's Annual Revenue Before Reimbursements (in billions of US$)	
2006	$16.65
2005	$15.57
2004	$13.67
2003	$11.82
2002	$11.57

As Indian companies reposition themselves for BPO, Accenture will have to stay one step ahead. The years to come will show whether the Accenture strategies of vertical growth, geographical positioning, and technological innovation will prevent Indian outsourcers from grabbing Accenture's market share.

Discussion Questions

1. What strategic advantage have Indian outsourcers used to capture market share away from Western IT leaders?

2. What risks did Accenture face when developing itself as a business process outsourcer?

3. Why have companies like DuPont chosen to outsource entire business processes?

4. How do Accenture's vertical assets give the company a strategic advantage over Indian competitors?

Endnotes

[1] Patrick Thibodeau, "GM Awards IT Outsourcing Contracts Worth $7B," *Computerworld*, 2 February 2006.

[2] Eli Lilly 2006 Annual Report, page 6.

[3] Eli Lilly 2006 Annual Report, pages 6–7.

[4] "Clinical data management system," Wikipedia.

[5] Paul McDougall, "Drug Giant Lilly Outsources Clinical Data to India," *InformationWeek*, 15 November 2006.

[6] "Tata Consultancy Services Enters into Engagement with Eli Lilly and Company," press release, Tata Consultancy Services Web site at *www.tcs.com*, 15 November 2006.

[7] Kirsty Barnes, "Eli Lilly Awards Major Indian Contract," *DrugResearcher.com*, 14 November 2006.

[8] "Tata Consultancy Services Enters into Engagement with Eli Lilly and Company," press release, Tata Consultancy Services Web site at *www.tcs.com*, 15 November 2006.

[9] Ibid.

[10] Paul McDougall, "GlaxoSmithKline Outsources Data Management to TCS," *InformationWeek*, 29 March 2007.

[11] Paul McDougall, "Drug Giant Lilly Outsources Clinical Data to India," *InformationWeek*, 15 November 2006.

[12] Richard Pastore, "CIO Hall of Fame: Katherine M. Hudson," *CIO*, 15 September 1997.

[13] Ed Frauenheim, "Sears Ends $1.6 Billion Deal with Computer Sciences," *CNet News.com*, 17 May 2005.

[14] "Galaxy Nutritional Foods Reports First Quarter Operating Results," 23 August 2006 (accessed at *www.lexdon.com/article/*, 5 June 2007).

[15] Theo Forbath and Peter Brooks, "Outsourcing's Next Wave," *CRM Buyer*, 9 April 2007.

[16] Gary Hamel and C.K. Prahalad, "The Core Competence of the Corporation," *Harvard Business Review*, Volume 68, no. 3, pages 79–93, May–June 1990.

[17] "TCS to Provide Complete Outsourcing Services to Banco Pichincha," *BPO Times*, 17 January 2007.

[18] "TCS Inks $140m Banco Pichincha Deal," *finextra.com*, 16 January 2007.

[19] Douglas Brown and Scott Wilson, "50 Best Managed Global Outsourcing Firms," *www.sourcingmag.com/content/c060712a.asp* (accessed 28 May 2007).

[20] Andrew Burger, "Outsourcing, Part 1: Is Resistance Futile?" *CRM Buyer*, 26 April 2007.

[21] "Pizza Inn, Inc., Reports Results for the Third Quarter Fiscal Year 2007," PRNewswire, 9 May 2007.

[22] "Accenture and AIG Europe Sign 10-Year Business Processing Outsourcing Agreement," press release (accessed at *www.accenture.tekgroup.com*, 18 January 2006).

[23] "Mumbai International Airport Ltd. Signs MoU with TCS," press release, Tata Consultancy Services, 4 February 2007.

[24] Ronald Schmelzer, "Outsourcing, SOA, and the Industrialization of IT," SearchWebServices.com, 16 October 2004.

[25] Paul McDougall, "In Depth: When Outsourcing Goes Bad," *InformationWeek*, 19 June 2006.

[26] Andrew Burger, "Outsourcing, Part 1: Is Resistance Futile?" *CRM Buyer*, 26 April 2007.

[27] David Barboza and Louise Story, "Train Wreck," *The New York Times*, pages C1 and C4, 19 June 2007.

[28] Paul McDougall, "In Depth: When Outsourcing Goes Bad," *InformationWeek*, 19 June 2006.

[29] Ibid.

[30] Paul McDougall, "The Importance of an Outsourcing Prenup," *InformationWeek*, 23 May 2005.

[31] Ibid.

[32] Ed Frauenheim, "Dell Drops Some Tech Calls to India," *News.Com*, 23 November 2003.

[33] Robert McMillan, "Offshore Outsourcing Cited in Florida Data Leak," *Computerworld*, 26 March 2006.

34 Nari Kannan, "Reducing Operational Risk in Business Process Outsourcing," *SourcingMag.com*, 5 June 2007.

35 Bob Brown, "FBI Special Agent Recounts Outsourcing Horror Story," *Computerworld*, 16 May 2006.

36 Thomas Koulopoulos, "Value Creation Through Smart Sourcing," *Optimize*, Issue 22, March 2006.

37 "Outsourcing for Business Growth," *www.hp.com/changeartists*, 7 June 2007.

38 Mary Hayes Weier, "Strike Shuts Down Outsourcing in India," *InformationWeek*, 12 February 2007.

39 Bob Violino, "Benchmarking Takes Root," *InformationWeek*, 3 October 2005.

40 Ibid.

41 "Delivering High-Performance Outsourcing: Best Practices from the Masters," Executive Survey Results, Accenture, 2004.

42 Paul McDougall, "EDS Wins $1 Billion Outsourcing Deal with German Retailer," *InformationWeek*, 7 May 2007.

43 Jennifer Mears and Ann Bednarz, "Take It All Outsourcing on the Wane," *InformationWeek*, 30 May 2005.

44 Paul McDougall, "In Depth: When Outsourcing Goes Bad," *InformationWeek*, 19 June 2006.

45 Nari Kannan, "Reducing Operational Risk in Business Process Outsourcing," *Sourcingmag.com*, 5 June 2007.

46 Bob Violino, "Benchmarking Takes Root," *InformationWeek*, 3 October 2005.

47 Susannah Patton, "Cutting Costs with Multiple Outsourcers," *CIO*, 16 January 2007.

48 Antony Sawas, "Swansea Strike is a 'PR Disaster' for Outsourcing," *ComputerWeekly*, 13 September 2004.

49 Antony Sawas, "Swansea Project to Cost 50% More," *ComputerWeekly*, 27 September 2004.

50 Antony Sawas, "Swansea IT Staff Put Their Case for Secondment," *ComputerWeekly*, 1 November 2004.

51 Lindsay Clark, "Swansea Awards Capgemini £119m IT Services Deal," *ComputerWeekly*, 4 January 2005.

52 Antony Sawas, "Swansea Council Approves Transfer of IT Staff to Capgemini," *ComputerWeekly*, 15 April 2005.

53 Tim Richardson, "Swansea IT Workers Lose Outsourcing Fight," *ComputerWeekly*, 14 April 2005.

54 Bill Goodwin, "Swansea and Capgemini Outsourcing Deal Delayed," *ComputerWeekly*, 1 November 2005.

55 Patrick Thibodeau, "GM Awards IT Outsourcing Contracts Worth $7B," *Computerworld*, 2 February 2006.

56 Susannah Patton, "Cutting Costs with Multiple Outsourcers," *CIO*, 16 January 2007.

57 John Moore, "Accenture," *www.eweek.com/article2/0,1759,1254083,00.asp*, 7 May 2001.

[58] Linda Tucci, "IT happens: The 'Big Bang' at DuPont and its aftermath," *CIO News*, *http://searchcio.techtarget.com/originalContent/0,289142,sid19_gci1180320,00.html*, 20 April 2006.

[59] Ibid.

[60] Christine Ferrusi Ross, John C. McCarthy, Carey Schwaber, "Accenture's Growth Strategy: Mostly On The Right Track," Forrester, *www.forrester.com/ER/Research/Brief/Excerpt/0,1317,32680,00.html*, 30 September 2003.

[61] Accenture annual report, *www.accenture.com/Global/About_Accenture/Investor_Relations/Annual_Report/AnnualRep06Overview.htm*, 2006.

[62] "Accenture to Expand Management Consulting Capabilities in India," Accenture Newsroom, *http://newsroom.accenture.com/article_display.cfm?article_id=4525*, 20 March 2007.

CHAPTER 5

CORPORATE GOVERNANCE AND IT

THE IMPORTANCE OF CORPORATE GOVERNANCE

"Research into IT management practices at hundreds of companies around the world has shown that most organizations are not generating optimal value from their IT investments. The most important factor distinguishing top-performing from substandard-performing organizations is the level of leadership by business and senior managers."[1]

—Luc Kordel, information systems author, trainer, and auditor

HARLEY-DAVIDSON

Why Managers Must Get Involved in IT Governance

Harley-Davidson, founded in 1903 in Milwaukee, Wisconsin, is the oldest and arguably most renowned producer of motorcycles in the United States. Harleys are high-quality, heavy bikes designed for cruising on the highway and are known for the distinctive rumbling sound of their exhaust. The firm employs about 10,000 workers and is a reliable source of steadily increasing revenue, with recent annual revenue exceeding $5 billion.

Like other publicly held U.S. corporations, Harley-Davidson is subject to many federally mandated regulations, including the Sarbanes-Oxley Act, the Health Insurance Portability and Accountability Act (HIPAA), and Gramm-Leach-Bliley. (See Table 5-1 for a summary of these regulations and others.)

TABLE 5-1 Regulations that influence corporate and IT governance

Regulation	Goal(s) of the Regulation
Sarbanes-Oxley Act (SOX)	Ensures that internal controls are in place to govern the creation and documentation of financial statements. Section 404 of the act requires a signed statement by the CEO and CFO attesting that the information in any of their firm's SEC filings is accurate, with stiff penalties for false attestation.
Health Insurance Portability and Account-ability Act (HIPAA)	Requires employers to use national standards for electronic healthcare transactions with insurers and healthcare providers, and requires that employers ensure the security and privacy of employee health data.
Gramm-Leach-Bliley	Requires all financial-services institutions to communicate their data privacy policies and honor customer data-gathering preferences.
Foreign Corrupt Practices Act	Prohibits corrupt payments to foreign officials for the purpose of obtaining or keeping business.
Basel II	An international set of risk and capital management requirements designed to ensure that banks hold capital reserves commensurate to the risk banks assume through their lending and investment practices. The less the exposure, the less capital the bank needs to safeguard its solvency and overall economic stability.

In the early 2000s, Harley-Davidson had only rudimentary IT internal controls in place. Further-more, its managers were not well versed in implementing and managing internal controls. In recognition of this relatively poor state of affairs and in response to an increased emphasis on effective management controls, Harley-Davidson formed a new IT compliance department whose purpose was to improve internal controls and mitigate the risk of noncompliance to federal mandates. The group also was instructed to work effectively but to avoid hurting product quality or slowing down production.

Initially, the group began to implement a vendor's general computer controls model. However, after a key member of the department attended a Control OBjectives for Information and Related Technology (COBIT) user convention, the member persuaded the group to convert the effort to implement the

COBIT control framework. Harley-Davidson decided to convert to the COBIT control framework for several reasons:

- COBIT is an internationally accepted standard for IT governance and control practices.

- The use of COBIT provides a common language, terminology, and processes that can be used by business managers, end users, auditors, and IT professionals.

- Acquisition of the COBIT framework, including tools and templates, is essentially free as a download.

- COBIT is a set of decades-old standards that are highly regarded by auditors because they provide specific guidelines and are completely independent of any computer hardware or software platforms. Thus, the company easily gained agreement with its external auditors to use the COBIT framework and control objectives.[2]

The IT compliance team began by comparing the firm's existing controls to those recommended by the COBIT framework. A key decision was made to expand the scope of work beyond Sarbanes-Oxley compliance to provide a broad control framework that encompassed the full range of IT-related activities. Many gaps in the existing control procedures were uncovered:[3]

- There was no standardized user process to access data and IT applications, which made life difficult for end users and created potential opportunities for hackers.

- There was no defined change management process to capture information about who made changes to IT infrastructure components or why.

- There was no impact analysis of proposed changes; even a seemingly trivial change could unexpectedly cause a chain reaction and affect several other components.

- The backup and recovery processes were not fully tested and 100 percent reliable.

- To meet the even more stringent control requirements of Sarbanes-Oxley, HIPAA, and Gramm-Leach-Bliley, the firm had a long way to go.

Sobered by their findings, the team prioritized the work to be done and made plans to eliminate the gaps over a reasonable period of time. The team recognized that a key to the successful introduction and adoption of more effective internal controls was ensuring that managers understood *why* they needed to care about improved internal controls.

As changes in internal controls and processes were being designed and implemented, it was a constant challenge to ensure that control owners truly understood the concept of effective control. Often they would assume that "more controls are better." The team had to prove time and again that having one or two good controls is better than having six or seven ineffective controls. Once control owners understood the value of expending fewer resources and less time for an equal or better control, they became enthusiastic about the internal controls improvement project.

The team also identified tracking and reporting as important components of Harley-Davidson's ongoing IT governance activities. They developed a Microsoft Access issues tracking database to enable managers and the internal audit staff to view information about known control weaknesses. The database enabled team members to learn about carryover and repeat audit findings, and to follow up with management action plan owners to ensure continued effort to address the issues.

Harley-Davidson has achieved many benefits by implementing an improved control model:[4]

- The company passed its Sarbanes-Oxley one-year compliance review.

- COBIT provides Harley-Davidson management with an objective gauge of where the company is positioned regarding controls and what must be done to improve. Previously, it had been difficult for Harley-Davidson to find internal controls from other manufacturers for the purpose of benchmarking.

- The firm gained external audit agreement on the company's control position, providing a foundation for all future internal and external audits.

- Prior to implementing the COBIT framework, business areas and processes that were audited externally were chosen randomly or based on loose justifications. Areas selected for auditing now are based firmly on recognizing the area's business value and need for controls.

LEARNING OBJECTIVES

As you read this chapter, ask yourself:

- What is governance and what are the key elements of an effective governance process?
- How can an effective governance program improve the likelihood of organizational success?

This chapter defines the goals of IT governance and clarifies the importance of good governance in terms of achieving organizational objectives and managing risk.

WHAT IS IT GOVERNANCE?

Corporate governance is the set of processes, customs, rules, procedures, policies, and traditions that determine how to direct and control management activities. The primary people involved in corporate governance include the board of directors, CEO, senior executives, and shareholders. Corporate governance addresses issues such as:

- Preparation of the firm's financial statements
- Monitoring the choice of accounting principles and policies
- Establishment of internal controls
- Hiring of external auditors
- Nomination and selection of people to the board of directors
- Compensation of the chief executive officer and other senior managers
- Management of risk
- Dividend policy

Interest in corporate governance has grown due to recent accounting scandals resulting in bankruptcies at companies such as Arthur Andersen, Enron, Global Crossing, Tyco, and Worldcom. In addition, board members who are responsible for paying executives have been challenged as a result of several scandals in which executives received compensation perceived by some critics as overly generous. For example, Richard Grasso at the New York Stock Exchange was fired by the board of directors after his $139 million annual salary became public. These examples show that governance decisions are only as good as the people who make the decisions. Senior executives must have character and integrity to avoid improper conduct.

IT governance is a decision-making process that involves investments in IT. Governance includes defining the decision-making process itself, as well as defining who makes the decisions, who is held accountable for results, and how the results of decisions are communicated, measured, and monitored.[5]

An organization's executives and board of directors are responsible for governance. They carry out this duty through committees that oversee critical areas such as audits, compensation, and acquisitions. Enlightened organizations recognize that IT governance is not the responsibility of IT management but of executive management, including the board of directors.

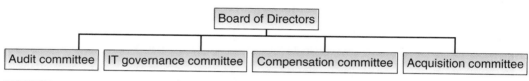

FIGURE 5-1   Board of directors and various subcommittees

There are two primary goals of effective IT governance: ensuring that an organization achieves good value from its investments in IT and mitigating IT-related risks, as shown in Figure 5-2. IT governance is similar to financial portfolio management, in which a manager weighs the rate of return and balances it against the risks associated with each investment. The manager then makes choices to achieve a good rate of return at an acceptable level of risk. Achieving good value from IT investments requires a close alignment between business objectives and IT initiatives. Mitigating IT-related risks means embedding accountability and internal controls in the organization.

Ensuring that an Organization Achieves Good Value From its Investments in IT

At one time, IT was viewed simply as a support function that was separate and distinct from a business. Today, however, IT infrastructure and applications are so integral to various business lines and functions that many parts of the organization could not operate without IT. This is especially true for organizations that electronically integrate partners and customers into their business processes.[6] If IT is integral to a business and business managers must take a key role, then the means by which managers discharge their responsibilities—governance—must be applied to the management of IT. Senior executives must take leadership for creating an effective partnership between the IT organization and the rest of the organization.

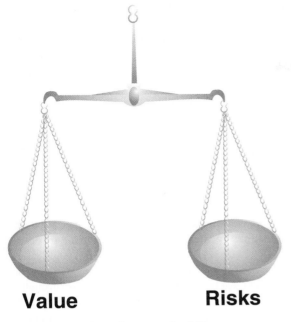

Value   **Risks**

FIGURE 5-2   Two primary goals of IT governance

An effective IT strategic planning process, as discussed in Chapter 2, can help an organization achieve good value from IT investments by ensuring close alignment between business goals and objectives and IT project goals and objectives. Only IT projects that are consistent with the business strategy and that support business goals and objectives should be considered for staffing and funding. Such projects will deliver the organization's strategic goals, whether they are increased revenues, decreased costs, improved customer service, or increased market share. Other projects should be rejected. This process ensures that IT is being effective and is working on the right projects.

For IT projects to be aligned with business goals and properly staffed, funded, and executed, the projects must deliver expected business results on time and within budget. This process involves applying good project management principles, as discussed in Chapter 3, to ensure that work is done efficiently and that results can be achieved with a high degree of predictability.

Mitigating IT-Related Risks

IT-related risks include the failure of IT systems and processes to meet government rules and regulations (e.g., Sarbanes-Oxley), security risks from hackers and denial-of-service attacks, privacy risks from identity theft, and threat of business disruption due to a disaster or outage. Good internal controls and management accountability must be embedded in the organization to avoid IT-related risks.

Internal control is the process established by an organization's board of directors, managers, and IT systems to provide reasonable assurance for the effectiveness and efficiency of operations, the reliability of financial reporting, and compliance with applicable laws and regulations. For example, one internal control might be that a company's purchasing department, accounts payable department, and IT systems do not allow the same person to authorize a major purchase and then approve its payment. Internal controls play a key role in preventing and detecting fraud and protecting the organization's resources.

Improper conduct of senior managers and failure to hold managers accountable can circumvent even a good system of internal controls. For example, Siemens is one of the world's largest electrical engineering and electronics firms, with 475,000 employees worldwide. The firm makes everything from light bulbs to high-speed trains. Serious allegations have surfaced that Siemens managers set aside funds to bribe potential customers abroad in return for orders. From the 1990s through 2004, Siemens may have paid as much as $540 million in bribes to major customers in China, the Middle East, Nigeria, and Russia.[7] Critics question why these bribes were not detected earlier. No one identified that senior managers were violating established business practices and circumventing internal controls.

As another example of improper conduct among senior managers, Computer Associates, the fifth-largest software provider in the world, had a series of accounting problems through the 1990s that led to the conviction of its former executives, including ex-CEO Sanjay Kumar. Kumar and other executives allegedly instructed salespeople to complete deals after a quarter had closed and then concealed the lateness of the sale by removing timestamps from faxes and changing other documents. Again, no one identified that senior managers were violating established business practices and circumventing internal controls.

Various rules and regulations such as the Foreign Corrupt Practices Act (FCPA) and the Sarbanes-Oxley Act have been established to hold senior management accountable for the integrity of their organization's financial data and internal controls. The goal is to prevent problems like those at Siemens and Computer Associates.

There is increased corporate demand for help in the arena of risk mitigation and compliance. Many accounting, consulting, and software firms can provide products and services. For example, IBM announced a new service called IBM Data Governance Maturity Model Assessment in 2006 to help companies evaluate their data governance practices.[8]

Figure 5-3 shows the five key activities needed for effective IT governance. IT value delivery and risk management are the goals. Strategic alignment and IT resource management are the methods for achieving these IT governance goals. Performance measurement is the means by which management tracks how well its IT governance efforts are succeeding.

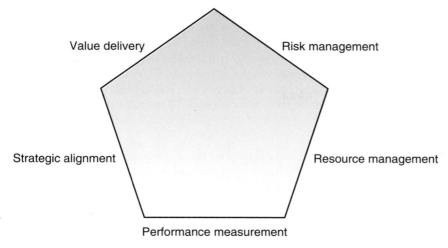

Value delivery

Risk management

Strategic alignment

Resource management

Performance measurement

FIGURE 5-3 Five key activities needed for effective IT governance

WHY MANAGERS MUST UNDERSTAND IT GOVERNANCE

Leveraging IT to transform an enterprise and create value-added services, increased revenue, and decreased expenses has become a universal goal for businesses. Successful managers seek opportunities to deliver the potential benefits promised by IT. However, IT-related initiatives are seldom simple and straightforward. They are influenced by many factors: the vision, mission, and values of the organization; community and organizational ethics and values; a myriad of laws, regulations, and policies; industry guidelines and practices; changing business needs; and the values of the IT stakeholders and company owners. Thus, successful managers need a process that can help them achieve high value from their investments in IT, manage associated risks, and deliver IT-related solutions that comply with increasing regulatory compliance demands. IT governance is just such a process.

In organizations that have good IT governance, the IT organization is better aligned and integrated with the business, risks and costs are reduced, and IT helps the company gain a business advantage. Organizations in which IT governance is lacking have inadequate direction and leadership, lack of accountability, and no measurement of the outcome of IT-related decisions. IT governance is an important tool to ensure the delivery of real value from IT expenditures and to mitigate IT-related risks.

IT GOVERNANCE FRAMEWORKS

Organizations can use a number of frameworks as a basis to develop their own governance model. The two best known are the IT Infrastructure Library (ITIL) and the Control OBjectives for Information and Related Technology (COBIT). ITIL and COBIT are not really competing frameworks but complements to each other. ITIL provides best practices and criteria for effective IT services such as help desk, network security, and IT operations. It is a useful tool to improve IT operations efficiency and IT customer service quality. COBIT provides guidelines for more than 30 processes that span a wide range of IT-related activities, including planning and organization, acquisition and implementation, delivery and support, and monitoring. COBIT is a useful tool to improve the quality and measurability of IT governance or to implement a control system for improved regulatory compliance.[9] Organizations adopt one of these frameworks to get a "jump start" on improving IT-related processes that are of most concern. The Plan-Do-Check-Act problem-solving approach is used in quality improvement and can be applied to improve IT-related processes.

IT Infrastructure Library (ITIL)

The **IT Infrastructure Library (ITIL)** is a set of guidelines initially formulated by the UK government in the late 1980s and widely used today throughout Europe and the United States to standardize, integrate, and manage IT service delivery. ITIL provides a proven and practical framework to plan and deliver IT operational services based on a synthesis of ideas from international practitioners. The framework defines "best practices" in 24 IT-related disciplines, and continues to be updated and improved; as of July 2007, ITIL was in its third version. More than 15,000 organizations have adopted ITIL, including Barclays Bank, British Airways, Guinness, Hewlett-Packard, IBM, Microsoft, and Procter & Gamble.

ITIL consists of five distinct volumes: ITIL Service Strategy, ITIL Service Design, ITIL Service Transition, ITIL Service Operations, and ITIL Continual Service Improvement. The ITIL Service Strategy volume covers the core principles of ITIL itself. It addresses strategy and value planning, roles and responsibilities of key players, planning and implementing service strategies, business planning and IT strategy linkage, and risks and critical success factors for implementing ITIL. There is also an IT Service Management Forum, an independent forum of ITIL users that promotes the exchange of information and experience among IT service providers worldwide. People can receive training and become certified in ITIL at three different levels: Foundation, Practitioners, and Managers.

Control OBjectives for Information and Related Technology (COBIT)

Control OBjectives for Information and Related Technology (COBIT) is a set of guidelines whose goal is to align IT resources and processes with business objectives, quality standards, monetary controls, and security needs. These guidelines are issued by the IT Governance Institute. They provide metrics, best practices, and critical success factors for COBIT-defined IT-related processes. COBIT's best practices represent the consensus of experts. You can download the guidelines at *www.isaca.org/cobit.htm*.

The initial set of COBIT guidelines was published in the mid-1990s. Since then the framework has been refined and improved several times; the current version, 4.1, was released in the summer of 2007. The IT Governance Institute, through its COBIT Steering Committee, intends to continually evolve the guidelines. The COBIT framework provides guidance for more than 30 IT-related processes grouped into four major categories, as shown in Table 5-2.

TABLE 5-2   The COBIT processes grouped into four major categories

Category	Process
Plan and Organize	Define a Strategic Plan and Direction
	Define the Information Architecture
	Determine Technological Direction
	Determine the IT Processes, Organization, and Relationships
	Manage the IT Investment
	Communicate Management Aims and Direction
	Manage IT Human Resources
	Ensure Compliance with External Requirements
	Assess and Manage Risks
	Manage Projects
	Manage Quality
Acquire and Implement	Identify Automated Solutions
	Acquire and Maintain Application Software
	Acquire and Maintain Technology Infrastructure
	Enable Operation and Use
	Procure IT Resources
	Manage Changes
	Install and Accredit Solutions and Changes
Deliver and Support	Define and Manage Service Levels
	Manage Third-Party Services
	Manage Performance and Capacity
	Ensure Continuous Service
	Ensure Systems Security
	Identify and Allocate Costs
	Educate and Train Users
	Manage Service Desk and Incidents
	Manage the Configuration
	Manage Problems
	Manage Data
	Manage the Physical Environment

TABLE 5-2 The COBIT processes grouped into four major categories (continued)

Category	Process
	Manage Operations
Monitor and Evaluate	Monitor and Evaluate IT Processes
	Monitor and Evaluate Internal Control
	Ensure Regulatory Compliance
	Provide IT Governance

Each of the processes is described in terms of:

- *The process inputs*—Define what the process owner needs from others
- *The process description*—Describes what the process owner has to deliver
- *The process outputs*—Lists specific deliverables the process must produce
- *The goals and metrics*—Describe how the process should be measured
- *The RACI chart*—Defines what has to be delegated and to whom (see Table 5-3)
- *The maturity model*—Describes how the process can be modified for improvement

TABLE 5-3 RACI roles and descriptions

Role	Description
R	The person who owns the problem
A	The person to whom "R" is accountable; this person must approve the work
C	People who should be consulted must have information necessary to complete the work
I	People who must be informed of the decision, but need not be consulted

For each of the COBIT processes, the "maturity level" of management processes can be evaluated on a scale of 0 to 5. The scale is defined as follows:

- 0 Non-Existent—Management processes are not applied at all
- 1 Initial—Processes are ad hoc and disorganized
- 2 Repeatable—Processes follow a regular pattern
- 3 Defined—Processes are documented and communicated
- 4 Managed—Processes are monitored and measured
- 5 Optimized—Best practices are followed and automated

Organizations can use the scale for each process to evaluate a number of items:

- Determine the organization's current maturity level
- Define the maturity level the organization needs to achieve
- Identify the maturity level that is considered "best practice" in their industry
- Identify the maturity level achieved by their strongest competitor

Organizations can then use this information to choose which processes have priority for improvement and which can be addressed later.

Using PDCA and an IT Governance Framework

The Plan-Do-Check-Act (PDCA) model, shown in Figure 5-4, is a tried and proven method that can be applied to a specific targeted process that has been identified for improvement. Each step in the model has the following specific objectives:

- The Plan step requires the improvement team to identify its target improvement area, analyze how things work currently, and identify opportunities for improvement.
- In the Do step, the change decided in the Plan step is implemented, often on a pilot or limited basis to assess the potential impact of the proposed change(s).
- In the Check step, the results of the change are measured. Were the results achieved? Were there unexpected negative side effects? Is further improvement needed?
- In the Act step, the improvement team considers whether it is worth continuing the process with the recently implemented change. If the change is too complicated for people to follow or it led to insignificant improvements, then the change may be aborted. At this point the team would go back to the Do step and start over. Thus, the completion of one cycle of improvement flows into the beginning of the next cycle.

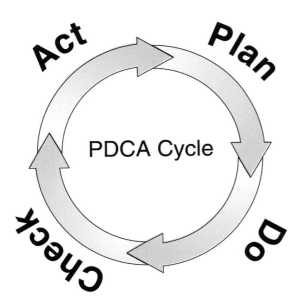

FIGURE 5-4   PDCA cycle for implementing process improvement

The ITIL or COBIT governance frameworks provide an excellent set of best practices for various IT-related processes. A process improvement team can use these best practices in the

Plan step to assess their organization's current practices and identify areas of improvement. The best practices also can provide improvement ideas for the Do step and measures for the Check step. Thus, many organizations combine the use of PDCA and an ITIL governance framework to get excellent results in their process improvement projects. Figure 5-5 depicts how the PDCA model can be applied to improve an IT-related process using the COBIT framework as a benchmark. Read the following special feature to see an example.

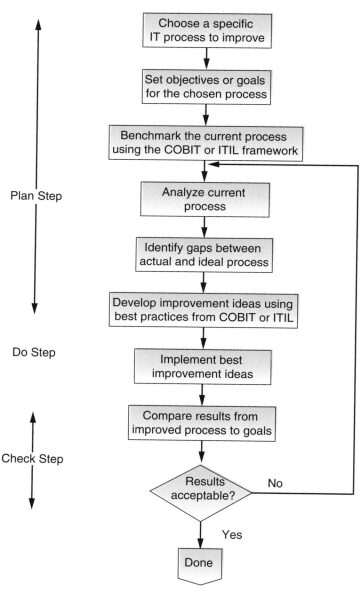

FIGURE 5-5   Process improvement using PDCA and COBIT or ITIL

Audatex Uses PDCA and ITIL to Improve Its Service Offerings

Audatex is a UK-based subsidiary of the Soleara Company. It operates as a service provider for body shops and insurance companies. Audatex offers an integrated suite of software to support auto insurance collision repair shops in automating claims processing, repair job estimating, and shop management. The Audatex software enables insurance firms to increase customer satisfaction by shortening repair times, reducing loss adjustment expenses, and improving claims processing control and efficiency. The system includes support for such actions as ordering courtesy cars, sending text messages to customers to provide them with progress updates on their repair job, ordering parts from suppliers for repairs, invoicing insurance companies, and matching invoices to repair estimates.[10]

The firm must invest heavily in product development, new technology, and improved products and services to remain competitive. To that end, Audatex recently converted from a stand-alone software package that ran on a user's personal computer to a Web-based service that users access over the Internet. The new infrastructure allows product specifications and other information to be updated once at a central point rather than multiple times on a site-by-site basis.

Ross McEleny, IT services director at Audatex, recognized that running the Web-based system would introduce new, demanding requirements for security, regulation, and service levels. Audatex must not only meet these requirements but avoid passing unnecessary costs along to its customers. Also, because Audatex stores and manages data centrally on behalf of its many clients, it has a legal level of responsibility under regulations such as the Data Protection Act, the main legislation that covers the protection of personal data in the UK.[11] From an insurer's point of view, Audatex has to make data both highly available and highly secure, as well as keep it safe for seven years in case of legal or fraud-related action. With all these issues in mind, McEleny formed a process improvement team. McEleny said, "It's no good having high-performance systems that aren't available, or highly available systems that don't perform."[12]

During the Plan step, the team defined the goal as being able to deliver a consistent and measurable service to Audatex clients. The analysis began with an audit of existing systems and processes. The ITIL framework was used as a standard for comparison, which allowed McEleny to check the efficiency of processes and identify gaps where new systems and standards needed to be introduced. The team identified the need for additional staff training and capturing useful data from each customer call.

During the Do step, the team introduced staff training in two separate areas: service support and service delivery. They also implemented the ITIL program best practices for call logging and problem resolution.

During the Check step, the team introduced useful charts (dashboards) to measure up-to-date progress on problem resolution.

continued

Not completely satisfied with the results, Audatex established a continuous improvement loop that uses many of the techniques built into ITIL to capture calls from customers and feed them into the firm's incident management and system development processes. Audatex also uses customer surveys to measure progress. The company emphasizes that it is just beginning a long journey of continuous improvement.

Discussion Questions:

1. What role did the ITIL guidelines play in this process improvement project?
2. Can you identify areas at Audatex other than customer service in which ITIL guidelines and the PDCA process could be brought to bear and achieve significant benefits?
3. At what point would a team stop this ongoing process improvement?

Now that we have discussed basic IT governance frameworks, let's examine one of management's key governance responsibilities: business continuity planning.

BUSINESS CONTINUITY PLANNING

A disaster is an unplanned interruption of normal business operations for an unacceptable period of time. Unfortunately, the list of potential business-disrupting events seems to be getting longer (see Table 5-4). Whether it affects a broad geographical area and thousands of organizations or is confined to one floor of a building of one organization, a disaster can result in many negative consequences:

- Loss of staff through death or injury
- Unavailability of staff due to a disruption in their ability or willingness to travel
- Adverse psychological effects on staff, including stress and demoralization
- Damage to buildings, equipment, raw materials, and finished products
- Inability to run time-sensitive processes such as order processing, payroll, accounts payable, accounts receivable, and inventory control
- Loss of data processing capability
- Loss of voice and data communications
- Loss of essential electronic and manual records
- Disruption to customers and dependent organizations
- Damage to an organization's reputation
- Loss in stock price and increased difficulty in borrowing money

TABLE 5-4 Examples of disasters

Type of Incident	Examples
Widespread natural disasters	Hurricane, flood, earthquake, tsunami, forest fire
Localized natural disasters	Tornado, wind damage, landslide
Isolated incidents	Fire, power outage, death of key personnel, backhoe severs an electrical or communications line
Deliberate attacks	Denial-of-service attack, terrorist attack, civil unrest

Examination of recent major disasters reveals certain key planning assumptions that must be built into an organization's business continuity plan (see Table 5-5).

TABLE 5-5 Hard lessons learned from recent major disasters

Disasters are not always contained to a limited geographical area.
The most basic essentials (including potable water, electricity, and passable roads) may not be available.
Essential police and firefighting services may not be available.
Employees may not be able to reenter their former place of work for months, or ever.
The ability to recover may be limited by lack of building material, equipment, and workers.
The impact of a disaster may linger for months.
Key members of an organization, including members of the disaster recovery team, may be lost.
Suppliers and key vendors on which an organization depends also may be struggling to recover.
Certain cities and even entire countries may not be safe locations for a major corporate facility.
Organizations need to consider carefully the trade-off between the efficiency and cost savings of placing all operations within a small geographical area and the added safety of distributing operations across multiple, dispersed locations.
When building new facilities or expanding existing ones, use fire-retardant material, smoke detectors, sprinkler systems, wide stairwells, and safety floors.

A **business continuity plan** defines the people and procedures required to ensure timely and orderly resumption of an organization's essential, time-sensitive processes with minimal interruption. **Due diligence** is the effort made by an ordinarily prudent or reasonable party to avoid harm to another party. Failure to make this effort may be considered negligence. Being able to show a written, tested business continuity plan is considered part of due diligence. Indeed, many laws and regulations specify requirements for business continuity planning. The requirements vary by country and by industry. However, regardless of any legal requirements, there is a growing sense of urgency across the global business community to be prepared for a disaster by implementing a comprehensive business continuity plan. Each week seems to bring a disaster that affects some part of the globe, but every day there are hundreds of isolated incidents that hurt the ability of an organization somewhere to operate. The directors of an organization are responsible for preparing to deal with a disaster or some lesser incident that hurts their organization's ability to function.

The scope of a full business continuity plan addresses the health and safety of all workers; minimizes financial loss, including damages to facilities, critical data, records, finished products, and raw materials; minimizes the interruption to critical business processes; and provides for effective communications to customers, business partners, and shareholders. A well-considered business continuity plan can mean the difference between your organization's survival and failure in a disaster.

A **disaster recovery plan** is a subset of the business continuity plan, and focuses on keeping components of the IT infrastructure functioning during a disaster or recovering them quickly afterward. The COBIT process known as "Ensure Continuous Service" describes how to create an effective disaster recovery plan. The next section describes the process shown in Figure 5-6 to develop a business continuity plan.

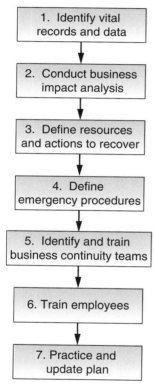

FIGURE 5-6   How to develop a business continuity plan

Process for Developing a Business Continuity Plan

Identifying Vital Records and Data

Every company has key electronic records and hard copy data that are essential to manage and control the cash flow and other tangible assets of the organization. These records include customer data, contracts, current order information, accounts payable data, accounts receivable data, inventory records, and payroll information. Companies must identify vital records and data and then determine where and how they are being stored and backed up. Then, considering various disaster scenarios, companies must assess the adequacy of the current data storage plan.

Some approaches are not recommended, but unfortunately have been widely practiced. One bad approach is to have employees take backup copies of vital data home at the end of the work day. Such data is easily stolen or lost. Another bad approach is to have backup data stored in a building across the street: a disaster that affects the local area could wipe out the primary data and the backup data.

Other approaches are recommended and widely implemented. For example, as online databases are updated, companies can have these changes mirrored on a backup database hundreds of miles away. This approach is expensive, but it provides rapid access to current data in the event of a disaster. Another approach is to copy online databases every night to high-volume, inexpensive magnetic storage devices and ship them off-site to a data storage facility in another state. This low-cost solution minimizes the potential for losing more than one day of data.

Conducting a Business Impact Analysis

An effective business continuity plan can be developed only after an organization's unique requirements are identified. The needs of a small manufacturing facility clearly are not the same as those of a financial institution with thousands of employees around the world.

Even within the same organization, different parts of the business have different needs. Functions that control the cash flow of the company and provide essential services to customers are considered the most critical. A useful way to classify business functions is shown in Table 5-6. The classification is based on identifying and quantifying financial, operational, and service impacts associated with a business function becoming inoperable. Important to this analysis is determining how soon the impact will be felt. The time within which a business function must be recovered before an organization suffers serious damage is called the **recovery time objective**. Based on this data, each business function can be placed in the appropriate category. This triage approach to business continuity planning enables people to focus on meeting the recovery needs of the most essential functions.

TABLE 5-6 Business function classification

Business Function Classification	Recovery Time Objective	Example
AAA	This business function is extremely critical to the operation of the firm and cannot be unavailable more than a few *minutes* without causing severe problems.	Order processing
AA	This business function is critical to the operation of the firm and cannot be unavailable more than a few *hours* without causing severe problems.	Accounts receivable Accounts payable
A	This business function, while significant, can be unavailable for up to a few *days* without causing severe problems.	Payroll
B	This business function can be unavailable for several days in times of a major disaster without causing major problems.	Employee recruiting

Defining Resources and Actions Required to Recover

For all AAA priority business functions, document all the resources needed to recover the business function within the recovery time objective: number of people, phones, files, desks, office space, faxes, computers, software, printers, and so on. Establish a notification list that identifies who needs to be notified in the event of a disaster, including key suppliers, customers, and members of the media. Many organizations have established a continual voice conference to be used in a disaster so employees can call in, report their status, and get up-to-date information and instructions.

Next, identify the sequences of steps that must occur to recover from a disaster. Several scenarios need to be addressed, from a relatively isolated incident that affects one floor of a single building to a complete and widespread loss of facilities, people, and equipment. Specific features to consider for inclusion in the recovery of a AAA priority business function include:

- Use of an emergency generator to replace lost public utility services.
- Contingency plans to relocate operations to another site or run IT equipment from a backup facility. Many organizations pay for use of an alternate site to house employees, store their backup files and data, and operate backup equipment if employees cannot return to the workplace.
- Consideration of backup warehouse, production, and distribution capabilities to enable a company to continue making its product and get it to market.
- Intelligent switching capabilities and backup networks for voice and data communications. Intelligent switches can recognize when a portion of the network has been lost and automatically reroute voice and data communications over alternate communications paths to locations that still work.

When all the preceding tasks have been completed for the AAA priority business functions, repeat the process for all the AA priority business functions, and then for all the A priority business functions.

Defining Emergency Procedures

Emergency procedures define the steps to be taken during a disaster and immediately following it. A little planning and practice of such procedures can minimize loss of life and injuries as well as reduce the impact on the business and its operations. It is best to develop these plans in conjunction with professional first responders such as fire departments, police departments, and civil defense organizations. To the greatest extent possible, computer, data, and equipment backup processes should be triggered automatically or with a minimum of human intervention. For example, a file server should automatically detect that the primary power source has been lost so that the equipment can run on a battery or alternate power source. This detection should trigger an automatic procedure to back up key files over the network to servers located elsewhere.

Identifying and Training Business Continuity Teams

The three business continuity teams are the control group, the emergency response team, and the business recovery team. The members of these teams should be carefully selected based on their areas of expertise, experience, and ability to function well under extreme pressure. More members should be selected and trained than are actually needed, in case personnel are lost in a disaster. For the same reason, it is wise to cross-train people.

The control group provides direction and control during a disaster and operates from a secure emergency operations center equipped with emergency communications gear. The group gathers and analyzes data needed to make decisions and direct the work of the emergency response team and business recovery team. Communications must be maintained between the control group, the emergency response team, and the disaster recovery team.

For most organizations, the emergency response team includes members of the fire department, police department, and other first responders. Some large organizations have their own emergency firefighting department. Their role is to help save lives and contain the impact of the disaster.

The recovery team includes employees and nonemployee specialists who assess the situation once it is safe to do so. They assess the extent of the damage and decide if or when it may be safe to reenter the affected work area. They recommend whether the disaster recovery plan needs to be put into effect, depending on the impact of the disaster or incident.

Training Employees

All employees should be trained to recognize and respond to various types of disaster warnings, such as those for fires, tornados, gas releases, and so on. Employees must know whether to evacuate their office, seek shelter in a basement, stay put, or take some other action. In addition, it is a good practice to identify "floor wardens" who are responsible for evacuating a given floor or work area. These floor wardens receive additional training in crowd control, first aid, CPR, operation of defibrillators, and helping handicapped workers evacuate. They provide leadership and direction for other workers in time of disaster. Most organizations conduct one or two disaster drills per year to ensure that employees know what to do and that wardens can operate effectively. The drills may simulate dealing with injured workers and helping handicapped workers.

Practicing and Updating the Plan

It does no good to develop a plan and never practice it. The business continuity plan needs to be tested to ensure that it is effective and that people can execute it. Many companies practice the business continuity plan for at least one AAA priority system once per year. The practice may be announced in advance to avoid excessive alarm, but employees are expected to exercise the business continuity plan and restore operations within the desired recovery time. It is essential to capture problems or issues not addressed by the plan and revise it to incorporate solutions. In this manner, the plan is continually upgraded to become more effective.

Organizations are in a constant state of change, so the plan must be continually updated to account for changes such as:

- Changes in personnel and their roles within the organization
- The acquisition, divestiture, and merger of organizational units
- Relocation of people, organizational assets, and locations where functions are performed
- Upgrades in software, hardware, and other equipment
- Changes in key suppliers and customers
- Changes in membership and contact information for the control team, emergency response team, and disaster recovery team

The manager's checklist in Table 5-7 provides a set of recommended actions for business managers to improve the success rate of their organization's projects. The appropriate answer to each question is yes.

TABLE 5-7   Manager's checklist

Actions	Yes	No
Does your organization have an effective corporate governance function?		
Does your corporate governance function include IT governance?		
Is there a good understanding of the IT-related risks that must be mitigated?		
Do business managers lead and direct the IT governance function?		
Does your organization have a documented business continuity plan?		
Does the business continuity plan take into account hard lessons learned from recent disasters?		
Is the business continuity plan developed using a triage of business functions based on their potential negative impact if they become inoperable?		

Chapter Summary

- IT governance is a decision-making process that involves investments in IT. Governance includes defining the decision-making process itself, who makes the decisions, who is held accountable for results, and how the results of decisions are communicated, measured, and monitored.

- Enlightened organizations recognize that IT governance is not the responsibility of IT management but of executive management, including the board of directors.

- Good internal controls and management accountability must be embedded in the organization to avoid IT-related risks.

- The five central themes of IT governance are: (1) IT value delivery, (2) risk management, (3) strategic alignment, (4) IT resource management, and (5) performance measurement.

- IT governance can help managers achieve high value from their investments in IT while enabling them to manage associated risks to deliver IT-related solutions that comply with an increasing level of regulatory compliance demands.

- Organizations can use a number of frameworks as a basis to develop their own governance model. The two best known are the IT Infrastructure Library (ITIL) and the Control OBjectives for Information and Related Technology (COBIT).

- ITIL and COBIT are not really competing frameworks but complements to each other. ITIL provides best practices and criteria for providing effective IT services. COBIT provides guidelines for more than 30 processes that span a wide range of IT-related activities, including planning and organization, acquisition and implementation, delivery and support, and monitoring.

- Each organization must perform an objective assessment of its unique risks, develop a comprehensive plan, and be prepared to train, test, maintain, and enhance the plan as needs change.

Discussion Questions

1. Discuss fully: Should Harley-Davidson, as described in the chapter's opening vignette, be criticized for using the COBIT framework rather than ITIL to improve its internal controls and mitigate the risk of nonconformance to various federal mandates?

2. Provide a strong argument for creation of an IT governance committee that reports to a board of directors.

3. Identify and briefly discuss the five central themes of IT governance.

4. Do you think that organizations should try to develop and use a system of internal cross-charging? Why or why not?

5. What are the goals of corporate governance and what issues does it address? What are the goals of IT governance and what issues does it address?

6. What is the goal of an organization's system of internal controls? Provide several examples of good internal controls and several examples of poor internal controls.

7. In what ways are ITIL and COBIT similar? How are they different?

8. Is it important for business managers to understand and be involved in IT governance? Why or why not?

9. Have you ever used the PDCA model? Briefly describe how it was used and the results achieved.

10. What is the scope of a business continuity plan? How is it different from a disaster recovery plan?

11. Should suppliers and customers have any role in defining the business impact of various business functions? Explain why or why not.

Action Memos

1. You are a senior manager for your firm, and are responsible for leading the IT governance subcommittee. You just received a text message from a young IT project manager who you met last week. "We are at an off-site meeting with IBM, and following a review of their new service called the IBM Data Governance Maturity Model Assessment, we are about to sign a contract for this service. We'd like your input. Please call my cell phone at 555-3686." You were not aware of any effort in this area. How do you respond?

2. You were appointed by senior management to be the project leader for the organization's business continuity planning effort. No one in the company has even looked at the plan in more than three years, let alone tried to execute the plan. Senior management appointed you to be in charge of "dusting off and freshening up" the plan. You have just read an e-mail from another appointed member of the team. He has challenged you to tell him why he should "waste his time" on a meaningless effort. How do you respond?

Web-Based Case

Do research to find detailed documentation of your school's business continuity plan and which school officials are responsible for creating and maintaining it. After reading the plan, write a brief memo assessing the thoroughness and effectiveness of the plan.

Case Study

Goldman Sachs Plans for Business Continuity

Goldman Sachs was founded in 1869 and is one of the world's leading investment banking and securities firms. Its headquarters is in the Lower Manhattan area of New York at 85 Broad Street. It has offices in major U.S. cities, including Chicago, Los Angeles, and San Francisco. It also has offices in major financial centers around the world, including, Beijing, Hong Kong, London, Moscow, Sao Paulo, Tokyo, and Zurich.

The firm has a diversified client base of corporations, financial institutions, and governments, as well as individuals from some of the wealthiest families in the world. Goldman Sachs provides a broad and comprehensive set of products and services through the following organizations:[13]

- Investment banking helps governments to raise money by issuing and selling securities, aids public and private corporations in raising funds in the equity and debt capital markets, and provides strategic advisory services for mergers, acquisitions, and other types of financial transactions.

- Trading and principal investments involves fixed income and equity products, currencies, commodities, and derivatives on these products. In addition, Goldman Sachs engages in specialist and market-making activities on equities and options exchanges, and clears client transactions on major stock, options, and futures exchanges worldwide.

- Asset management provides investment advisory services and financial planning services and offers investment products to a diverse group of institutions and individuals worldwide. It also provides brokerage services, financing services, and securities lending services to institutional clients, including hedge funds, mutual funds, pension funds, and foundations, and to investors with high net worth worldwide.

- Global investment research provides in-depth analysis of markets, companies, industries, and currencies worldwide. Clients can gain online access to market intelligence and trading opportunities through the firm's institutional portal.

Goldman Sachs believes that "as open markets and global finance transform economies, capital markets will play an increasingly vital role in connecting capital to ideas necessary for growth. Goldman Sachs helps allocate capital and manage risk, and through this process fosters entrepreneurship and innovation, drives efficiency, and encourages economic reform."[14]

TABLE 5-8 Financial highlights

Key Measures ($ amounts in millions)	2006	2005	2004
Net revenues			
Investment banking	$5,629	$3,671	$3,374
Trading and principal investments	$25,562	$16,818	$13,728
Asset management and securities services	$6,474	$4,749	$3,849
Total net revenue	$37,665	$25,238	$20,951
Net earnings	$9,537	$5,626	$4,553
Total assets	$838,201	$706,804	$531,379
Total number of employees	26,467	23,623	21,736

Goldman Sachs is moving its world headquarters to the 90-acre Battery Park City, a planned community at the southwestern tip of Manhattan. Battery Park City includes the World Financial Center, the Goldman Sachs world headquarters, The New York Mercantile Exchange, and several condominium and apartment complexes. The new headquarters will be housed in an impressive 43-story tower that is expected to open in 2009, and it will provide room for more than 9,000 employees in 2.1 million gross square feet of office space. The building will house six high-tech trading floors, three floors of meeting and amenity spaces, and 30 floors of office space.

Terrorist attacks around the world, as well as recent natural disasters, have revealed the inadequacy of the business continuity planning of many organizations. As a result, new regulatory mandates and business pressures place increased emphasis on the need for strong business continuity measures, especially within the financial service industry. For example, following the terrorist attacks in the United States and the subsequent four-day closing of the financial markets in 2001, the National Association of Securities Dealers (NASD) conducted a survey of 150 member firms to assess their ability to survive significant business disruptions. They discovered a number of serious shortcomings:[15]

- Many firms simply did not have business continuity plans
- Many small and mid-size firms failed to store backup data and systems away from their primary records and systems
- Less than half the member firms had back-up facilities capable of handling the same volume of business as their primary facilities
- Less that half the larger member firms maintained back-up systems to support investor communications

As a result of these findings, the New York Stock Exchange (NYSE) and the National Association of Securities (NASD) established new measures addressing business continuity plans. These measures were put in place to ensure that firms are able to protect vast volumes of data, as well as communicate with clients and the financial markets during a disaster. NYSE Rule 446, which went into effect in 2004; requires member organizations to develop, maintain, review, and update business continuity and contingency plans that establish procedures to follow in an emergency or significant business disruption. The NASD developed a virtually identical set of measures covering business continuity and contingency plans, known as Rules 3510 and 3520. Both of these rules define a set of ten items that must be covered in the business continuity plan of member organizations: [16]

1. Books and records backup and recovery (hard copy and electronic)
2. Identification of all mission-critical systems and backup for such systems
3. Financial and operational risk assessments
4. Alternate communications between customers and the firm
5. Alternate communications between the firm and its employees
6. Alternate physical location of employees

7. Critical business constituent, bank, and counterparty activity

8. Regulatory reporting

9. Communications with regulators

10. How member organizations will assure customers prompt access to their funds and securities if member organizations cannot continue their business

In addition to these items, both the NYSE and NASD guidelines advise that business continuity planners consider several scenarios that could occur including: the loss of an entire building or a business interruption that impacts an entire business district, city, or geographic region. While these rules apply only to NYSE and NASD member firms, many planners agree that these rules can provide worthwhile guidelines for all organizations.

The Federal Financial Institutions Examination Council, an interagency council that prescribes uniform standards for the U.S. financial industry, has also provided business continuity guidelines that financial institutions should strongly consider. For example, because a pandemic event differs from other disaster events in terms of scale and duration, financial institutions are being asked to incorporate plans for such an event into their business continuity plan. If an institution does not implement the recommendations, examiners will likely inquire why and push for implementation. Further, institutions that play a critical role in financial markets may be asked by the agencies to implement the standards, in an effort to monitor and control the risk of one financial firm significantly impacting the entire market.

Here is an example of how following a business continuity plan can provide real business benefits. First State Bank is in the small town of Parkersburg in northeast Iowa. In May 2008, Parkersburg was hit by a category 5 tornado with winds in excess of 200 mph that devastated many of its schools, homes, and businesses. Roads near the town were covered with fallen trees and debris, and were impassable. The bank was unable to operate out of its facility. The tornado hit on the Sunday before Memorial Day, which gave the bank two extra days to recover operations before the work week started. The bank followed its business continuity plan and within five days it was operating out of two double-wide trailers complete with LAN networks, teller windows, customer service desk, drive up facilities, and a cash vault.[17]

The Goldman Sachs business continuity plan also provides reasonable assurance that the firm's business operations could continue even if normal operations are disrupted at firm's critical facilities. This plan guarantees that the firm can protect not only the employees and assets of Goldman Sachs, but importantly its clients and their assets. It was developed in accordance to the NYSE rule 446 and the NASD rules 3510 and 3520.

You can find "The Goldman Sachs Business Continuity Program for Disaster Recovery: Overview" at the Goldman Sachs Web site at *http://www2.goldmansachs.com/our-firm/about-us/continuity.pdf*. Read this plan and then answer the discussion questions.

Discussion Questions

1. Visit the Goldman Sachs Web site and review recent press releases and SEC filings, including the annual report. Can you find evidence that the firm continues to evolve and improve its business continuity plan? Can you find any information that could adversely affect the firm's business continuity plan? Write a paragraph that summarizes your findings.

2. Goldman Sachs' movement to a new world headquarters is cause for the firm to rethink and improve the business continuity plan. Identify several general areas of the plan that may need revisions and updates. Identify specific areas of the plan that should be changed.

3. Identify two business functions that should be classified as AAA for Goldman Sachs. For each business function, define its inputs, processing, and outputs. Discuss how Goldman Sachs might provide business continuity for these two functions.

Endnotes

[1] Luc Kordel, "IT Governance Hands-on: Using COBIT to Implement IT Governance," ISACA, 2004 (accessed at *www.isaca.org*, 3 August 2007).

[2] Mike Rothman, "COSO and COBIT: The Value of Compliance Frameworks for SOX," *SearchSecurity.com*, 25 July 2007.

[3] "COBIT and IT Governance Case Study: Harley-Davidson," accessed at *www.isaca.org/cobitcasestudies*, 27 July 2007.

[4] "Harley-Davidson: Using COBIT to Simplify Compliance," *COBIT Focus*, December 2006.

[5] Craig Symons, "IT Governance Framework," Forrester, 29 March 2005.

[6] Luc Kordel, "IT Governance Hands-on: Using COBIT to Implement IT Governance," ISACA, 2004 (accessed at *www.isaca.org/*, 3 August 2007).

[7] Christoph Hammerschmidt, "Siemens Stock Rides Out 'Black Money' Scandal," *InformationWeek*, 5 February 2007.

[8] Thomas Calburn, "IBM to Offer Data Governance Consulting," *InformationWeek*, 13 December 2006.

[9] John Morency, "Best Practice, Practice, Practice," *Network World*, 10 January 2005.

[10] "Audatex, About Us," *www.audatex.us/about_us/index.htm*, accessed 25 July 2007.

[11] Wikipedia.

[12] "ITIL to the Motor Rescue," BCS Service Management, accessed at *www.bcs.org*, 22 July 2007.

[13] Goldman Sachs 2006 Annual Report, page 26.

[14] Goldman Sachs 2006 Annual Report, page 1.

[15] Nick Benvenuto, Brian Zawada, "Complying with New Business Continuity and Contingency Plan Rules," *Information Management Magazine*, October 1, 2004.

[16] Nick Benvenuto, "Complying with New Business Continuity and Contingency Plan Rules," *DM Review Magazine*, October 2004.

[17] "Iowa Bank Continues to Serve Community and Support Rebuilding Efforts," *Business Wire*, June 25, 2008.

CHAPTER 6

COLLABORATION TOOLS AND WIRELESS NETWORKS

THE IMPORTANCE OF NETWORKING AND COLLABORATION

"Motto for the 21st century: Network or Die."

—Steve Hamm, in *BusinessWeek*'s Information Technology's section[1]

IBM'S INNOVATION FACTORY

The telecommunications industry is highly competitive and companies are scrapping to gain a competitive edge through the introduction of next-generation services. The capability to identify and launch new services quickly is a critical success factor. Forward-thinking companies require a highly agile technical infrastructure to conceive, test, launch, and commercialize new voice and data services.

For example, the Innovation Factory is a collection of software and services from IBM that is designed to support the innovation process from an idea's incubation through commercialization. It employs social networking and information discovery capabilities, including blogs, wikis, social tagging, surveys, and pools. The goal is to accelerate innovation by enabling collaboration among employees, partners, and customers through online communities.[2]

Sprint Nextel is a global provider of voice, data, and Internet services. While it is already recognized for its highly innovative services, such as mobile data, walkie-talkie capabilities, and wireless networks, the firm is an early adopter of the IBM Innovation Factory. Sprint Nextel knows that to compete, it must accelerate its time to market for new offerings and further strengthen its reputation as an innovative company.[3]

Sprint Nextel is using the Innovation Factory to launch new service trials in a way that brings service providers closer to the subscribers, and to spot market trends through early and direct subscriber interaction.[4] Sprint Nextel employees can enter the Innovation Factory and quickly connect with other subject-matter experts inside and outside the company to collaborate on developing new products or services. They can request the software and analysis needed to develop and test their ideas. Sprint Nextel can offer selected customers a free trial on innovations that are not yet available to the general public and invite them to interact directly with the Sprint Nextel innovators. They can provide online support and documentation for trials as well as capture feedback from trial participants. All this speeds up the innovation cycle and positions Sprint Nextel even more closely to its customers, enabling it to create products that more accurately address customer and market needs.[5]

LEARNING OBJECTIVES

As you read this chapter, ask yourself:

- How do managers determine which collaboration tools are most effective for meeting organizational needs?
- How can managers effectively employ the new capabilities of emerging wire-less networks?

This chapter discusses the variety of collaboration tools that organizations can use to improve communication and enhance productivity. It also covers how wireless communications are used to meet the needs of mobile workers.

WHY MANAGERS MUST UNDERSTAND NETWORKING AND COLLABORATION TOOLS

Collaboration is essential to the success of every human endeavor, from building great cities, to waging war, to running a modern organization. Over time, the way people collaborate has evolved: We began with face-to-face sign language and verbal communications, then advanced to communications over distances via smoke signals, telegraph, and telephone. Today we use electronic messaging and networking tools over wireless networks. These technologies are shrinking the world and enabling people anywhere to communicate and interact effectively without requiring face-to-face meetings.

- As we saw in the opening vignette, organizations are using collaboration tools to get closer to customers by allowing them to try new products, identify issues, and make recommendations for improvement—all before the new product even goes into a formal test market.
- Today's field workers, traveling executives, salespeople, and service workers at customer locations must be able to access pertinent corporate data and critical messages, regardless of the time of day or where they are. Their organization's customers and business partners have come to expect this instant access as well. For example, Northrop Grumman Corp. issued 5,500 Black-Berry units to its employees. The BlackBerry, manufactured by Research in Motion, makes mobile e-mail an option and has changed the lives of busy workers by breaking down the walls between work and home and eroding the nine-to-five work schedule. Effective use of these units is considered so essential to running global operations that the company tests each new release of its Research in Motion software in a corporate jet to ensure that executives will have mobile services while in transit.[6]
- Organizations use networking tools to hold virtual meetings and deliver training to employees, business partners, shareholders, and customers around the world.
- Collaboration tools, including instant messaging, Web conferencing, and desktop sharing, will continue to help virtual teams stay connected and work collaboratively and productively.

COLLABORATION TOOLS

This section discusses some of the tools being employed for electronic collaboration. These tools are summarized in Table 6-1. While you may have used many of these tools in your personal life, pay particular attention to the examples of how organizations use collaboration tools, both internally and across organizational boundaries. These examples help explain why collaboration tools are widely used to meet organizational needs.

TABLE 6-1 Frequently used collaboration tools

Tool	Common uses
Bulletin boards	Support threaded discussions among employees, customers, and business partners Capture comments, issues, and suggestions
Blogs	Provide informal updates on products and projects Capture customer opinions and feedback
Calendaring	Determine resource availability Schedule people for meetings
Desktop sharing	Provide technical support and product demos for customers Support virtual meetings among employees and business partners
Instant messaging	Provide timely updates Link key resources to address a problem
Podcasts	Enable digital content such as training material, company announcements, and music to be distributed over the Internet for playback on portable media players and personal computers
RSS feeds	Enable people to receive automatically new postings to their favorite blogs or Web sites
Shared workspaces	Provide ease of access to documents
Web conferencing	Provide presentations and training to employees, customers, and business partners
Wikis	Allow users to create and edit Web page content freely using any Web browser

Bulletin Board

An electronic **bulletin board** allows users to leave messages or read public messages that provide information or announce upcoming events. Organizations often add bulletin board capabilities to their Web site to attract a community of users and increase site traffic. Often the bulletin board is used to keep visitors informed about current events and developments associated with the organization. Posters must be extremely careful in what they add to public bulletin boards, even if a pseudonym is used. In early 2007, Whole Foods Markets, an operator of natural foods stores and farmer's markets in North America, announced that it wanted to acquire rival Wild Oats for $565 million. The Federal Trade Commission decided to block the proposed takeover, however, in part because it learned that the Whole Foods Markets CEO had bashed Wild Oats for years under a pseudonym on a Yahoo! bulletin board.[7]

Blog

A **blog** is a Web site in which contributors ("bloggers") provide ongoing commentary on a particular subject. A blog often is used as a personal online diary or to address current issues and local news. Images and links to other blogs, Web pages, and other media related to the topic of the blog may be incorporated within the text. With most blogs, readers are able to leave comments. Several dedicated search engines can search blog contents, including blogdigger, Feedster, and Technorati.

Increasingly, organizations are using corporate blogs externally for branding, marketing, or public relations purposes. Often, executives or public relations people write the posts with the goal of improving a firm's public image, its products, and its services. Corporate blogging can be a good way for a corporation to make itself appear more personable and appealing. The key is to ensure that the blogs allow realistic discussions about issues that are important and relevant to readers, including topics that are potentially problematic for the firm. Failure to be open and objective can cause the organization to appear biased and self-serving, thus creating a worse public image than if the firm had never tried blogging.[8] Company executives and public relations people should "only write blogs if they actually can write in an interesting way; blog postings that reek of 'marketing speak' largely will be ignored or, worse, ridiculed."[9] Corporate bloggers must also recognize that people who respond to a blog posting have a strong need to feel that someone is listening to them. Furthermore, bloggers immediately are discouraged by any response they consider dismissive or insensitive.

Pitney Bowes Inc. launched its first blog with posts written by its executive chairman. The posts covered a wide range of topics that were not related to the company, including public funding for research on Alzheimer's disease. Manpower Inc., a major provider of temporary employees around the globe, maintains an employment law blog (or blawg) at *www.manpowerblogs.com*. Written by company attorney Mark Toth, the blog can be searched for employment topics such as wrongful termination, sexual harassment, and privacy in the workplace. These blogs require an investment of resources, time, and money with no guarantee of a tangible return on investment. Such thinking runs counter to traditional marketing approaches.[10]

A growing number of organizations allow employees to create their own personal blogs. Great care must be exercised—as such blogs provide an outlet for uncensored commentary and interaction. Under the best of conditions, individual employees use their own blogs to ask for help, to transfer information in a manner that invites conversation, or to invite other people to refine or build on a new idea. Of course, employees can also use their blogs to criticize corporate policies and decisions.[11]

Calendaring

Calendaring software allows people to capture and record scheduled meetings and events. The software enables you to check the electronic calendar of team members for open time slots and notify or remind meeting and event participants by e-mail. Calendaring products include Google Calendar, IBM Lotus Notes, Microsoft Exchange, and Now Up-to-Date & Contact from Novell Group Wise. With most of these packages, you can choose who has access to see your calendar, which details they can view, and whether they can "book" your available time. You can also create automatic event reminders, including mobile phone notifications.

In 2006, the Pennsylvania state senate adopted the use of group calendaring software to enable them to interact with staff, colleagues, and constituents. The software can be accessed from any type of computer, via BlackBerry devices from the office, or while the user is traveling or working away from the office. The software provides multiple calendars that enable users to view and manage their personal and business calendars separately.[12]

Desktop Sharing

Desktop sharing includes a number of technologies and products that allow remote access and remote collaboration on a person's computer. Remote log-in and real-time collaboration are the most common forms of desktop sharing.

Remote log-in allows users to connect to their office computer while they are away from the office. For example, users might be at home and need to print an important document or update a spreadsheet for an unexpected early-morning meeting. Products such as GoToMyPC from Citrix allow access to your office computer via the Internet from any computer at any location. LogMeIn, Inc. offers a suite of remote access and support solutions that provide instant, secure connections between remote PCs over the Web. The service has applications for desktop remote control, data backup, file sharing, remote system administration, and on-demand customer support.

Remote log-in makes it possible for certified technicians to access users' computers distantly to perform setup, training, diagnostics, and repair. All of these services are provided with no need for travel by the user or the technician, and without requiring the shipment of computers back and forth.[13] Also, there is no need to set up an appointment for a visit by a technician. Such service minimizes the users' downtime and enables them to return to productive work as quickly as possible.

Real-time collaboration is discussed in the "Web Conferencing" section later in this chapter.

Instant Messaging (IM)

Instant messaging (IM) offers real-time, informal communications based on the often rapid exchange of typed messages. IM is less formal than e-mail and is used primarily in a synchronous communications mode, with all parties sending and receiving messages in real time. IM systems allow their users to set an *online status* or *away message* so their contacts know when the user is available, busy, or away from the computer. Because IM users need not respond to incoming messages, IM is considered less intrusive than phone calls. IM also makes it easy to send and receive information that is not conveyed easily via phone, including URLs and electronic attachments such as documents or spreadsheets. Some IM systems support additional features such as Webcams or the retention of a "conversation" for later use.

IM has become an accepted communications tool in nearly 90 percent of all organizations.[14] In those organizations, IM is all but eliminating the use of interoffice "snail mail" and dramatically cutting the number of phone calls. Unfortunately, unsecured IM applications are easy prey for hackers who want to take advantage of an organization's vulnerabilities. Security professionals warn that businesses must pay careful attention to their employees' use of instant messaging. Worms and viruses, such as Bropia, Kelvir, and MyDoom, have been launched specifically to breach IM tools. Security firm Akonix measured a 79 percent increase in malicious code attacks over IM networks from January 2007 to June 2007 versus the same time period in the previous year.[15]

In addition to potential hacker problems, IM raises the issue of unintended release of private information. For example, Mark Foley represented Florida in the U.S. House of Representatives from 1995 until 2006. Although known as a champion in the fight against child abuse and exploitation, he was forced to resign from Congress after it was revealed that he had sent sexually explicit instant messages to underage boys serving as congressional pages. The episode underscores the fact that instant message conversations can be just as damaging as e-mail.

The first step to reducing risks associated with IM is to create a policy that states who can use it, for what purposes, for what kinds of data, and whether file attachments are allowed. Organizations must decide whether and how to archive conversations and for how long. Key to the success of this policy, companies must communicate how the policy will be monitored and enforced.[16]

A second step is to establish an enterprise server that functions as a hub for all IM messages. Businesses must install security software on this hub to monitor and inspect all IM traffic. The use of content filters and keywords can help safeguard important data from leaving the network. In addition, software can be installed to check for malware signatures, like IM-specific worms, and to prevent spam from getting through.

Podcast

A **podcast** is a digital media file distributed over the Internet using syndication feeds for playback on portable media players and personal computers. Users simply download a podcast to their computer and then transfer it to an iPod or other player device for listening at their convenience. The method by which a podcast is syndicated is called *podcasting*. Podcasting differs from other types of online media distribution; it is based on a subscription model and uses automatic feeding mechanisms such as Really Simple Syndication (RSS) or Atom to deliver files to audiences.

Podcasters create audio programs, usually in the form of MP3 files, which they upload to Web sites. Anyone with a computer and a microphone can now create a podcast. Podcasts are available from many commercial broadcast and publishing concerns, including newspapers, television networks, National Public Radio, the BBC, magazines, and other informational Web sites. Numerous Web sites, such as PodcastAlley, and Podcast.com, index and facilitate the finding of and subscription to podcasts according to subject matter, source, and other criteria.

Really Simple Syndication (RSS)

Really Simple Syndication (RSS) is a family of data formats that help people automatically receive feeds anytime there are new postings to their favorite blog sites, updates in the news headlines, or new information posted at specified Web sites. RSS content is read using software called an RSS reader. Content distributors syndicate a Web feed, allowing users to subscribe to it by entering the feed's link into the reader or by clicking an RSS icon in a browser that initiates the subscription process. The RSS reader is programmed to check automatically all subscribed feeds on a regular basis to look for new content and download any updates that it finds. This process allows users to stay current on topics of interest.

An Internet aggregator makes a number of Web feeds accessible at one URL. The number of sites that offer RSS feeds is growing rapidly, and includes BlogStreet, myRSS, NewsIsFree, and Purple Pages. Syndic8.com, a syndicator Web site, has hundreds of feeds on a variety of subjects, including 136 feeds on 32 business-related topics from accounting to transportation and logistics. The capabilities of RSS feeds and aggregators make them attractive to business managers who want to stay informed.

For example, Phoenix Technologies is a maker of electronics components for PC motherboards. CIO Cliff Bell noticed that many e-mails at his firm went unread. So, he decided to test RSS as a tool to keep employees, customers, and business partners current by pushing relevant information to them via subscription rather than counting on them to find pertinent information in the hundreds of e-mails they receive each month. The Phoenix Technologies legal team now uses RSS for legal review, combining RSS subscriptions with internally generated RSS feeds for specific projects. With this approach, the attorneys no longer must search through their e-mail or the intranet to get information about particular projects. The use of RSS ensures that if something new happens, company attorneys will receive an RSS feed about it. CIO Bell envisions the use of additional departmental or project-based RSS feeds across the company to keep people well informed and on top of matters. Another application of RSS feeds is to create a common channel for company-wide announcements.[17]

RSS feeds can also be used to increase revenue flow by helping companies develop strong relationships with consumers and create brand loyalty. RSS feeds that present interesting and worthwhile information can attract existing customers and prospective clients by notifying them of specials, discounts, product announcements, technical support tips, news, and industry studies.[18]

Shared Workspace

A **shared workspace** is an area hosted by a Web server in which project members and colleagues can share documents, models, photos, and other forms of information to keep each other current on the status of projects or topics of common interest.

One organization that uses shared workspaces is the Center for Scientific Review (CSR), which is responsible for reviewing more than 80,000 grant applications received each year by the National Institute of Health (NIH). Each application is reviewed thoroughly to ensure that the NIH funds only the most promising research. CSR recruits more than 18,000 external experts to conduct the necessary reviews. In the past, NIH incurred considerable travel expenses to bring these experts together and review the applications. Recently CSR has been using shared workspace technology to provide an effective and efficient way to eliminate the need to travel. Reviewers access a shared team workspace set up on a restricted server that ensures privacy and controls access to the applications and associated data. Reviewer comments are posted in the discussion threads of the shared virtual workspace server where they can be viewed and commented on by other reviewers. Over the course of a few days, the reviewers read the applications, interact via IM, submit comments and rebuttals to each of the grant application discussion forums, and vote on the grant applications. Not only does this approach dramatically cut travel expenses and speed up the review process, it also enables NIH to recruit reviewers who are willing to participate in the process but cannot afford to spend the time traveling to a single location.[19]

Web Conferencing

Web conferencing is a way to conduct live meetings or presentations over the Internet. In a Web conference, each participant sits at his or her own computer, and is connected to other participants via the Internet. Web conferencing is possible using either a downloaded application on each of the attendees' computers or a Web-based application that requires attendees to enter an online address to join the conference. An important capability of Web conferencing software is application sharing, the ability of one person to share a document or spreadsheet on his desktop and pass the control of the application to someone else in the meeting. Web conferencing products include GoToMeeting, Live Meeting, Netviewer, SkypePro, WebEx, and Yugma. Most of these products provide support for the following features:

- Rich media presentations
- Live video via Webcam or digital video camera
- Panoramic video
- Active speaker indicator
- Public events page
- Personal recordings
- Virtual breakout rooms
- Whiteboard with annotation, which allows the presenter and attendees to highlight or mark items on the slide presentation
- Text chat for live question-and-answer sessions
- Polls and surveys, which allow the presenter to pose questions with multiple-choice answers to the audience

A *Webcast* is a presentation of information in one direction only, like watching a video on the Internet. A *Webinar* is a type of Web conference in which the direction of the presentation is primarily one way from the presenter to the audience; however, it can also be interactive between the presenter and audience.

Web conferencing often is sold as a service that is hosted on a Web server and controlled by the vendor. The service may be offered on a per-minute usage basis or for a fixed monthly fee. Some vendors make their conferencing software available as a licensed product, allowing organizations that make heavy use of conferencing to install the software on their own servers. Adobe Acrobat Connect, Genesys Conferencing, GoToMeeting, Lotus Sametime, Microsoft Office Live Meeting, Unyte, and WebEx are all examples of Web conferencing services.

American Lighting manufactures a wide range of innovative and high-quality lighting products to meet residential, commercial, and specialty needs. The firm places great emphasis on developing new products and using trendsetting technology, as continued innovation is a core business strategy. With a widespread base of operations, including manufacturing facilities in China and Taiwan, warehousing in North America, and offices in Hong Kong, ShenZhen, Mexico, and Canada, rapid and effective communications can be a challenge. American Lighting uses Web conferencing to conduct effective meetings with employees around the globe. The technology has helped American Lighting achieve substantial savings in conducting overseas sales and company meetings, as well as greatly reduce expenses related to quality assurance and diagnosing issues.[20]

Wikis

A **wiki** is a collaborative Web site that allows users to create and edit Web page content freely using any Web browser. A wiki supports hyperlinks and has a simple text syntax for creating new pages and cross-links between internal pages. Wikis can be constructed from all types of corporate data, such as spreadsheets, Microsoft Word documents, PowerPoint slides, PDFs, and anything else that can be displayed in a browser. One of the best-known wikis is Wikipedia, the largest free-content encyclopedia on the Internet, with more than 7 million articles written in more than 200 languages.

Wikis allow users to determine the relevancy of content rather than depending on a central document control group. After the wiki is established, users control it, not the document control group. The wiki has built-in version control so that no changes can be made without creating a record of who made the change. If necessary, it is relatively easy to revert to an earlier version of the wiki.[21] The following list presents a few benefits of corporate wikis:

- Corporate wikis are easy to link to useful corporate information systems such as phone books and people directories.
- The use of corporate wikis reduces the amount of e-mail traffic within a company by enabling all relevant information to be shared by people working on a given project. In addition, wikis reduce the use of lengthy distribution lists that burden recipients with excessive and irrelevant messages.
- Wiki access rights and roles can be established based on a user's role or organization so that unauthorized people cannot view or edit certain pages.
- Wikis provide a tool for building consensus, as they enable people to express their views on specific topics.
- Wikis allow users to build and organize useful new sources of data for a variety of projects, issues, and ideas. For example, you can build a wiki that defines acronyms in common use within your firm.

Motorola has more than 3,300 wikis that are used by 69,000 employees and another 8,000 business partners. The total wiki content measures 5 terabytes of data. A wide range of topics are covered, from six-sigma, to key engineering efforts, to presales factbooks for various products. Toby Redshaw, a corporate vice president, views wikis as a tool to capture and take advantage of the brain power Motorola gains through acquisitions.[22]

The following feature illustrates how one company makes effective use of several collaboration tools.

A MANAGER TAKES CHARGE

Ryan Companies Enables Collaboration

Founded in the 1930s as a small lumber company in northern Minnesota, Ryan Companies US, Inc. is now a commercial real estate firm that provides integrated design, development, and real estate management services to customers in more than 150 cities around the country. Recent annual revenues exceeded $750 million. Roughly 45 percent of the firm's 1,000 employees work from remote offices and job sites. CIO John Leeper recognized that the firm must employ cutting-edge technology to enable employees to deliver high-quality products, on time and on budget. He led the effort for Ryan Companies to successfully adopt and implement Web-based collaboration services to support its employees.

continued

The firm's construction managers carry laptops to access and maintain complex project schedules. If the managers have a problem with their laptops, such as a virus, they need help in a hurry. They cannot afford to be without their laptops for the three days it would take to ship them to the help desk, have them repaired, and shipped back. So the IT organization's help desk people use GoToAssist to fix problems remotely and get the user back up and running, often within minutes.

Collaboration among a project's engineers, architects, and project managers is critical, but these employees are spread out across the country. It is impossible to get them together in one place on short notice, and even if such meetings were possible, the travel costs and time wasted in airports would be prohibitive. Much of the project work involves the use of AutoCAD (software for creating and updating blueprints), and e-mailing these large files has been a problem. Ryan Companies employs GoToMeeting to enable quick, simple collaboration among all the people involved on a project. The ease of setting up a GoToMeeting session actually has increased the amount of collaboration, with resulting benefits in on-time and on-budget projects, not to mention the savings in travel expenses.

Ryan Companies also employs GoToMyPC to enable mobile workers to have "anytime, anywhere" access to applications, files, and data. Thus, an architect working in Phoenix easily can access AutoCAD drawings stored on a computer at the firm's headquarters in Minneapolis.

Discussion Questions:

1. Discuss how Ryan Companies might have rolled out these collaboration tools to ensure that they were widely used.
2. What issues do you think might arise in the use of these collaboration tools? How might these issues be addressed?
3. Identify and briefly discuss how two other collaboration tools could help Ryan Companies.

Now that we have discussed many of the collaboration tools, let's see how organizations use wireless communications to meet the needs of their mobile workers.

WIRELESS COMMUNICATIONS

The increasing use of wireless communications is an important business trend of the twenty-first century. Wireless communications is used to keep in touch with employees, customers, and business partners; access important corporate data and business applications; and to interact over the Internet and Web. In many cases, the collaboration tools just discussed are employed in a wireless network environment. Such applications and the need for wireless access are driving the development of very high-speed, wide-area Internet access for mobile devices.

Figure 6-1 shows the future trends expected in the area of wireless networks along two dimensions: the geographic area covered by wireless networks and the capabilities of those networks. The coverage of networks can range from a local area network—for example, a wireless network that connects your mouse or printer to your personal computer—to global networks that span the world. The most desired network characteristics include secure communications that are free from intrusion by unauthorized people or devices, extremely high-speed communications that enable unimagined new services, high-quality service that does not allow a dropped or degraded connection, robust applications that support real business needs, and "always, anywhere" availability by all users.

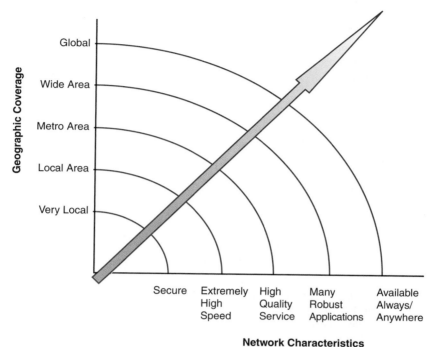

FIGURE 6-1   Future trends in wireless networks

Staying abreast of the rapidly changing developments in wireless technology is difficult. While vendors announce new products, features, and enhancements at a dizzying pace, the IEEE and other standards-setting groups seem to ratify a new and improved wireless standard every few months. The next section provides a few basics to enable IT and business managers to put the changing communications landscape into perspective.

Communications Fundamentals

A **communications channel** is a path that carries a signal from sender to receiver. The channel may support simplex, half-duplex, or full-duplex communications. A *simplex channel* supports communications in only one direction, like the circuit in your home that connects the doorbell to a ringer. A *half-duplex channel* supports communications in both directions, but only in one direction at a time. CB radio and walkie-talkies are examples of half-duplex communications. A *full-duplex channel* supports communications in both directions simultaneously. The telephone and cell phone are examples of full-duplex communications because both parties can speak at the same time.

Transmission media are used to propagate the communication signal. These media either may be guided, in which signals travel along a solid medium, or wireless, in which signals are broadcast over airwaves as a form of electromagnetic radiation (see Table 6-2). Common guided transmission media include: twisted pair wires, which are used in telephone lines; coaxial cable; fiber optics; and even power transmission lines. Examples of wireless communications include cell phone and satellite TV.

TABLE 6-2 Summary of transmission media types

Media type	Description	Common use
Twisted pair wires	Shielded or unshielded pairs of copper wire	Conventional land-line telephones
Coaxial cable	Inner conductor wire surrounded by insulation	Cable TV
Fiber-optic cable	Thousands of tiny strands of glass bound together in a sheathing; uses light beams to transmit signals	High-speed LANs
Broadband over power lines	Data is transmitted over standard high-voltage power lines	Internet access
Microwave (terrestrial and satellite)	High-frequency radio signal (300 MHz to 300 GHz) sent through the atmosphere or space; often involves communications satellites	Satellite TV, alternative to land-line telephone
Radio	Operates in the 30-300 MHz range	Cell phone communications
Infrared	Signals sent over the airwaves as light waves	Wireless keyboards and mouses

The frequency at which a signal is transmitted is measured in cycles per second called Hertz, which is abbreviated as **Hz** (see Table 6-3). **Bandwidth** is the range of frequencies that an electronic signal occupies on a given transmission media. The greater the bandwidth, the more information per unit of time you can convey. Bandwidth can be measured in bits per second (bps).

TABLE 6-3 Speed of transmission in Hz

Abbreviation	Meaning
Hz	Cycles/second
KHz	Thousands of cycles/second
MHz	Millions of cycles/second
GHz	Billions of cycles/second

The electromagnetic spectrum is divided into frequency bands, as shown in Figure 6-2. Use of radio frequency bands is regulated by the governments of most countries to avoid frequency interference and to ensure usability. The regulation must not cause complications in transmitting information across national boundaries. For example, in the United States, the 700-MHz spectrum (698 MHz to 806 Mhz) used by UHF TV channels 52 through 69 will be vacated once the transition from analog to digital TV transmission occurs in February 2009. Of this newly available spectrum, thirty MHz probably will be reserved for communications to support public safety. Another 60 MHz of this spectrum is up for auction beginning in January 2008. Winners will be able to create new services that operate in this frequency range once the digital TV transition is complete.[23]

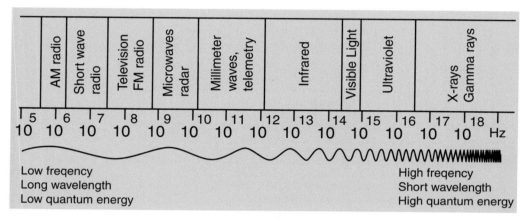

FIGURE 6-2   Electromagnetic spectrum

Cell Phone Services

The cell phone has become ubiquitous and is an essential part of life in the twenty-first century. The scope of its capabilities has grown far beyond a voice communications device. It is now used for Web browsing, e-mailing, listening to audio, watching video, computing, keeping track of contacts and to-do lists, text messaging, and taking and sending photos.

Cell phones operate in the radio frequency range of the electromagnetic spectrum. A cell phone uses two frequencies to support each call, thus creating a full-duplex communications link. The cell phone communications area is divided into hexagonal cells (see Figure 6-3). Sometimes hundreds of cells are required to cover a large city or geographical area. Each cell includes a base station that consists of a tower rising hundreds of feet into the air, and radio transmitters and receivers that connect with the antennae on the tower through a set of thick cables. Both the base station and cell phone transmitter operate at relatively low power levels. This ensures that transmissions from one cell do not disrupt communications by "spilling over" into a neighboring cell. It also minimizes the battery power required to operate the cell phone.

In the United States, the government requires the presence of two cell phone service providers in every market, with each carrier assigned a non-overlapping range of frequencies to be used for operations. In the United States, the largest wireless license-holders run

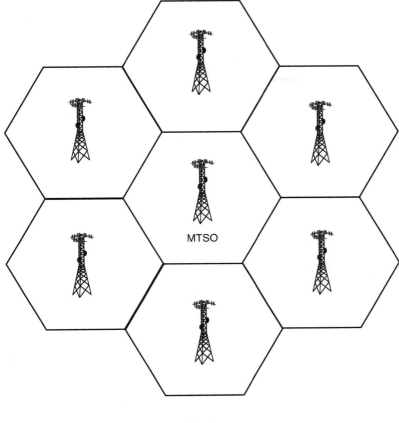

MTSO

Cluster

FIGURE 6-3   Cellular network

mobile networks that are in near-constant transition; these companies include AT&T, Sprint Nextel PCS, T-Mobile, and Verizon. Each cell phone service provider establishes its own Mobile Telephone Switching Office (MTSO) to manage communications across a number of cells. Service providers often share base stations, so you may see two or more sets of antennae on a cell tower.

The cell base station continuously monitors the signal strength of each call within its range. As the user moves away from the base station in one cell, the base station detects that the signal strength is weakening for a particular call. At the same time, the base station in an adjacent cell will detect a strengthening of the signal strength for the same call. The two base stations coordinate through the MTSO to handoff the call from one cell to the next. The cell phone receives a signal from the MTSO to change frequencies as the call passes from one cell to the next. Cell phone technology has evolved through three generations and is about to enter its fourth generation, as described in the following section.

1G

The first generation of cell phones operated in the 800-MHz range of the electromagnetic spectrum and employed analog technology. The frequency used in analog voice channels is 30-KHz wide, which is sufficient to provide a conversation of the same quality as a wired telephone. With such technology, each cell covers about 10 square miles and can support about 56 concurrent voice communication links.

2G

The first digital cell phones represent the start of the 2G services. Such cell phones also are called personal communication service (PCS) in the United States. 2G cell phones operate in the 1900-MHz frequency range and perform millions of calculations per second to convert your voice (an analog signal) into a digital format of 0s and 1s. The use of digital signals in a 2G network increases system capacity so that more concurrent calls can be made within a single cell. Digital signals also enabled the introduction of new digital data services such as short messaging service (SMS) for text messaging and e-mail. 2G networks use three different approaches to encode and transmit voice and data in a digital format:

- Time-division multiple access (TDMA) assigns each call to a certain portion of time on the designated frequency. TDMA systems operate in either the 800-MHz or 1900-MHz frequency band and support three times the number of calls as the analog 1G system.
- Frequency-division multiple access (FDMA) assigns a separate frequency for each call.
- Code-division multiple access (CDMA) assigns a unique sequence code to each call and transmits the call in small pieces over a number of discrete frequencies available for use in the specified range. CDMA systems can provide between 10 to 20 times the call carrying capacity of analog equipment.

Various 2G cell phone service providers have built services based on one or more of the following mobile communications standards:

Global System for Mobile Communications Service (GSM) is the most widely adopted digital cellular technology in use today. GSM uses time and frequency division techniques (TDMA and FDMA) to optimize the call-carrying capacity of a wireless network. In addition to voice services, GSM also provides a number of carefully standardized and broadly supported capabilities such as Short Message Service (SMS), circuit switched data (CSD), and General Packet Radio Services (GPRS).

General Packet Radio Service (GPRS) is a type of mobile data service available to users of GSM mobile phones that provides fast data transfers over a very large area. It provides moderate-speed data transfer by employing unused TDMA channels in the GSM network.

Enhanced Data Rates for Global Evolution (EDGE) is a type of network connection that provides faster data transfer rates than GPRS over a similar-sized area. EDGE defines new signal modulation schemes that will permit much higher data rates to be achieved in GSM and TDMA networks. Rates up to 384 Kbps are expected, which will enable the delivery of multimedia and other broadband applications to mobile phone and computer users.

3G

3G is a digital cell phone service that operates at an increased bandwidth to enable data transfer rates of up to 3 Mbps. Such a high transfer rate makes it feasible to download data from the Internet, surf the Web, and send and receive large, multimedia files. 3G phones typically have considerable processing power and can support advanced services such as video conferencing, receipt of streaming video from the Web, and the downloading of e-mails with attachments, all in a mobile environment. In many countries, 3G networks do not use the same radio frequencies as 2G, so 3G service providers must build entirely new networks and license entirely new frequencies. In the United States, carriers operate 3G service in the same frequencies as other services.

4G

4G is a term that describes the next evolution in wireless communications beyond 3G. Currently, no clear standards exist that exactly define 4G. This does not stop various telecommunications providers and consumers from trying to define what 4G means to them, however.

The World Wireless Research Forum includes Ericsson, Huawei Technologies, and Motorola, three companies that helped define 3G wireless. This group states that 4G is likely to be similar to 3G, in that it will be a collection of technologies and standards, not just one single standard. They also believe that a 4G system will offer a comprehensive IP solution in which voice, data, and streamed multimedia can be provided to users on an "anytime, anywhere" basis, at higher data rates (100 Mps to 1 Gbps), higher quality, and greater security than previous generations.[24]

CEO Gary Forsee of Sprint Nextel states that 4G technology must meet four criteria: It must allow Sprint Nextel to be the first to market; it must provide economical performance; it must let the carrier create an environment in which multiple suppliers exist for the network; and it must develop, offer, and support high-speed services. To achieve this goal, Sprint Nextel committed $1 billion in 2007 and $2 billion in 2008 to build a 4G network that will offer transmission speeds of 2 Mbps to 4 Mbps for mobile users. Sprint Nextel expects to have 100 million users of its 4G service by 2008.[25]

Wi-Fi Solution for Local Area Networks

Wi-Fi is a wireless communications technology brand owned by the Wi-Fi Alliance, which includes more than 300 technology companies such as AT&T, Dell, Microsoft, Nokia, and Qualcomm in 20 countries. The goal of the alliance is to improve the interoperability of wireless local area network products, which are based on the 802.11 communications standards developed by the Institute of Electronic and Electrical Engineers (IEEE). Wi-Fi operates at 2.4 GHz or 5 GHz, considerably higher frequencies than the ones used for cell phones. The higher frequency allows the signal to carry more data per unit time (bps). Wi-Fi is based on the set of 802.11 networking standards, which come in many flavors, as shown in Table 6-4.

TABLE 6-4 Common 802.11 communications standards

Characteristic	802.11a	802.11b	802.11g	802.11n
Frequency used	5 GHz	2.4 GHz	2.4 GHz	2.4 GHz & 5 GHz
Data transmission rate	54 Mbps	11 Mbps	54 Mbps	140 Mbps
Uses regulated frequency spectrum?	Yes	No	No	No

The area covered by one or more interconnected wireless access points is called a **hot spot**. Free hot spots often are provided by businesses or governments that seek to offer a worthwhile service to customers, citizens, and visitors. Thus, many people use Wi-Fi networks in airports, coffee shops, college campuses, fast food restaurants, and hotels to connect to the Internet. In a Wi-Fi wireless network, the user's computer or personal digital assistant has a wireless adapter that translates data into a radio signal and transmits it using an antenna. A wireless access point, which consists of a transmitter with an antenna, receives the signal and decodes it. The access point then sends the information to the Internet over a wired connection (see Figure 6-4). When receiving data, the wireless access point takes the information from the Internet, translates it into a radio signal, and sends it to the computer's wireless adapter. Wi-Fi uses unlicensed spectrum to provide access to a network, typically covering only the network operator's own property.

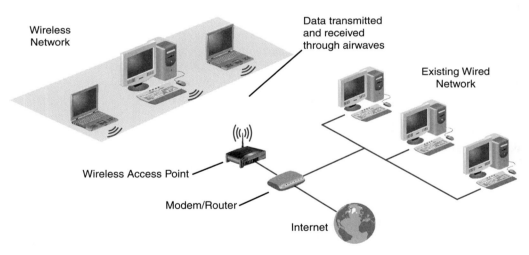

FIGURE 6-4 Typical Wi-Fi network

If the router fails or if too many people try to use the Wi-Fi network at one time, the users can experience interference or dropped connections. Most laptop computers come with built-in wireless transmitters and software to enable computers to discover automatically the existence of a Wi-Fi network.

City officials have been attracted to the idea of establishing wireless networks for use by meter readers and other municipal workers, as well as providing low-cost Internet access to its citizens and visitors. For example, EarthLink has a contract to build municipal Wi-Fi networks in Anaheim, New Orleans, Philadelphia, San Francisco, and nine other cities. The success of such efforts is far from assured, as the incumbent telecom and cable service providers have countered with alternative technologies that provide improved capabilities, albeit at higher costs to individual users.[26]

By no means are all Wi-Fi networks operated free of charge. Indeed, high costs can be a drawback that limits the use of Wi-Fi networks. Although many hot spots support Wi-Fi communications, a charge often is associated with use of the network. Wi-Fi access charges of more than $7 per day are not uncommon by hot spot operators. For example, suppose you are flying from Cincinnati to Miami with an intermediate stop in Atlanta, and you check your e-mail using a Wi-Fi network at each airport. Three separate Wi-Fi operators will charge a total of $21!

Security is another important issue for Wi-Fi users. Because communications are broadcast, anyone within "listening distance" of the wireless access point can intercept your Wi-Fi communications. Every time you log on to an unsecured Web site via a public Wi-Fi access point, your logon name and password are sent over the open airwaves. There is a high risk that someone might eavesdrop on your electronic communications. In addition, there is a high potential for criminals to steal and reuse the IP addresses of network users or hop on an unprotected Wi-Fi network to commit criminal acts. For example, police used the IP address to identify the home address of a suspected pedophile who traded child pornography online. They went to the address prepared to make an arrest but instead found an elderly woman who they quickly determined was not the real suspect. Someone had hopped onto her wireless network to conduct their nefarious business.[27] Another potential danger is logging onto a hot spot set up in a public place by a cybercriminal to capture logon IDs, passwords, credit card numbers, and other vital information.

All Wi-Fi equipment supports some form of encryption technology, which scrambles messages so that they cannot be read easily. Public Wi-Fi networks usually default to nonencrypted communications, however. It is highly recommended that you encrypt the data you send over any wireless network. Unfortunately, the most common wireless encryption standard, Wired Equivalency Privacy (WEP), is easily broken. Use the Wi-Fi Protected Access (WPA) and IEEE 802.11i (WPA2) encryption methods, which are much stronger than WEP.

Access points and routers all use a network name called the service set identifier (SSID). The SSID is a 32-character unique identifier attached to the data packets sent over the wireless network. The SSID acts as a password when a mobile device tries to connect to the network. Manufacturers typically ship their products with the same default SSID name. (The network operator should change the default name to discourage network break-ins.) To avoid connecting to a malicious hot spot, you should know which providers operate in your area and check the list of displayed SSIDs to ensure you are connected to the right one. Do not set your wireless card to connect automatically to any available network.

WiMAX, a Solution for Metropolitan Area Networks

Worldwide Interoperability for Microwave Access (WiMAX) is the common name for a set of 802.16 wireless metropolitan-area network standards that support different types of communications access. This access can be mobile (similar to that via a cell phone) or the location of the end user's wireless termination point can be fixed. In many respects, WiMAX operates like Wi-Fi, only over greater distances and at faster transmission speeds. Just as with Wi-Fi, a WiMAX tower connects directly to the Internet via a high-bandwidth, wired connection. A WiMAX tower can also connect to another WiMAX tower using a line-of-sight, microwave link. A single WiMAX tower can provide up to 3,000 square miles of coverage, so only a few towers are needed to cover an entire city. WiMAX can support data communications at a rate of 70 Mbps.

WiMAX can provide two basic forms of wireless service. The first is a non-line-of-sight service in which a small antenna on a PC connects to a WiMAX tower, which transmits in the 2-GHz to 11-GHz frequency range where transmissions are not disrupted easily by physical obstructions. The distance between the WiMAX tower and the PC antenna must be less than six miles. This form of service is similar to Wi-Fi, but again has coverage area and speed advantages. Some cities have experimented with setting up WiMAX base stations in key areas for business and then allowing people to use them for free. Fewer WiMAX base stations are required to cover the same geographical area than when Wi-Fi technology is used.

The second form of WiMAX service is a line-of-sight option in which a fixed-dish antenna points at the WiMAX tower from a rooftop or pole. Transmissions are in the 10-GHz to 66-GHz frequency range, making this option much faster than non-line-of-sight service. The distance between the WiMAX tower and antenna can be as great as 30 miles. Some companies are experimenting with this approach and charging users a fee for unlimited monthly access or a "pay-as-you-go" fee based on the amount of usage. This form of WiMAX may have a dramatic impact on today's cable and DSL Internet service providers. Because a physical connection is not needed from the WiMAX tower to each subscriber, the WiMAX service provider would have a potential cost advantage over cable and DSL providers. WiMAX also might be used to deliver voice-over Internet service and thus compete with traditional phone service providers.

Some cellular companies are evaluating WiMAX as a means of increasing bandwidth for a variety of data-intensive applications. Sprint Nextel and Clearwire are collaborating on a $12 billion joint venture to build a nationwide WiMAX network that will reach about 140 million users by the end of 2010. To supply the necessary phones, computer chips, and back-end equipment, Sprint Nextel is working with heavyweights like Intel, Motorola, Nokia, and Samsung to provide WiMAX-capable PC cards, gaming devices, laptops, cameras, and even phones by the end of 2008.

WiMAX is a key component of Intel's broadband wireless strategy to deliver innovative mobile platforms for "anytime, anywhere" Internet access. Intel is betting that over time, WiMAX will achieve universal acceptance just as Wi-Fi has. Intel is active in bringing this vision to fruition. The company is making its Centrino laptop computers WiMAX-enabled and is partnering with Clearwire to push the technology even further ahead.

Samsung is active in the mobile WiMAX market, having won contracts to provide networks in Brazil, South Korea, and Venezuela. It also supplies WiMAX equipment to Sprint Nextel and is working with the U.S. Army's Communications Electronics Research and Defense Engineering Center to study the use of WiMAX equipment in a military environment.[28]

This chapter discussed the use of collaboration tools, especially their use in an organizational setting, and it covered the basics of wireless networks. Table 6-5 is a manager's checklist that summarizes how organizations can use collaboration tools and wireless networks. It will be exciting to see how these technologies evolve to provide new capabilities and services.

TABLE 6-5 A manager's checklist for using collaboration tools and wireless networks

Recommended management actions (appropriate response is "Yes")	Yes	No
Does your organization use collaboration tools to get closer to customers?		
Is your organization using collaboration tools and wireless networks to enable executives, salespeople, and service workers to access pertinent corporate data and critical messages?		
Is your organization using networking tools to hold virtual meetings and deliver training to employees, business partners, shareholders, and customers around the world?		
Is your organization using collaboration tools to help virtual teams stay connected and to work collaboratively and productively?		
Does your organization have standards and guidelines to ensure safe usage of wireless networks?		
Is your organization actively following the evolution of wireless network technology and seeking opportunities to capitalize on new capabilities?		

Chapter Summary

- Collaboration is essential to the effective operation of modern organizations, and enables people anywhere to communicate and interact effectively.
- Bulletin boards support threaded discussions among employees, customers, and business partners. They can also be used to capture comments, issues, and suggestions.
- Blogs provide informal updates on products and projects, and capture customer opinions and feedback.
- Calendaring software is used to determine resource availability and to schedule people for meetings.
- Desktop sharing can provide technical support and product demos for customers. It can also support virtual meetings among employees, business partners, and customers.
- Instant messaging is a way to provide timely updates and link key resources to solve a problem.
- A podcast is a way to distribute digital media files over the Internet to allow playback on portable media players and personal computers.
- RSS makes it easy to receive updates automatically from Web sites and blogs.
- A shared workspace can provide a workgroup with an easy way to access documents, spreadsheets, and other information of interest to the group.
- Web conferencing is used to conduct live meetings or give presentations over the Internet.
- A wiki is a collaborative Web site that allows users to freely create and edit Web page content using any Web browser.
- The future trends in wireless networks are proceeding along two dimensions: the geographic area covered and the capabilities provided.
- Communications channels, transmission media, bandwidth, and the division of the electromagnetic spectrum into bands are important concepts to grasp in order to understand developments in wireless communications.
- The scope of capabilities provided by the cell phone has grown far beyond a simple voice communications device. Cell phones are now used for Internet browsing, e-mail, listening to audio, watching video, computing, keeping track of contacts and to-do lists, text messaging, and taking and sending photos.
- Cell phone technology has advanced from 1G, which are strictly analog voice communications, through 2G (digital communications) to 3G (higher-speed services) to 4G (extremely high-speed communications and advanced services).
- Wi-Fi is a wireless communications technology based on a set of 802.11 standards defined by the IEEE. Wi-Fi is broadly available, but users must be wary of potential high costs and security issues.
- WiMAX is a wireless communications technology based on a set of 802.16 metropolitan-area network standards established by the IEEE. In many ways, WiMAX operates like Wi-Fi, only at greater speeds and over longer distances.

Discussion Questions

1. How would you define collaboration? Describe an example of effective collaboration from your own experience.
2. How would you differentiate between a bulletin board and a blog?
3. Identify three recommendations that should be followed when using a corporate blog to improve an organization's image.
4. Identify two potential issues associated with the use of a corporate calendaring system to schedule meetings.
5. What are the advantages and disadvantages associated with using instant messaging at the corporate level?
6. How might a project team use a shared workspace to improve communications among project team members and other project stakeholders?
7. What is the difference between a communications channel and transmission media?
8. What is the name of the primary frequency bands used for wireless communications?
9. Briefly explain how a cell phone network works.
10. Briefly describe the primary differences among the four generations (1G to 4G) of cell phones.
11. What is Wi-Fi? What is WiMAX? In what ways are they similar? In what ways are they different?
12. What are two potential problems associated with the use of public Wi-Fi networks?

Action Memos

1. The CEO e-mails you: "I've been thinking about this for some time, and I believe that it would be a great idea to enable our employees to create their own personal blogs as a way to improve openness and communications within the firm." As the vice president of human resources, how would you reply?
2. Draft an instant message corporate usage policy that will encourage the use of IM and provide useful guidance to avoid potential problems.

Web-Based Case

Subscribe to RSS feeds on three topics of your choice. Write a brief memo that summarizes your experiences and thoughts on the pros and cons of RSS.

Case Study

Wireless Technology: Saving Time and Lives

Since the Institute of Medicine released its 2000 landmark report "To Err is Human," which showed that thousands of people die each year as a result of medical error, hospitals have come under increasing pressure to reduce error through computer intervention. Electronic medical record (EMR) systems, which allow medical personnel to enter and view patient data electronically, might help hospitals prevent a surgeon from amputating the wrong leg. Computer physician order entry (CPOE) systems, which require doctors to communicate treatment, prescription, lab, and radiology orders electronically, might help hospitals prevent a nurse from administering an accidental overdose. Yet, despite indications that these systems reduce error and error-related casualties, hospitals have been slow to adopt the technologies. A 2006 survey by the Greater New York Hospital Association found that, of the 74 hospitals and 11 healthcare systems surveyed, only 30 percent had fully operational CPOE systems and only 22.5 percent had fully operational EMR systems. An even smaller 12.5 percent had adopted clinical decision-support systems.[29]

Reasons for the industry's sluggish approach might be understood from such examples as the early failure of Los Angeles' Cedars-Sinai Medical Center's $34 million CPOE system, which was launched in 2002. The system had been developed in-house, but only a handful of the hospital's doctors had been involved in the process. The resulting change to the workflow proved unwelcome. For each visit, a doctor would have to find a computer, log in, and check box after box to record medications, allergies, symptoms, diagnoses, and tests. The process took considerably longer than scribbling the information in shorthand. "Who's got five extra hours in a day?" asked Andrew Wachtel, a pulmonary specialist. "As it is, we work 80 hours a week." In January 2003, several hundred doctors confronted the administration and forced it to scrap the system after only two months.[30]

Apart from early experiences with difficult-to-use systems, several other barriers typically are cited regarding the adoption of these technologies. Doctors, who are used to unconstrained work environments, typically resist initiatives that force them to follow very controlled procedures, particularly if they clash with an established workflow. In addition, few reports until recently have demonstrated economic benefits from using these relatively expensive systems.

All this has begun to change, however, with the rise of one new technology: mobile and wireless systems. Hospitals can now do what they couldn't do with stationary desktops: Save time.

A number of new wireless devices and systems are making this improvement possible. COWS (computer on wheels) or WOWS (wireless devices on wheels) are rolled on battery-powered trolleys to a patient's bedside, where nurses can record or chart patient data, access doctors' orders, or check drug compatibility. Using notebooks and other portable computers, physicians can check electronic medical records, radiology reports, or lab tests from wherever they happen to be. Handless communication devices worn around the neck save nurses from going to the nurse's station to call a doctor about changing symptoms or ask about an order.

"It's like Star Trek. You push a button, say a name, and you're connected," said Nicholas Christiano, CIO of Health Quest, the parent organization of Vassar Brothers Medical Center in Poughkeepsie, N.Y. The medical center deploys 600 Vocera communication badges across its two-building campus.[31]

Hospitals have even begun using powered, attachable RFID tags that can be read by Wi-Fi access points to track mobile medical equipment, physicians, and even wandering patients.

New studies have begun to report significant time savings. A recent study at George Eliot Hospital in England tracked time savings during a pilot project in which the hospital provided 20 notebook or tablet PCs to a cross-section of nurses and mobile clinicians (see Table 6-6). The 440-bed acute care hospital employed manual, paper-based processes in delivering clinical services. Nurses could access only computerized information at centralized desktop computers where staff would often line up. The trial measured time saved and reported a significant increase in efficiency for some tasks. Other tasks, however, such as entering handwritten notes into the system or printing lab results for physicians, were eliminated entirely from the workflow. [32]

TABLE 6-6 Results of mobile technology productivity test at George Eliot Hospital

Task	Time saved
Finding pathology results	5 minutes per patient visit
Scanning pre-op patients	45 minutes per shift
Recording surgery notes	20 minutes per surgery
Locating surgery notes for follow-up visit	10 minutes per patient visit
Monitoring chronic patients	45 minutes daily
Charting inpatients	20 minutes per shift

The study projected an amazing return on investment. It found that the mobile technology system would deliver ROI in just 15 months. If senior medical staff and consultants were given access to the system, the ROI would be met in five months.[33]

As mobile IT catches on, more and more hospitals are using it to implement health information systems that save lives and improve the quality of care. In 2004, Piedmont Hospital in Atlanta began a 15-month deployment of its CPOE system using mobile technology. "When we first started out with CPOE, the thought was that physicians would use Toughbooks [laptop computers]," said Sandy Wiggs, Manager of Clinical Services at Piedmont Hospital.

The problem was that there was no place to put the Toughbook computer down. In the emergency room, for example, the physician might put the Toughbook on a patient's bed. When the patient moved, the computer would fall to the floor. The majority of physicians did not want to use them. So, Piedmont procured a variety of mobile computers: thin devices stored in cabinets mounted on the wall, monitors on swing arms, COWS, notebooks, and tablets.

"People have their own preferences. That's why you have to offer a variety of devices," said Wiggs.

The effort paid off. By March 2006, the hospital had achieved 100 percent physician adoption of the CPOE system.

"Now our physicians can log on to a computer, place an order prior to coming in, so they don't have to call a nurse. That's a big time saver," explained Wiggs. "Physicians have to sign their orders. They would have hundreds and hundreds of orders. Now, they can do that electronically and they can do it from home."

During the paper-based era, hospital staff would take a copy of a prescription down to the pharmacy, wait, and then take the prescription back up to the patient. Today, automation saves time, both for the staff and for the patient who urgently may need the medication. For example, the hospital cut by half the time to treatment for patients with critically low potassium. Piedmont's CPOE deployment now serves as an astounding success for others—not just in achieving 100 percent adoption or saving time, but in vastly improving the quality of care and saving lives. The hospital reduced mortality by 6 percent and medication errors from 5.5 to 0.9 per 10,000 doses.[34]

Even with the explosion of wireless technologies, health information systems face challenges, and chief among them is establishing standards for portability and interoperability. Establishing these standards is much more complicated for health care than it is for other industries. The Office of the National Coordinator for Health Information Technology (HIT) in Washington, D.C. has spearheaded the Nationwide Health Information Network, a project that eventually will allow a physician in Baltimore to access the records of a patient electronically who was last seen in Los Angeles. Nancy Szemraj, manager of communications and outreach, points out that the amount of data that needs to be standardized for medical information systems is vastly greater than the amount that must be standardized for financial information systems.

Wireless technology is not making the process any simpler. Once medical data is standardized so outpatient systems can communicate with in-patient systems and one hospital can access data from another, the issue of privacy is still a concern. Suppose a physician is in a hotel in Florida and needs to access his patient's files. Is it okay to send data through the Wi-Fi that the patient is HIV positive? Yet hospitals are overcoming these hurdles to jump on the wireless bandwagon. Today, new paperless hospitals are being built in the United States and abroad. Established institutions are eliminating paper-based processes, replacing stationary hardware, and integrating existing proprietary systems. Successful projects frequently incorporate a development phase in which consultants study the paper-based workflow of the medical staff.

While CPOE and EMR systems are not yet widespread, many experts believe that hospital adoptions will achieve a critical mass soon. As standards and wireless technology enable adoption the resulting changes will transform the patient experience and the cost and quality of care.

Discussion Questions

1. Are the requirements imposed on wireless communications systems for hospitals more stringent or different from the requirements placed upon such systems in other industries? Support your response with examples.

2. Identify three opportunities to use some of the collaboration tools discussed in this chapter in a hospital setting. The medical industry has been slow in adopting such technologies. Why do you think this is true?

3. Why did early adoptions of CPOE and EMR systems fail?

4. How does wireless technology help overcome the obstacles that many hospitals face in convincing their staff to adopt new IT systems?

Endnotes

[1] Steve Hamm, "Lessons from IBM's Innovation Factory," *BusinessWeek*, September 2007.

[2] "Made in IBM Labs: IBM Opens 'Innovation Factory' Using Collaboration to Acceleration of New Products, Services," IBM Press Room, 28 March 2007.

[3] "The Business Case of IBM Lotus Connections," IBM Lotus News, 26 March 2007.

[4] "Made in IBM Labs: IBM Opens 'Innovation Factory' Using Collaboration to Acceleration of New Products, Services," IBM Press Room, 28 March 2007.

[5] "The Business Case of IBM Lotus Connections," IBM Lotus News, 26 March 2007.

[6] "A Workforce on the Move," The Untethered Worker, *Computer World Executive Briefings*, 12 July 2007.

[7] Michael Goldberg, "When a Pseudonym Doesn't Work," Blog: Information Collective, 13 July 2007.

[8] Heather Havenstein, "Corporate Blogs Take On a New Edge," *Computerworld*, 30 July 2007.

[9] "Blogs and Wikis: Technologies for Enterprise Applications?" *The Gilbane Report*, Volume 12, Number 10, March 2005.

[10] Heather Havenstein, "Corporate Blogs Take On a New Edge," *Computerworld*, 30 July 2007.

[11] C.G. Lynch, "Seven Reasons for Your Company to Start an Internal Blog," *CIO*, 20 June 2007.

[12] "Group Calendaring at the Pennsylvania Senate," Novell Web site, 11 June 2006.

[13] Jennifer LeClaire, "Consumers on the Hunt for Quality Tech Support," *TechNews World*, 19 April 2006.

[14] Jack M. Germain, "IM at Work, Part 2: Tools for Locking Down," *E-commerce News*, 6 July 2007.

[15] Ibid.

[16] Eileen Kennedy, "Corporate IM Still Lacks Security, Policies," *SearchWinIT.com*, 4 June 2007.

[17] Galen Gruman, "Really Simple Syndication as a Knowledge Management Tool," *CIO*, 1 September 2006.

[18] Sharon Housely, "Future of RSS is Not Blogs," accessed at *www.feedforall.com/future-rss-not-blogs.htm*, 12 October 2007.

[19] CSR Web site, accessed at *http://cms.csr.nih.gov/AboutCSR/Welcome+to+CSR/*, 29 October 2007.

[20] "Who Uses Megameeting-Retail and Service," accessed at *www.MegaMeeting.com*, 10 November 2007.

[21] Ezra Goodnoe, "How to Use Wikis for Business," *InformationWeek*, 8 August 2005.

[22] Dan Bricklin blog, "Podcast with Toby Redshaw of Motorola on their Continued Wiki Use," accessed at *danbricklin.com*, 20 March 2007.

[23] Eric Bangeman, "FCC Readies 'For Sale' Sign on Beachfront 700 MHz Property," *ars technica*, 25 April 2007.

[24] Denise Pappalardo, "What You Need to Know About 4G," *NetworkWorld*, 21 May 2007.

[25] Denise Pappalardo, "Sprint Nextel Backs WiMAX," *NetworkWorld*, 15 August 2006.

[26] Ronald Grover, "A Failure to Communicate," *BusinessWeek*, 21 May 2007.

[27] Preston Gralla, "Sound Off: Why You Need Wireless Protection," *Computerworld*, 25 September 2007.

[28] Martyn Williams, "U.S. Army to Evaluate Mobile WiMAX Use," *Computerworld*, 26 April 2007.

[29] "Greater New York Hospital Association Hospital Information Technology Survey Report," January 2006.

[30] Ceci Connolly, "Cedars-Sinai Doctors Cling to Pen and Paper," *Washington Post*, 21 March 2005.

[31] David Haskin, "Costs down, quality up: A hospital goes wireless," 14 February 2007, *Computerworld*, accessed at *www.computerworld.com/action/article.do?command= viewArticleBasic&taxonomyName=wireless_trends_and_ technologies&articleId=9011258&taxonomyId=78*, 22 August 2008.

[32] "Trial of mobile technology at George Eliot Hospital shows significant benefits," MTP Europe, Technology for Healthcare, *www.mtbeurope.info/content/ft60101.htm*.

[33] "Trial of mobile technology at George Eliot Hospital shows significant benefits," MTP Europe, Technology for Healthcare, *www.mtbeurope.info/content/ft60101.htm*.

[34] Blair, Robin, "Passing the 'Yo' Mamma Test," *Health Management Technology*, June 2006.

E-BUSINESS

EDMUNDS.COM INC.

Why Managers Must Get Involved in E-business

Edmunds began business in 1966 as a publisher of pricing guides designed to help car shoppers make purchasing choices. The firm operated under this business model for nearly 30 years. A few visionary employees recognized that the Internet represented a tremendous opportunity and created a Web site as an experiment in 1995. Had they not ventured onto the Internet, Edmunds would not be in business today.

The Edmonds.com Web site soon became one of the most popular Web sites for shoppers looking to compare, price, and locate new and used vehicles. Along the way, Edmunds had to adapt its business model to generate revenue by selling advertising on its Web site to auto manufacturers and makers of auto products, as well as by selling sales leads to auto dealers.

Researching a new car using online resources really caught on with potential car buyers. As a result, numerous auto-oriented Web sites sprung up—auto manufacturer sites; industry expert sites such as Consumer Reports; and third-party sites, such as CarGurus, Yahoo! Autos, and Edmunds; which incorporated user-generated content.[1] Edmunds management recognized that in order to be successful against all of this competition they needed to do more than attract highly motivated one-time visitors. Therefore, Edmunds created a set of auto-themed Web sites designed to keep consumers returning even after they purchased a vehicle. Inside Line was launched in 2005 and currently has a monthly readership of 3.5 million who appreciate its photos, videos, and columns on road tests, future vehicles, and auto show news. CarSpace was created in 2006 as an automotive social networking site that lets visitors come together and discuss the ins and outs of car buying and car ownership. AutoObserver was established in 2007 to provide insightful automotive industry commentary and analysis. Edmunds even makes its content available to mobile-device users through its Edmunds2Go family of wireless Web sites. As a result of its expanded line of Web sites, Edmunds has increased its potential audience from the 16 million people who buy cars in a given year to a much larger crowd of auto enthusiasts who are intensely loyal.[2]

Management vision and leadership enabled Edmunds to make a highly successful leap from a business model based on the use of print to one based on the use of interactive media. While earnings figures are not publicly disclosed, Edmunds' revenues have more than tripled in the past seven years. Its Web sites are so successful that Edmunds got completely out of the print business in 2006.[3]

LEARNING OBJECTIVES

As you read this chapter, ask yourself:

- What sort of benefits can arise from well chosen e-business opportunities?
- How can business managers recognize and capitalize upon these opportunities?

This chapter provides several examples of organizations making effective use of e-business and highlights the essential role of managers in recognizing and leading the implementation of appropriate e-business opportunities. After discussing why it is important for managers to understand e-business, this chapter will continue by discussing several forms of e-business, identifying e-business critical success factors, and defining many of the advantages and issues associated with e-business.

WHY MANAGERS MUST UNDERSTAND E-BUSINESS

IBM defines **e-business** as "the transformation of key business processes through the use of Internet technologies."[4] E-business enables organizations and individuals to build new revenue streams, to create and enhance relationships with customers and business partners, and to improve operating efficiencies. E-business is critically important to today's business. As we saw with Edmunds in the opening vignette, conversion to an e-commerce business model is essential to the survival of some organizations. For many other organizations, the revenue associated with e-business is substantial and growing.

During the late 1990s, many poor ideas for Web-related businesses were proposed and funded in a wave of "irrational exuberance" for all things associated with the dot-com economy. In many cases, these new businesses ignored traditional business models built on delivering fundamental value for customers, achieving operational excellence, and generating revenues in excess of costs. Instead many companies placed an unhealthy emphasis on increasing market share with little regard for bottom-line profits. With their focus on the wrong things, it really was not a surprise when hundreds of the dot-com companies failed. It is estimated that the bursting of the dot-com bubble wiped out $5 trillion in market value of technology companies from March 2000 to October 2002.[5] While many of the early start-up dot-com organizations vanished, many established firms went on to incorporate e-business elements into their business operations.

To succeed, business managers must understand their customers and the fundamentals of the markets in which they operate. They must then run their businesses on the basis of those fundamentals. If they are to incorporate e-business into their business, business managers need a clear understanding of how the Internet differs from the traditional venues for business activity, and they must employ business models appropriate to the Internet. The business-to-business, business-to-consumer, consumer-to-consumer, and e-government models of e-business will now be discussed to help you gain an understanding of the broad scope of e-business.

Business-to-Business (B2B) E-business

There are several forms of e-business including business-to-business (B2B), business-to-consumer (B2C), consumer-to-consumer (C2C), and e-government (e-gov). B2B is the exchange of goods and services between businesses via computer networks. The revenue generated via B2B transactions greatly exceeds B2C revenue by a factor of more than 6 to 1.[6] There are several forms of B2B Web sites in operation today.

Private Stores

Many organizations have established Web sites that function as private stores for each of their major customers. Access to the **private store** requires that the buyer enter a company identification code and password to make a purchase from a selection of products at pre-negotiated prices typically based on an established annual minimum purchase quantity. For example, the Sprint Private Store Web site (*www.mycompanyrates.com*) shown in Figure 7-1 enables employees of companies who have an agreement with Sprint to shop for Sprint equipment, rate plans, and accessories at exclusive corporate discount pricing.

FIGURE 7-1   Sprint private store

Customer Portals

These are private stores that offer additional customer services beyond simply placing an order. Goodrich is a global supplier of systems and services to the aerospace, defense, and homeland security markets. It offers an extensive range of products, systems, and services for aircraft and engine manufacturers, airlines, and defense forces around the world. Goodrich generates annual sales of over $6 billion.[7] Goodrich built a customer portal (*https://customers.goodrich.com/portal/site/public*) that consolidates Goodrich commercial aftermarket products and services into a single Web site accessible online from anywhere in the world at any time. Goodrich customers and employees can use the customer portal to search for parts, place orders, check order status, and inquire about lead times for items from Goodrich business unit e-commerce sites. Visit the Goodrich Web site shown in Figure 7-2 to view a demo of how a customer portal Web site operates.

FIGURE 7-2 Goodrich enterprise customer portal

Private Company Marketplaces

Today companies rarely manufacture all the components of increasingly complex pieces of equipment, such as appliances, aircraft, automobiles, computers, engines, motor homes, and televisions. Instead, such items are made up from component parts that are then used to build subassemblies that go together to create the final product. A high percentage of B2B transactions take place between companies that supply parts and components (Original Equipment Manufacturers) and the companies that sell the final product. Some of the companies that do business with OEM suppliers include General Motors, Ford, and Toyota in the automobile industry; Boeing and Cessna in the aircraft industry; Dell and HP in the personal computer industry; and Sony, Phillips, and Mitsubishi in the television industry. Each of these companies deals with dozens, even hundreds, of OEMs whose parts go into the final product.

Often large manufacturers that purchase goods and services from many small suppliers build a **private company marketplace** to manage their purchasing functions through a Web site. Suppliers are required to bid on providing goods and services by publishing a schedule of prices at which they would sell each of their various items to the manufacturer. The manufacturer compares that pricing to bids from other providers to select the winning supplier for each item. The selected supplier must then provide product price and description information in an electronic format suitable for loading the data into the manufacturer's e-procurement system.

E-procurement software allows a company to create an electronic catalog with search capability. Authorized purchasers within the manufacturing firm then use the catalog to identify needed products and services. E-procurement software can also automate key functions of the purchasing process including creating, reviewing, and approving purchase orders and transmitting these purchase orders electronically to the supplier. More advanced e-procurement systems can support the use of negotiated prices for the purchase of goods and services. The negotiation may be done through some form of reverse auction process (suppliers compete to submit the lowest bid for a set of products or services) and/or a request for quotation process (the buyer describes a business need to be met and invites potential suppliers to submit creative, low-cost solutions).

United Technologies Corporation (UTC) is a diversified company whose products include Carrier heating and air conditioning, Hamilton Sundstrand aerospace systems, Otis elevators and escalators, Pratt & Whitney aircraft engines, and Sikorsky helicopters. UTC has 225,000 employees, does business in approximately 180 countries, and has annual revenues over $54 billion.[8] With its wide diversity of products, UTC must deal with many suppliers. A key to UTC's success has been its ability to develop a supply chain of highly competitive, global suppliers who work closely with the firm to deliver world-class products and services. In the mid-1990s, the firm created a private company marketplace that allows the company to purchase over $10 billion in goods with an estimated savings of $2 billion through more competitive prices and lower transaction costs.[9]

Industry Consortia-Sponsored Marketplaces

In many cases, companies are not large enough or do not have sufficient purchasing power to require suppliers to deal with them through a private company marketplace. In such a situation, several companies in a particular industry may join forces to create an **industry consortia-sponsored marketplace** to gain the advantages of the private company marketplace for all members of the consortia.

Avendra is an industry consortia-sponsored marketplace serving hospitality-related industries. It was founded in 2001 by ClubCorp USA, Fairmont Hotels & Resorts, Hyatt Hotels, Intercontinental Hotels Group, and Marriott International. Avendra offers its customers a wide range of purchasing programs with over 900 suppliers providing items such as food and beverages, uniforms, linens, soaps and shampoos, office supplies, janitorial supplies, kitchen equipment, and golf course maintenance. Avendra's programs cover over $2.5 billion of annual purchases and generate considerable cost savings for the buyers.[10] Avendra also provides many benefits to suppliers, including:[11]

- Improved method of communicating product descriptions and availability
- Access to new customers in the hospitality and related industries
- Enhanced customer service through better reporting and improved information access
- Standardized and simplified business processes
- Increased sales

Business-to-Consumer (B2C) E-business

Business-to-consumer (B2C) e-business is the exchange of goods and services between business organizations and individual consumers. One of the first and most successful B2C retailers is Amazon.com, which began its online bookstore in 1995 and had recent annual net income of $476 million on sales of $14.8 billion. Today, the majority of large brick-and-mortar retailers have at least experimented with some level of B2C.

B2C sales in the United States are growing at a rate of nearly 40 percent per year as shown in Figure 7-3. B2C sales represented about 2 percent of overall 2006 retail sales, but are expected to increase to about 16 percent by 2011.[12] For all of 2007, U.S. consumers spent $122.7 billion on retail e-commerce. In the first quarter of 2008, U.S. retail e-commerce sales represented 3.3 percent of all U.S. retail sales.[13]

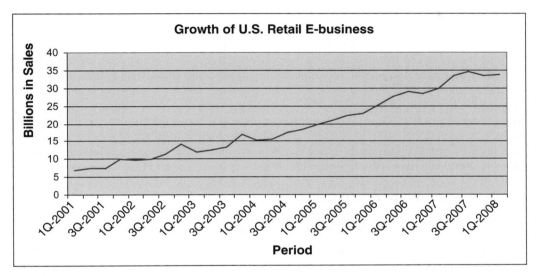

FIGURE 7-3   Growth of U.S. retail e-business

B2C Web sites must focus on attracting prospects, converting them into customers, and retaining them to capture additional future sales. These have long been necessary objectives of brick-and-mortar retailers as well. Now, however, shoppers use online tools and data to become better informed shoppers. Many shoppers research products online before going to a store to make a purchase. Many also look at an online peer review before making a purchase.

Brick-and-mortar retailers are finding that they must modify their in-store operations and procedures to meet shoppers' new expectations that are based on online shopping experiences. Now when one store location is out of an item, consumers expect salespeople to simply walk over to a computer and find a store where it is in stock. Many consumers no longer have the patience to search around large stores looking for a specific item, so retailers like Barnes & Noble are installing kiosks that allow people to search inventory, locate merchandise, and order out-of-stock items. AMR Research estimates that retailers will spend nearly $800 million on providing Web-like technology in their stores over the next few years.[14]

Over the past decade, many big retailers have built effective and efficient online Web sites. Part of their e-commerce strategy is to lure online shoppers into their brick-and-mortar store by allowing customers to pick up their purchases at a local store rather than wait for it to be shipped. Getting the customer into the store provides an opportunity for more sales. Circuit City guarantees online purchases will be available for pickup within 24 minutes at a local store or the customer gets a $24 gift card. As a result, nearly 50 percent of its online orders are picked up in a store.[15]

A brick-and-mortar store can only stock so many items based on the size of the store. With the use of an electronic catalog on the Web and large, highly efficient distribution centers, the amount of products that can be offered grows substantially, allowing customers many more choices. This new electronic catalog was propelled by a new value proposition known as "The Long Tail" first coined by Chris Anderson: "Here is what the idea says: Many of us see the same movies and read the same books because the bookstore can store only so many books and the movie theater can play only so many movies. There isn't enough space to give us exactly what we want. So we all agree on something we kind of want. But what happens when the digital age comes along, allowing the bookstore to store all the books in the world? Now, it doesn't sell 1,000 copies of one book that we all kind of want; it sells one copy of 1,000 books each of us really wants."[16]

Consumer-to-Consumer (C2C) E-business

Consumer-to-consumer (C2C) e-business is the exchange of goods and services among individuals, typically facilitated by a third party. Craigslist is an example of a third party that has established local classifieds and forums for 500 cities worldwide. It posts over 30 million new classified ads and receives over 2 million new job listings each month. It receives over 10 billion hits per month! Craigslist is used by both individuals and organizations.[17] Successful use of Craigslist requires that individuals or organizations place their ads under the Craigslist category that will best attract the target audience for their goods and/or services.

eBay is another online auction and shopping Web site from which people and organizations buy and sell millions of appliances, automobiles, collectibles, equipment, furniture, and other items on a daily basis. eBay has established localized Web sites in more than two dozen other countries in addition to the United States eBay supports auction style listings, in which the seller offers one or more items for sale by a specific date and time. The highest bidder whose bid exceeds any reserve price set by the seller purchases the item. eBay also supports a fixed price forum that lets the seller specify a "Buy It Now" price. A buyer who agrees to pay that price immediately purchases the item at that price with no bidding involved. The eBay business model generates revenue from fees paid by the lister/seller of an item. PayPal, a wholly owned subsidiary of eBay, can be used to send and receive payments via the Internet. PayPal also charges fees for these transactions.

Because the U.S. dollar is currently weak against many of the world's currencies, U.S. goods represent bargains for shoppers worldwide. This increases the demand for e-business between U.S. sellers and global buyers. While some U.S. e-commerce sites and eBay sellers don't ship outside the country—it requires a mountain of paperwork—foreign buyers have found a way around the problem. They use private forwarding services, which receive goods at a U.S. address and send them on to the purchaser.

E-government Applications

E-government (e-gov) involves the use of information technology (such as Wide Area Networks, the Internet, and mobile computing) by government agencies to transform relations between the government and citizens (G2C), the government and businesses (G2B), and among various branches of the government (G2G).[18] Table 7-1 lists many of the most popular e-gov G2C Web sites.

At last count, businesses and citizens spent approximately over 8 billion hours and more than $320 billion filling out paperwork and complying with government regulations. Users spent much of this time navigating complex government hierarchies and wading through millions of documents meant to help their businesses become compliant with laws and regulations.[19] One of the primary objectives of e-gov is to save time and money spent on regulatory compliance by providing quick and easy access to business laws, government regulations, forms, and agency contacts. Additional desired benefits include better delivery of government services to citizens, improved government interactions with business and industry, easier citizen access to information, and more efficient government management.[20]

Unfortunately, in the United States, citizens are not completely satisfied with the quality of e-gov G2C Web sites. The American Customer Satisfaction Index (ACSI) E-Government Satisfaction Index for 2007 showed that user satisfaction with government Web sites slipped for the third straight year. On average, the federal government Web sites scored 72.9 on the ASCI's 100-point scale for the July-August-September 2007 quarter, the lowest scores since April-May-June of 2005. The two highest-rated U.S. e-gov Web sites are Internet Social Security Benefits Application with a score of 88 and Help With Prescription Drug Plan Costs with a score of 87.[21]

TABLE 7-1 Frequently used E-gov G2C Web sites

E-gov Web site	Description
GovBenefits.gov	Provides single point of access for citizens to locate information and determine potential eligibility for government benefits and services
USAJobs.gov	Provides information regarding career opportunities within the Federal government
Business.gov	Provides a single access point to government services and information from the Small Business Administration to help the nation's businesses with their operations
Grants.gov	Functions as a central storehouse for information on over 1,000 grant programs and provides access to approximately $400 billion in annual awards
Forms.gov	Serves as the U.S. government's official hub site for various forms including tax forms, small business forms, social security forms, veteran benefits, and FEMA forms

TABLE 7-1 Frequently used E-gov G2C Web sites (continued)

E-gov Web site	Description
USCIS.gov	Provides information on the services provided by U.S. Citizen and Immigration Services regarding citizenship, lawful permanent residency, family- and employment-related immigration, employment authorization, and inter-country adoptions
IRS.gov	Enables tax filers to download tax forms, obtain answers to frequently asked questions about filing, and electronically file a tax return

County and local governments also have made attempts at implementing e-gov Web sites. For example, the city of Chicago Web site at *http://egov.cityofchicago.org* provides easy access to information for residents of the city and for people who plan to visit the city.

The U.S. General Services Administration is the managing partner for several e-gov G2B initiatives directed at improving the efficiency and effectiveness of government operations through programs such as the following:[22]

- E-Gov Travel is a collaborative, interagency program whose goals are to deliver cost-savings and increased services associated with an automated and integrated approach to managing the travel function of the federal government's civilian agencies. There are thousands of civilian employees who travel for business every day. This service will replace more than 250 travel-booking practices at various government agencies and reduce travel management expenses by 50 percent over the next 10 years.
- Federal Asset Sales is an effort to develop a secure, effective, and efficient one-stop online environment that provides clear information and a marketplace for buyers and sellers of federal assets.
- The Integrated Acquisition Environment (IAE) Project will create a platform to support the acquisition of $200 billion/year of goods and services. The goal is to transform the way government agencies interact and transact with their business partners to reduce costs and streamline business processes while improving customer service.

Mobile Commerce

Mobile commerce(m-commerce) is the buying and selling of goods and/or services using a mobile device such as a cell phone, smartphone, PDA, or other such device (Figure 7-4). Mobile commerce can be used to support all forms of e-commerce—B2B, B2C, C2C, and G2C. Mobile spending is expected to exceed $500 million in 2008 and grow to almost $2 billion by 2010 according to JuniperResearch. To put this in perspective, e-commerce exceeded $100 billion in 2007 according to comScore Networks.[23]

FIGURE 7-4 Smartphone
Image copyright Perry, 2008. Used under license from Shutterstock.com.

.Mobi

Worldwide, there are more digital mobile phones than personal computers and TVs combined. Most mobile phones have full Internet capabilities. However, these mobile phones have a number of limitations that make it difficult to view standard Web pages. The main limitation of course, is the size of the viewing screen. .Mobi (also known as dot-Mobi) is a top-level domain approved by the International Corporation of Assigned Names and Numbers (ICANN) and managed by the mTLD global registry. Its goal is to deliver the Internet to mobile devices. It works with mobile operators, handset manufacturers, and content providers to ensure that the .Mobi destinations designed for mobile phones work fast, efficiently, and effectively with user handsets. One means of doing this is by developing and publishing a set of style guides that contain mandatory and recommended best practices for developing mobile content and services.

Mobile Payments

There are many payment systems based on the use of mobile devices. One of the newer and more innovative systems is Mocapay, which allows people to pay for purchases without the use of cash or a credit card. Subscribers to Mocapay set up an account linked to their bank account and cell phone. To use the system in a store that accepts Mocapay, subscribers just text their four-digit pin to Mocapay. The Mocapay computers then verify the account number and determine the current balance in the account. Subscribers wait for the transaction number and balance to arrive via SMS message format and present this transaction number to the Mocapay merchant. The transaction number works one time only. The cashier then provides the subscribers with a receipt of payment. The amount is automatically deducted from the subscriber's bank account. Merchants like the fact that the transaction fee is a mere $.19 compared to up to 4 percent of each transaction for credit cards.[24]

Mobile Ticketing

Tickets can also be purchased via mobile devices. The tickets are sent to the mobile device, and users present their phones at the venue to gain entrance. The same approach can be taken to distribute vouchers, coupons, or loyalty cards as a virtual token that is sent to a mobile device. Customers can then present their mobile devices at the point of sale to gain the same privileges and benefits as customers with the actual physical voucher.

The Washington Nationals Major League Baseball team allows fans to purchase mobile tickets that are delivered to their phones via a text message. Mobile ticket purchasers must bring the phone to the game where the image on the phone is scanned to allow entrance. Research in Motion and Ticketmaster provide a joint service that enables BlackBerry smartphone users to browse, search, and purchase tickets available on Ticketmaster.com, TicketsNow.com, and Getmein.com. Continental Airlines is testing the use of electronic boarding passes that allow travelers to pass through security and board the plane without handling a piece of paper. Their boarding pass is an image of an encrypted bar code displayed on the BlackBerry's screen, which can be scanned by gate agents and security personnel.

Location-based Services

If a mobile device is equipped with GPS and appropriate software, the location of a mobile device user can be determined to a high degree of accuracy. The user can then request local maps and walking or driving directions to points of interest, as well as obtain local traffic and weather information. Where available, the mobile device user can receive offers for local goods and services (e.g., stop in for lunch today at Izzy's on Main Street and receive $2.00 off on each order).

Mobile Banking

Banks, brokerage firms, and other financial institutions are keenly interested in enabling customers to use mobile devices to access account information, withdraw and transfer funds among various accounts, and purchase stocks and bonds. With Mobile Banking from Bank of America, you can use your cell phone or smartphone to access balance information, pay your bills, transfer funds, and find nearby ATMs or banking centers.

Web 2.0 and E-commerce

Web 2.0 is a term describing changes in technology (see Table 7-2) and Web site design to enhance information sharing, collaboration, and functionality on the Web. The emergence of Web 2.0 is dramatically changing the ways companies interact with consumers. Indeed, business-to-consumer e-commerce Web site designers must take advantage of Web 2.0 to remain competitive. Consumers who visit Web sites such as eBay, which are full of recommendations, user reviews, and ratings, expect similar features from other e-commerce Web sites. According to Gene Alvarez, a Gartner Group analyst, "Web sites don't get a second chance to impress. Sites have a one-shot, one-visit time to win. If you don't get them the first time, you have to win them back."[25] While business-to-consumer organizations clearly see how to employ Web 2.0, business-to-business organizations are racing to figure out how to take advantage of these capabilities.

TABLE 7-2   Partial list of Web 2.0 capabilities

Web 2.0 Capability	How Used
Blogs	Enables the customer to get to know your organization in a different way and allows a two-way dialogue.
Forums	Create open or moderated forums to enable discussions on your Web site.
Mashup	Combines content from a variety sources and in various forms to create multimedia messages for Web site visitors.
Multiple product comparisons	Provides valuable and highly desired information for Web site visitors.
Newsletters	Allow users to sign up online, create multiple subscriber groups, and manage newsletter issues.
Page notes	Allow visitors to comment on content you have published on your Web site.
Podcasts	Provide high-quality messages to customers.
Polls	Create instant polls to collect information from visitors and display the results.
RSS newsfeeds	Allow visitors to your Web site to subscribe to RSS newsfeeds to receive fresh, compelling content from your firm or third-parties.

Before simply adding Web 2.0 capabilities to your Web site, you must determine what you are trying to accomplish. Are you trying to create a more engaging online experience for your current users? Are you trying to acquire new users? Are you trying to learn more about visitors to your Web site? Do you wish to engage and reward your most loyal customers? In addition, you must realize that many Web 2.0 capabilities require that retailers let go of control and allow visitors to have their say—good, bad, or indifferent—about your organization and its products and services.

E-business Critical Success Factors

Now that various e-commerce models and examples have been discussed, the critical factors needed to make an organization's e-business operation successful will be outlined. There are numerous factors that contribute to making an e-business operation successful, including identifying appropriate e-business opportunities, acquiring necessary organizational capabilities, directing potential customers to your site, providing a good customer online experience, providing an incentive for customers to buy and return in the future, providing timely and efficient order fulfillment, offering a variety of easy and secure payment options, handling returns smoothly and efficiently, and providing effective customer service.

Identifying Appropriate E-business Opportunities

E-business initiatives can be risky and extremely challenging due to an organization's lack of e-business skills, uncertainty in regards to how business processes and policies must be changed, and the need to make new investments in IT-related hardware and software. Before embarking on such a dangerous journey, an organization must consider carefully how each potential e-business initiative fits into its overall business strategy. Just like any other business initiative, specific, achievable objectives and time-based measures need to be defined. An example of a specific, achievable objective with a time-based measure is "Reduce the cost of direct advertising by 5 percent within 12 months of start-up." Initiatives whose objectives and goals do not match those of the organization or that do not seem feasible either should be rejected or redefined.

Acquiring Necessary Organizational Capabilities

Many organizations lack the skills and experience to succeed in their initial e-business initiatives, or the organization culture may be such that people harbor a strong resistance to change. Senior management must make an objective assessment as to whether or not the organization has adequate skills, sufficient experience, and the corporate culture necessary to succeed in its e-business initiatives. Often, organizations will elect to hire or contract with experienced resources to help evaluate and lead the implementation of their early e-business projects rather than proceed on their own.

Directing Potential Customers to Your Site

Successful e-commerce Web sites must be able to attract prospects in order to convert them into customers. The effective use of a search engine is critical to attracting prospects to the Web site. A search engine is software that maintains an index of billions of Web pages and uses that index to quickly display the URLs of those pages that "best match" the user's search term. To perform the matching process, many search engines such as Google, Yahoo!, and MSN use software called crawlers to score Web sites. The score of a Web site is based on how relevant the site is to the search term depending on things such as link popularity, density, frequency of keywords in the page content, number of Web sites referencing the site, and numerous other factors. In addition, Web site designers can specify other key words to be associated with the Web page. The search engine lists the URLs of those pages that "best match" the user's term in descending order of score. The user can then click on the displayed URLs to visit those sites.

Numerous studies have shown that top placement in the results returned by search engines can provide a higher return on investment that spending on mail campaigns or radio and TV advertising. Thus many organizations invest great amounts of time and money in **search engine optimization** to ensure that their Web site appears at or near the top of the search engine results whenever a potential customer enters search terms that relate to their products or services. If an organization understands how the crawler ranks its findings, it can attempt to raise its ratings by modifying the text on its Web pages or specifying more or different key words to be associated with the Web page.

An **organic list** is a type of search engine result in which users are given a listing of potential Web sites based on their content and keyword relevancy. Web sites can also bid on keyword phrases to have their site appear among the results listed. The higher the bid, the higher their ad will appear on the results page. The Web site owners then pay an additional small fee each time the ad is clicked on. Search engine results that appear because of the payment of fees are called **paid listings**. Critics of paid listings complain that the practice causes searches to return results of little relevancy to search engine users.

Google attempts to quickly return highly relevant results based on the content of the page, the relevancy of links pointing to that page, and other criteria. Google also allows companies to pay for their Web sites to appear at the top of the results page, but it clearly separates paid listings from organic listings. For a fee, Overture.com will display your Web site in a full-screen pop-under window on Web sites in its publisher network. Unlike a pop-up window that loads over a Web page, a pop-under window invisibly loads under the Web page. Web site visitors don't even see the pop-under window until they are finished at the Web site and close the window.

An organization can also attract potential customers to its site through the use of Web page banner ads that display a graphic and include a hyperlink to the advertisers' Web site. Some companies participate in a banner exchange network that coordinates ad sharing so that other sites show one company's ad while that company's site shows other exchange members' ads. Another approach is to find Web sites that appeal to the same target audience and pay those sites to allow placement of your banner. Companies can also work with banner advertising networks, such as Google or ValueClick. The banner advertising network acts as a broker between Web sites and advertisers.

See Table 7-3 for a summary of the pros and cons of different strategies for directing potential customers to your Web site.

TABLE 7-3 Strategies to direct potential customers to your site

Strategies	Pros	Cons
Search Engine Optimization	No additional out of pocket cost.	Requires special expertise and there is stiff competition for placement in the list of results.
Paid Listings	Can ensure your Web site appears on the results page for specified search terms.	Additional advertising cost and users may complain if your Web site has little relevancy to their search term.
Banner Ads	Several options for placing banner ads: join a banner exchange network, pay for ad rights on sites that appeal to same target audience, use a banner advertising network.	Users can become oblivious to banner ads.

Providing a Good Customer Online Experience

The ultimate goals of most Web sites are to increase sales as well as to improve customer satisfaction and loyalty to an organization. To accomplish these goals, a company must create a Web site that will compel customers to return time and time again. Usability focus groups and testing with typical consumers should be conducted throughout the process of designing a Web site to ensure that these goals are met. Several steps must be taken to provide a good customer online experience. A few of the key steps are listed:

- Design the home page to be informative and visually appealing to your target customer.
- Ensure that the navigation is highly intuitive.
- Provide a simple search tool that returns search results with thumbnails of actual products.
- Provide product and service comparison tools so customers can become better informed about competitive products and suppliers.
- Use available profile data on customers to make appropriate product and service recommendations.
- Prominently feature a mix of up-sells and cross-sells as well as hot items and clearance items.
- Use simple, plain language—no jargon.
- Use bold and italic text sparingly.
- Allow sufficient white space so that the pages are not too dense with text and graphics.

Providing an Incentive for Customers to Purchase and Return in the Future

According to a 2005 survey by Bain & Company, "eighty percent of companies believe they deliver a superior customer experience, but only eight percent of their customers agree."[26] Researchers James Allen, Frederick Reichheld, and Barney Hamilton found that the companies that truly provided a great customer experience "pursued three imperatives simultaneously:

1. They design the right offers and experiences for the right customers.
2. They deliver these propositions by focusing the entire company on them with an emphasis on cross-functional collaboration.
3. They develop their capabilities to please customers again and again—by such means as revamping the planning process, training people how to create new customer propositions, and establishing direct accountability for the customer experience."[27]

Compare the previously discussed actions of companies that deliver superior customer service to these all too frequently encountered Web site shopping experiences:

- You visit a Web site only to be disappointed that the Web site inventory is quite limited compared to the brick-and-mortar store.
- Your in-store credit from previously returned items cannot be applied to purchases on the company's Web site.
- You visit a brick-and-mortar store to return an item purchased at the company's Web site and are told that online purchases cannot be returned in the store.
- You visit a brick-and-mortar store and talk to a salesperson who is completely uniformed about what is available on the Web site and is unwilling or unable to help answer any questions relating to Web site items or prices.

Providing Timely, Efficient Order Fulfillment

A number of components and processes must be considered in designing a timely, efficient order-fulfillment system. Adequate storage must be secured for inventory. Items must be stored safely and accessed easily for fast order fulfillment. Products might be stored on pallets, bins, racks, or simply on the floor. Systems and processes must be capable of receiving fast and accurate deliveries from suppliers. Accurate inventory counts and the ability to do sales forecasting with some degree of accuracy are also critical. This enables management to minimize inventory levels (and the associated costs) while still providing a high percentage of order fulfillment. Distribution processes must be capable of meeting customer expectations for delivery times and costs. Often this means that several different delivery solutions may be offered ranging from one-week ground transportation to air overnight. Of course, all of this begins with an accurate capture of the customer order and delivery information.

Borders Group, Inc., is a global retailer of books, music, and movies. It employs over 30,000 people and operates 1100 stores plus its Web site, Borders.com. When it came time to upgrade its Web site, Borders implemented Sterling Order Management software to manage its complex cross channel (stores, Web site, and call center) selling and order fulfillment processes. The Sterling Order Management system determines which Borders distribution location is the most efficient and least costly to manage each order's fulfillment. The system is also able to provide inventory information across channels to improve inventory utilization, reduce excess safety stock, and minimize lost sales due to stock-outs. The end result is timely, efficient order fulfillment that meets or exceeds the customer's expectations.[28]

Offering a Variety of Easy and Secure Payment Methods

Have you ever shopped at a store that will only accept cash—no checks and no credit cards? For most of us, such a restriction would limit our purchases and discourage us from returning in the future. You probably would have a similar reaction if you walked into a store that accepted only one type of credit card (especially if it wasn't one that you have). Web sites need to accept a variety of easy and secure payment methods to increase sales and encourage repeat business.

Credit cards are used for payment for over 85 percent of worldwide consumer Web purchases.[29] Other forms of payment cards such as debit and charge cards are used less frequently. However, there is a high risk of credit card fraud with Web purchases. While less than 5 percent of all credit card transactions are completed online, those transactions account for a disproportionate 50 percent of the total dollar value of credit card fraud.[30] It is estimated that fraudsters will divert approximately $3.6 billion from U.S. e-commerce in 2007, representing about 1.4 percent of online sales.[31]

Most Web sites require user verification information in addition to the payment card number to help ensure that the person using the card actually possesses the card and that the card account is legitimate. The additional information might include one or more of the following: customer's billing address, card expiration date, or the Card Verification Value (a three-digit code on the back of Discover, MasterCard, and Visa cards or a four-digit code on the front of the American Express card). The trade-off here is that if too much information is demanded of consumers, they may consider the Web site "too difficult" or "too invasive" to do business with.

A secure Web site uses encryption and authentication to protect the confidentiality of Web transactions. By default, the most commonly used browsers (including Internet Explorer, Netscape, Mozilla Firefox, and Safari) will inform you when you are entering or leaving a secure Web site. However, if you have turned these notifications off, the browsers also provide visual clues—typically a locked padlock will appear in a bottom corner of the browser window.

The most commonly used protocol for Web security is the Secure Sockets Layer. The **Secure Sockets Layer (SSL)** can be used to verify that the Web site to which the consumer is connected is indeed what it purports to be. SSL also encrypts and decrypts the information flowing between the Web site and the consumer's computer. Thus any hacker who may be eavesdropping on the "conversation" will only receive unintelligible gibberish.

When interacting with a secure Web site, the biggest risk for a consumer is not that credit card data will be intercepted in transit, but that the retailer databases on which this data is stored may be compromised. Each year for the past several years, there have been dozens of incidences in which a large amount of credit card data has been stolen from the databases of retail organizations. For example, TJX operates more than 2500 stores worldwide under such brand names as Bob's Stores, Marshalls, and TJ Maxx. In the largest computer data breach in corporate history; more than 65 million Visa account numbers and 29 million MasterCard numbers were stolen from the company's database. Some of the numbers were used to make fake credit cards and buy millions of dollars of merchandise from various retailers. In August 2007, TJX said that the total cost of dealing with the breach (including fixing the company's computer systems and dealing with lawsuits and investigations) would exceed $250 million. Other experts outside TJX estimated the costs could go as high as $1 billion![32]

One approach to securing credit card data is being taken by the PCI Security Standards Council founded by American Express, Discover Financial Services, JCB, MasterCard Worldwide, and Visa International. Its **Payment Card Industry (PCI) data security standard** is a multifaceted security standard that requires retailers to implement a set of security management policies, procedures, network architecture, software design, and other critical protective measures to safeguard cardholder data. It also requires retailers to store certain card data for up to 18 months in the event of a dispute with the cardholder. Retailers can be fined for failure to meet the various implementation deadlines of this standard.

Unfortunately, implementation of the PCI standard has taken a long time and has been costly for retailers. As a result, many retailers have not fully implemented the standard. In addition, the National Retail Federation insists that credit card companies take more responsibility for storing card data, not retailers. Their point of view is that retailers need store only minimal information such as the authorization code provided at the time of a sale to validate a charge, plus a receipt with truncated credit card information to handle refunds and returns. Limiting the data stored by retailers would virtually eliminate any motivation for hackers to gain access to their credit card databases.

Another approach to enabling secure online transactions is through the use of "smart cards." A **smart card** resembles a credit card in size and shape, but it contains an embedded microchip that can process instructions and store data for use in various applications such as telephone calling, electronic cash payments, storage of patient information, and providing access to secure areas. The microchip can store the same data as the magnetic stripe on a payment card and more. Thus no name or card number need appear on the smart card, making it more difficult for thieves to use. Smart cards are used heavily in Europe in banking and healthcare applications; in the United States their use is quite limited primarily because of the significant investment in an extensive magnetic stripe-based infrastructure.

The international payment brands Europay, MasterCard, and Visa jointly developed the EMV standard specifications for debit and credit cards, the corresponding card acceptor devices (terminals), and the applications supported by them in order to perform debit or credit payments using smart cards. The objective is to ensure that multiple-payment systems interface properly by ensuring that they all employ terminals and card approval processes that are compliant with the EMV specifications.

"Contact" smart cards have a contact area on the front face of the card to interface with a payment terminal. Contactless smart cards do not have a contact area, but have an embedded inductive loop aerial, which allows them to work in proximity to a contactless card reader without physically making contact. Although not EMV compliant, these types of cards are already used by several toll systems and mass transit operators including the London Underground.

EMV financial transactions are considered more secure against fraud than traditional credit card payments due to the use of advanced encryption algorithms to provide authentication of the card to the processing terminal and the transaction processing center. Unfortunately, smart card processing takes longer than an equivalent magnetic stripe transaction, partly due to the additional processing to decrypt messages. Furthermore, many implementations of EMV cards and terminals confirm the identity of the cardholder by requiring the entry of a Personal Identification Number (PIN) rather than signing a paper receipt. "Chip and PIN" is a U.K. government-backed security measure that requires customers to present both a four-digit PIN and a bank card containing a smart chip in order to complete a purchase. In the United States, many banks and financial services companies have been reluctant to impose additional requirements for authentication because they don't want to add additional steps (and time delays) to the checkout process.[33] In the future, systems may be upgraded to use biometrics (technology that measures and analyzes human physical characteristics such as eye retinas, fingerprints, or voice patterns for security purposes); however, this approach is not currently considered economical for retail applications.

Handling Returns Smoothly and Efficiently

Online retailers should devote considerable attention to minimizing returns by providing sufficient information about a product so that consumers have a clear idea of what to expect when they make a purchase. Well-written product descriptions, thumbnail (or larger) photos, and customer-written product reviews can not only increase product sales, but can also go a long way toward eliminating returns.

Online retailers need to ensure that they do not upset customers with return policies that include "punitive" restocking fees or that offer only a limited choice of reimbursement methods. Strict handling of returns can result in temporary savings but at the expense of long-term customer loyalty and future sales.

"Brick-and-click" retailers should strongly consider allowing consumers to return online purchases to a brick-and-mortar store. A majority of consumers expect to be able to make returns and exchanges through any channel, no matter how they bought the product, yet only 42 percent of retailers make that possible.[34] However, it is estimated that over 25 percent of consumers returning a product purchased online to a store will purchase another item while they are in the store.[35]

Some retailers allow customers to return online purchases via a "preferred" package delivery service (e.g., U.S. Postal Service, United Parcel Service, FedEx, etc). The customer follows a streamlined process to contact the service and arrange for prompt pick-up and return.

Providing Effective Customer Service

Because a Web site is open 24 hours a day, many online customers expect to be able to receive customer service at any time of the day or night. If an organization cannot provide some level of customer service 24 hours a day, it may lose business to competitors. Often some form of automated system is employed to provide at least some level of service around the clock. For example, if customers need order delivery status information, they can be directed automatically to the Web site of the firm providing the delivery service. Once at that site, customers simply enter a bill or order number to obtain information on the current delivery status.

For click-and-mortar organizations, which sell from both physical locations and e-commerce Web sites, the call center customer service reps must have accurate and current information about all in-store and Web transactions so they are able to answer questions and provide help.

Many Web sites promote their capability to accept customer e-mail queries about such things as order status, after-sale information, or product information. It is critical that such queries are handled in an accurate and timely manner in order to maintain consumer interest and loyalty.

Often Web sites will provide several methods for customers to contact the organization for customer service—e-mail, instant messaging, live Web chat, automated systems, direct phone calls, and even virtual meetings. InQ employs around 300 sales agents who represent 20 online retail clients in live online chats with customers. Their goals are to help customers get the information they need and nudge them along toward a purchase. The sales agents offer personal one-on-one assistance—answering questions, offering demos of how a product works, and helping customers identify which model best meets their needs. The aim is to keep customers online and to keep them from abandoning their shopping cart before making a purchase.

Read the special interest box to learn how one manager has been able to implement these ideas to gain outstanding business success.

Jeff Bezos Provides the Vision for Amazon.com

Amazon.com began its online bookstore in 1995 as one of the first retailers to sell exclusively via the Web. For many years it was unclear if the firm would succeed against major, well-established competitors (e.g., Barnes & Noble, Borders, Waldenbooks, etc.) with dozens of brick-and-mortar stores. Indeed, it was not until 2003 that the company finally achieved enough sales and reduced expenses sufficiently to become profitable.[36]

Over the years, Jeff Bezos, founder of Amazon.com, has judiciously diversified its business model from selling just books from a single U.S. Web site, to selling many products including apparel, CDs, DVDs, consumer electronic devices, and home and garden supplies from multiple international Web sites. Today, Amazon.com not only operates its own retail Web sites, it offers programs that enable third parties to sell products on its Web sites. It also provides services for third-party retailers as well as marketing, promotional, and Web services for developers. Sales and profits have continued to grow at a healthy pace with a net income of more than $476 million on sales in excess of $14.8 billion in 2007.

Bezos has placed an almost fanatical emphasis on providing outstanding customer service by consistently enforcing secure Web transactions, ensuring timely order fulfillment and shipping, offering a diverse choice of products, and emphasizing price discounts. This focus on customer service has been rewarded by gaining Amazon.com a high degree of consumer confidence and a high sales volume from its clients.[37]

Bezos devotes considerable resources to continual improvement of the Web site design. As a result, in a recent survey of 2200 customers by Keynote (a provider of on-demand test and measurement products for mobile communications and the Internet), Amazon.com and Best Buy were identified as the two retailers that provide online customers with the best shopping experience.[38] The Amazon.com Web site earns high marks for offering customers a clean and simple Web layout and providing intuitive navigation. In addition, it delivers a friendly and more personalized shopping experience by customizing customers' shopping pages based on their past Web site visits and purchases. Amazon.com also encourages customers to post product reviews that other customers find extremely helpful.

Discussion Questions

1. Visit the Amazon.com Web site and see if your own experience confirms what has been said here. What do you like best about the Web site? What do you like least?

2. What other key factors to the successful operation of a B2C Web business can you identify that were not mentioned here?

3. Is there a risk that some consumers might be alarmed by the customization of the shopping experience based on their previous visits and purchases? Why do you think this may be?

Advantages of E-business

There are many advantages that result from the use of e-commerce. Interestingly, these advantages are not one-sided; there are advantages that accrue to the seller (see Table 7-4), the buyer (see Table 7-5), and to society as a whole (see Table 7-6). Most of these benefits are possible because of the global exposure of products sold on the Web and the ability of e-commerce to reduce the time and costs associated with both selling and purchasing.

TABLE 7-4   E-commerce advantages for the seller

The global reach of the Web enables organizations to place their products and services in front of the entire world market.
The global reach of the Web also makes it possible for organizations to explore more easily new business opportunities and new markets.
Organizations can gain a competitive advantage by implementing build-to-order processes that enable inexpensive customization of products and services that precisely meet the needs of individual customers.
The use of online advertising enables organizations to reach target audiences in a much more cost-effective manner than traditional print media or TV commercials.
Organizations can extend their hours of operation and thus increase sales by establishing a Web site that is always accessible from any Internet-connected device.
Online sales can be increased through targeted, online promotions as buyers visit your Web site.
Organizations can capture valuable data about their customers, which can be used to reach targeted market segments and support customer relationship marketing.
Organizations have an opportunity to interact with their customers in a manner that allows them to build increased customer loyalty.
The direct cost-per-sale for orders taken through a Web site is lower than through more traditional means (face-to-face or paper-based orders).
A Web site can be used as an information tool to draw informed customers into stores, save money on marketing material, and attract suppliers.
Potential customers can do research and make comparisons online so that salespeople will be dealing with more informed customers.

TABLE 7-5   E-commerce advantages for purchasing organizations and consumers (buyers)

E-commerce offers buyers the capability to buy products and services from providers around the globe, thus providing a much wider range of choices in suppliers, cost, quality, and features.
Shopping comparison tools can make product comparison and evaluation easier and more efficient.
Instant quotes for shipping costs based on various delivery speeds can be obtained instantly from FedEx, UPS, USPS, etc.
Buyers can shop from the convenience of their own home or office and at any time of the day or night.

TABLE 7-5   E-commerce advantages for purchasing organizations and consumers (buyers) (continued)

Delivery costs and time are dramatically reduced for items that can be delivered over the Internet such as games, e-books, music, software, and videos.
Buyers can view their order history and order and delivery status.

TABLE 7-6   E-commerce advantages for society

Consumers can stay in their homes or offices rather than traveling to a store to make purchases. This reduces traffic congestion, fuel consumption, air pollution, and CO^2 emissions.
Consumers in developing countries have the opportunity to purchase services and products that were previously unavailable to them.
Consumers can choose from a wider range of sources, which encourages competition.

Issues Associated with E-business

While there are many advantages associated with the use of e-business, managers must understand that there are also many limitations and potential problems. Failing to recognize this can cause a company to have overly optimistic expectations of its e-business initiatives or to fail to put in place critical safeguards and measures.

Customers Fear that Their Personal Data May Be Stolen or Used Inappropriately

E-commerce Web sites can gather a wealth of data about prospects and customers through site registration, questionnaires, and the order-placement process. Consumers have long had concerns about whether online data is secured from access by unauthorized users or hackers. These concerns are rising based on the widespread publicity of recent consumer data breaches such as the one at TJ Maxx already discussed in this chapter and at organizations such as CardSystems Solutions, Inc., ChoicePoint Inc., Citibank, and Wachovia Corp. Organizations doing e-business must put in place powerful safeguards to protect their customers. They must demonstrate the ability to operate in a safe and reliable manner that builds the trust of their customers. Failure to do so can cause severe damage to the good name of established businesses.

Cultural and Linguistic Obstacles

Web site designers must avoid creating cultural and linguistic obstacles that make a Web site less attractive or effective for any sub-group of potential users. It is estimated that while roughly 60 percent of the content available on the Web today is in English, less than half of the current Web users read English.[39] Furthermore, people feel more comfortable buying your products and services if you speak to them in their own language. Thus designers of Web sites are increasingly allowing visitors to select their home country on an initial home page and then display a version of the Web site designed to accommodate people from that country with correct language or regional dialect, print characters, and culture appropriate graphics and photos. This design approach is often called "think globally, act locally." There are numerous companies that provide Web page translation services and software including Applied Language Solutions, Berlitz, BeTranslated, ScanSoft, SYSTRAN, and Worldpoint Interactive.

Difficulty Integrating Web and Non-Web Sales and Inventory Data

Organizations that do business over multiple channels often have difficulty seeing the entire scope of their business. This is because they use separate, non-integrated systems and databases to capture and record order and inventory information for each sales channel. A Web order may be rejected because an item appears to be out of stock when looking at the amount of stock allocated to Web sales. However, there might be plenty of inventory when the total inventory available for both Web and in-store purchases is considered. Considerable additional cost and effort is required to connect inventory and order status data from the Web and non-Web channels.

High Costs Associated with the Development and Operation of an Effective Web Site

Major corporations have spent in excess of $140 million to create their online retail Web site and additional ongoing operating and support costs in excess of $10 million per year.[40] Of course, the cost of a Web site varies considerably depending on the business requirements it is designed to meet. Small (less than 50 employees) and medium (less than 500 employees) businesses (SMBs) clearly cannot afford this level of spending. In many cases, SMBs opt to combine packaged e-commerce software with experienced third-party Web hosting services to keep their initial investment low and to control the annual operating and support costs. There are literally thousands of ISPs, Web-hosting service providers, and Application Service Providers to choose from. An SMB considering developing and operating a Web site is well advised to seek out a consultant familiar with the various options. The consultant can help choose the best options based not only on cost but also on the functionality to be provided, the desired level of reliability, the need for backup and disaster recovery, the level of security desired, and the volume of Web site traffic expected.

The manager's checklist in Table 7-7 provides a useful set of questions to review your organization's e-commerce activities. The appropriate answer to each question is "yes."

TABLE 7-7 A manager's checklist for reviewing your organization's e-commerce activities

Do your organization's Web development efforts focus on the essential activities?	Yes	No
Identifying appropriate e-business opportunities.		
Directing potential customers to your site.		
Providing a good customer online experience.		
Providing an incentive for customers to buy and return.		
Providing timely, efficient order fulfillment.		
Offering a variety of easy and secure payment options.		
Handling returns smoothly and efficiently.		
Providing effective customer service.		

Chapter Summary

- E-business enables organizations and individuals to build new revenue streams, to create and enhance relationships with customers and business partners, and to improve operating efficiencies.

- In order to incorporate e-business into their business, managers must understand their customers and the fundamentals of the markets in which they operate, have a clear understanding of how the Internet differs from the traditional venues for business activity, and employ business models appropriate to the Internet.

- There are several forms of e-commerce including business-to-business (B2B), business-to-consumer (B2C), consumer-to-consumer (C2C), and e-government (e-gov).

- There are several forms of B2B Web sites in operation today including private stores, customer portals, private company marketplaces, and industry consortia-sponsored marketplaces.

- U.S. business-to-consumer (B2C) sales are growing at a rate of nearly 40 percent per year and represented about 3.3 percent of overall 1Q-2008 retail sales

- B2C Web sites must focus on attracting prospects, converting them into customers, and retaining them to capture additional future sales. These have long been necessary objectives of brick-and-mortar retailers as well.

- Brick-and-mortar retailers are finding that they must modify their in-store operations and procedures to meet shoppers' new expectations that are based on online shopping experiences.

- Consumer-to-consumer (C2C) e-business is the exchange of goods and services among individuals, typically facilitated by a third party.

- E-government (e-gov) involves the use of information technology (such as Wide Area Networks, the Internet, and mobile computing) by government agencies to transform relations between the government and citizens (G2C), the government and businesses (G2B), and among various branches of the government (G2G).

- Mobile commerce (M-commerce) is the buying and selling of goods and/or services using a mobile device such as a cell phone, smartphone, PDA, or other such device.

- There are numerous factors that contribute to making an e-business operation successful including identifying appropriate e-business opportunities; acquiring necessary organizational capabilities; directing potential customers to your site; providing a good customer online experience; providing an incentive for customers to buy and return; providing timely, efficient order fulfillment; offering a variety of easy and secure payment options; handling returns smoothly and efficiently; and providing effective customer service.

- There are many advantages that result from the use of e-business. There are advantages for the seller, for the purchaser, and for society in general.

- There are several potential problems associated with the use of e-business including customers, fear of loss of personal data, cultural and linguistic obstacles, difficulty in integrating data from the various sales channels, and the high costs associated with developing and operating a Web site.

Discussion Questions

1. How do you define e-commerce? What is the difference between e-commerce and e-business? Develop your own list of the top three reasons an organization should get involved in e-business.

2. What were some of the common mistakes made by many Web-based companies that failed during the dot-com bubble burst? Why were the managers of those companies unable to see they were headed for problems?

3. Identify and briefly describe four types of B2B Web sites. In what ways does a B2B Web site need to operate differently than a B2C Web site?

4. What business functions are performed by e-procurement software? Why might an organization attempt to build its own e-procurement software rather than use existing software packages to meet these needs?

5. Do you think the percentage of U.S. online retail sales to total retail sales will continue to increase? Why or why not?

6. How and why do brick-and-mortar retailers need to modify their in-store operations and procedures to meet new expectations of shoppers?

7. What is the new value proposition known as "The Long Tail" first envisioned by Chris Anderson? Can you provide an example of this?

8. What effect would a weak U.S. dollar have on the demand for e-business between U.S. sellers and global buyers?

9. Identify three variations of e-gov Web sites. Visit three e-gov Web sites and identify which Web site best meets your needs. Justify your choice of Web sites.

10. What is m-commerce? Provide four examples of m-commerce. Use a Web-enabled cell phone to access an m-commerce Web site. Jot down your reactions.

11. Review the list of e-business critical success factors. Identify the three factors that you feel are the most critical. Defend your choices.

12. Identify several problems associated with the set-up and operation of an e-commerce Web site.

Action Memos

1. You are the senior marketing manager for a manufacturing firm that is getting ready to launch its first e-commerce B2C Web site. The goal for the new Web site is to attract new customers in new markets and to boost sales by at least 5 percent by the end of the first year of operation. You have been asked by the CEO to prepare a 10-minute talk for the Board of Directors about basic business operating principles for the new Web site. You have decided to present the principles in terms of what will change and what will stay the same. The CEO

has asked you to stop by her office this afternoon to provide a "preview" of your talk. Prepare a brief outline emphasizing what will stay the same and what must change.

2. Your organization's first Web site was launched just six months ago, but already management is calling it a complete disaster. The site has failed to stimulate additional sales and has proven to be unreliable, with frequent periods of service interruption. Things are so bad that consumers are frequently calling the customer service center to complain. You are the manager of customer service and are surprised when the manager of marketing calls at 10 a.m. to invite you to lunch. She would like to discuss your ideas on how the situation can be "turned around." How would you prepare for this meeting? What approach would you recommend to better define the problems with the existing Web site?

Web-based Case

Do a comparative analysis of three competing Web sites (e.g., Best Buy, Circuit City, and Sears). Identify the primary features and capabilities of each Web site. Which Web site best meets your needs and why?

Case Study

The Borders Group Implements a "New" Web Site

The original Borders bookstore was started in 1971 in Ann Arbor, Michigan, by the Borders brothers, Tom and Louis, who attended the University of Michigan. Today Borders is the second-largest bookstore chain in the U.S. and sells a variety of books, CDs, DVDs, newspapers, and magazines. It employs around 30,000 people and total revenue for the fiscal year ending February 2, 2008 was $3.8 billion (Table 7-8).

TABLE 7-8 Borders 5-year financial summary

	2008	2007	2006	2005	2004
Statement of Operations Data					
Domestic Borders superstore sales	$2,847.2	$2,750.0	$2,709.5	$2,588.9	$2,470.2
Waldenbooks Specialty Retail sales	562.8	663.9	744.8	779.9	820.9
International sales	364.8	269.9	221.4	163.9	108.7
Total Sales (1)	$3,774.8	$3,683.8	$3,675.7	$3,532.7	$3,399.8
Operating income (loss)	$ 6.6	$ 8.5	$ 170.4	$ 201.2	$ 186.5
Income (loss) before cumulative effect of accounting change	$ (18.5)	$ (13.0)	$ 96.5	$ 117.9	$ 108.4
Cumulative effect of accounting change (net of tax)	-	-	-	-	$ (2.1)

TABLE 7-8 Borders 5-year financial summary (continued)

	2008	2007	2006	2005	2004
Income (loss) from continuing operations	$ (18.5)	$ (13.0)	$ 96.5	$ 117.9	$ 106.3
Income (loss) from operations of discontinued operations	(13.2)	(138.3)	4.5	14.0	8.9
Loss from disposal of discontinued operations	(125.7)	-	-	-	-
Income (loss) from discontinued operations	$ (138.9)	$ (138.3)	$ 4.5	$ 14.0	$ 8.9
Net income (loss)	$ (157.4)	$ (151.3)	$ 101.0	$ 131.9	$ 117.3
Diluted (basic) earnings (loss) per common share	$ (2.68)	$ (2.44)	$ 1.42	$ 1.69	$ 1.46
Cash dividends declared per common share	$ 0.44	$ 0.41	$ 0.37	$ 0.33	$ 0.08
Balance Sheet Data					
Working capital	$ 38.2	$ 106.6	$ 287.5	$ 511.1	$ 529.2
Total assets	$2,302.7	$2,613.4	$2,572.2	$2,628.8	$2,584.6
Short-term borrowings	$ 548.4	$ 542.0	$ 206.4	$ 141.0	$ 140.7
Long-term debt, including current portion	$ 5.6	$ 5.4	$ 5.6	$ 55.9	$ 57.3
Stockholder's equity	$ 476.9	$ 642.0	$ 927.8	$1,088.9	$1,100.6

Notes:

(1) Excludes results of discontinued operations of BordersIreland Limited, Books, etc., and U.K. Superstores

(2) All figures in thousands of dollars

In an attempt to launch a successful book division, Kmart acquired Waldenbooks in 1984 and Borders in 1992. Kmart had had trouble managing Waldenbooks and hoped that Borders senior management could help improve the operations of its fledgling book division. Instead, many Borders senior managers left Kmart. By 1995, Kmart was experiencing its own financial difficulties. In its efforts to shed unprofitable assets, Kmart allowed Borders to buy itself out. The new company came to be known as the Borders Group.

The company opened its first international store in Singapore in 1997. International superstores operate under the Borders name and are between 13,500 and 38,400 square feet with 2007 average sales of $8.3 million per superstore. As of February 2008, Borders had a total of 32 international superstores including 22 in Australia, 5 in New Zealand, 3 in Puerto Rico, and 2 in Singapore.

Borders began operating its own Web site in 1998, but after three years of losses from online sales, outsourced its online operations to Amazon.com. Amazon.com had the technical knowledge and plenty of infrastructure capacity to operate both a Borders.com and a Waldenbooks.com Web site. Under the terms of the agreement, Amazon.com was the merchant of record for all Web site sales and determined all prices and other terms and conditions for such sales. Amazon.com was also responsible for the fulfillment of all products sold through the Web sites and kept all payments from customers. Borders received commissions on sales for products purchased through the Web sites.[41]

Borders had a net loss of $151 million for 2006 compared to a net income of $101 million the previous year (as shown in Table 7-8). As a result of the company's declining financial position, Borders made a number of strategic announcements in March 2007 aimed at turning operations around.

- By the end of 2008, 250 unprofitable Waldenbook stores would be closed to bring the total number of locations to around 300. These closings were in addition to some 124 stores that had been closed during 2006.

- It would shed its U.K. and Irish businesses including Books Etc. to enable it to focus on reestablishing its core U.S. operations. (By September 2007, the U.K. and Irish businesses of 42 Borders Stores and 28 Books Etc stores had been sold for £10m or around US$20 million).

- It would launch its own Web site during the spring of 2008 at which time the Amazon agreement would be terminated. The Web site had been under development since fall of 2006. The two Web sites would be consolidated into a single infrastructure to enable across-channel experience

- The company would name a new CIO as former CIO Fred Johnson had left the company.

Cedric "Rick" Vanzura, executive vice-president of emerging business and technology and chief strategy officer justified the change in strategy by saying, "Technology costs have gone significantly down [on] the Internet while capabilities have gone up, and the availability of solid people to run these operations have gotten stronger. We've been happy with the Amazon relationship; it definitely served its purpose from the time we struck that deal, but now it's a new era, a new opportunity and we plan on taking advantage of it."[42]

Surprisingly, in July 2007, Borders announced that it was eliminating Vanzura's position and that he would leave the firm in September. Borders CEO George Jones stated, "Now that the strategic plan is set and its initiatives are being executed within the individual units, Rick and I agree that this is a logical time for him to transition away from his strategic duties.[43] Vanzura was replaced by the firm's new CIO, Susan Harwood, former CIO at Books-A-Million, an online bookseller.

The Web site project was led internally by Kevin Ertell, vice president of e-business. In October 2007, after a month of beta testing its new Web site, BordersStores.com, the firm made the site available for viewing by customers. They could access the site and offer comments; however, they had to continue to access the Amazon.com Web sites to place actual orders. One of the innovative features of the new Web site is what Borders calls the Magic Shelf. Web site visitors see a 3-D shelf of real book covers displayed just as they look in Borders' stores. To see information about a specific book, the user just clicks on the book. Purchases made online can be delivered directly to the customer's home or to a nearby Borders store. The Magic Shelf will become personalized for repeat shoppers based on their previous purchases. The site employs improved search and navigation capabilities and enables visitors to create a "wish list" of items they would like to buy or receive as gifts. The Web site also includes a feature called Borders Media that enables visitors to order exclusive and original video programs created by Borders. There will also be a BordersReward.com site for the 26 million members of the firm's loyalty program to track and redeem their rewards online.[44]

In March 2008 Borders announced that it would undergo a strategic alternative review process. The review would include investigation of a wide range of alternatives including the sale of the firm or certain divisions for the purpose of maximizing shareholder value. Barnes & Noble let it be known that it put together a team to study the feasibility of a takeover of its competitor.[45]

In May 2008, Borders severed its relationship with Amazon.com and began running its own Web site. Visitors to Borders.com can make choices from roughly 2 million books, 100,000 new movies, and 400,000 new CDs. Free shipping is offered on some orders over $25 and on shipments to its stores. In-store kiosks introduce the Borders.com shopping experience to customers. CEO George Jones stated, "Today is another milestone for our company as we launch Borders.com, a site that is much more than just another place to shop on the Web—it is a source of information and entertainment that brings a real bookstore experience to life online."[46]

The firm expects that the new site will become profitable in 2009.[47] However, many financial analysts are skeptical that Borders can regain the market share it lost to both online retailers and to discounters such as Wal-Mart.[48]

In June 2008 Borders completed the sale of its Australia, New Zealand, and Singapore business for roughly US$90 million in a further effort to focus on its U.S.-based business. In an effort to further cut expenses, the firm announced that it would cut some 20 percent of the company's jobs.[49]

Discussion Questions

1. Do research on the Web to learn about the current status of Borders. Write a few paragraphs documenting your findings.

2. It took Borders over two years and more than $10 million to develop its own Web site. Do you think this was an appropriate tactic for the firm? Defend your answer.

3. Visit the Borders Web site as well as the Web sites of Amazon.com and Barnes & Noble.

Which of the three Web sites provides their customers with a truly superior customer shopping experience? Defend your response.

4. With the benefit of 20-20 hindsight, what might Borders have done differently to improve its business results?

Endnotes

[1] Langley Steinert, "Shifting Online Auto Shopping Into High Gear," *E-Commerce Times*, November 20, 2007.

[2] Matt Vella, "Online, Souped Up, and Making Tracks, *Business Week*, December 10, 2007.

[3] Ibid.

[4] The Dave Chaffey Blog at *http://www.davechaffey.com/E-business/C1-Introduction/E-business-E-commerce-defined* accessed on September 10, 2008.

[5] "Will dotcom Bubble Burst Again?" *The Los Angeles Times*, July 17, 2006.

[6] Tanuja Singh, Geoffrey Gordon, and Sharon Purchase, "B2B E-Marketing Strategies of Multinational Corporations: Empirical Evidence from United States and Australia," *Red Orbit*, April 15, 2007.

[7] About Goodrich, Goodrich Web site accessed at *http://www.goodrich.com/* on September 10, 2008.

[8] Facts & Figures, About UTC, UTC Web site accessed at *http://www.utc.com/* on September 10, 2008.

[9] Gary Schneider, *Electronic Commerce*, copyright 2007 Course Technology, p. 252.

[10] "Avendra Renews Contracts with Its Five Founding Customer Partners," *Hospitality Net*, June 12, 2007.

[11] "Avendra – For Suppliers," Avendra Web site at *www.avendra.com*, accessed December 20, 2007.

[12] Katherine Noyes, "Holiday Returns: Ready or Not, Here They Come," *CRM Buyer*, November 10, 2007.

[13] "Online Retail Spending Rises 20% to $122.7 Billion comScore Says," *Internet Retailer*, January 25, 2008.

[14] Nanette Barnes, "More Clicks at the Bricks," *BusinessWeek*, December 17, 2007.

[15] Ibid.

[16] Malcolm Gladwell, "Chris Anderson," *Time Magazine*, May 14, 2007.

[17] Craigslist Overview accessed at *http://craigslistt.us/* on December 27, 2007.

[18] "Definition of e-Government," The World Bank Web site accessed at *http://www.worldbank.org/* on December 23, 2007.

[19] "About Us" Business.gov at *http://www.business.gov/about/* accessed December 24, 2007.

[20] "Definition of e-Government," The World Bank Web site accessed at *http://www.worldbank.org/* on December 23, 2007.

[21] Linda Rosencrance, "Satisfaction with E-gov Sites Slips a Bit with Users," *Computerworld*, December 18, 2007.

22 Initiative Overview & News at the E-Gov Web site at *http://egov.gsa.gov*, accessed December 24, 2007.

23 Keith Regan, "Sprint Cuts Ribbon on Mobile Shopping Service," *E-Commerce Times*, September 13, 2007.

24 Cliff Peale, "Merchants Now Sell by Cell," *Cincinnati Enquirer*, p. B1-2, December 26, 2007.

25 Beal, Barney, "Web 2.0 Takes Center Stage at Gartner CRM Summit," *SearchCRM.com*, September 19, 2007.

26 James Allen, Frederick F. Reichheld, and Barney Hamilton, "The Three 'Ds' of Customer Experience," *Working Knowledge*, November 7, 2005.

27 James Allen, Frederick F. Reichheld, and Barney Hamilton, "The Three 'Ds' of Customer Experience," *Working Knowledge*, November 7, 2005.

28 "Sterling Commerce Provides Order Fulfillment Foundation for New Borders.com," *Business Wire*, May 28, 2008.

29 Gary Schneider, Electronic Commerce, 7th annual edition, Copyright 2007 Course Technology, p. 495.

30 Gary Schneider, Electronic Commerce, 7th annual edition, Copyright 2007 Course Technology, p. 501.

31 Erika Murphy, "Report: E-Commerce Fraudsters' Haul May Reach $3.6 B in 2007," *E-Commerce Times*, November 18, 2007.

32 Ross Kerber, "Cost of Data Breach at TJX Soars to $256 Million," *The Boston Globe*, August 15, 2007.

33 Keith Regan, "UK Researchers Hack Chip and PIN Security," *TechNewsWorld*, February 6, 2007.

34 Katherine Noyes, "Holiday Returns: Ready or Not, Here They Come," *CRM Buyer*, November 10, 2007.

35 Ibid.

36 Juan Carlos Perez, "Amazon Turns 10, Helped by Strong Tech, Service," *Computerworld*, July 15, 2005.

37 Jena McGregor, "The 50 Most Innovative Companies," *Business Week*, May 4, 2007.

38 Linda Rosencrance, "Survey, Amazon, Best Buy Treat Customers Best," *Computerworld*, August 2, 2007.

39 Gary Schneider, "Electronic Commerce," published by Thomson Course Technology, copyright 2007, p. 33.

40 Gary Schneider, "Electronic Commerce," published by Thomson Course Technology, copyright 2007, p. 545.

41 Lee Copeland, "Borders Turns to Amazon for Outsourcing," *Computerworld*, April 16, 2001.

42 Linda Rosencrance, "Borders Turns a New Page on E-Commerce After Posting Losses," *Computerworld*, March 22, 2007.

43 John Soat, "Borders Opens New CIO Chapter," *Information Week*, July 17, 2007.

[44] Linda Rosencrance, "Borders Previews New E-Commerce Site," *Computerworld*, October 5, 2007.

[45] "Borders Launches New Retail Website," *The Boston Globe*, May 27, 2008.

[46] Linda Rosencrance, "Borders Launches New E-commerce Site," *Computerworld*, May 27, 2008.

[47] Corey Lorinsky, "Desperate Borders (BGP) Severs Amazon Deal in Futile Attempt to Go It Alone," *Clusterstock*, May 27, 2008.

[48] "Borders Launches New Retail Website," *The Boston Globe*, May 27, 2008.

[49] "Key Developments for Borders Group Inc," accessed at *www.reuters.com/finance/stocks/keyDevelopments?symbol=BGP.N* on July 11, 2008.

CHAPTER 8

ENTERPRISE RESOURCE PLANNING

THE CHALLENGE OF ERP IMPLEMENTATION

"Enterprise system implementations can be invasive, disruptive, and even counter-productive, causing considerable expense, possibilities of wrenching business-process change, and gnawing uncertainty in the minds of employees. Happily, while no magic pill guarantees an implementation will be quick, painless, and successful, there are steps manufacturers can take to secure ERP value without risk of catastrophic failure."

—Jim Fulcher, "Five Big Improvements in Just Five Months," *Manufacturing Business Technology*, August 2007.

B W A   W A T E R   A D D I T I V E S

Why Managers Must Understand ERP

While it might seem illogical, the best way to clean dirty water is by putting a lot of chemicals into it. BWA

Water Additives manufactures industrial water treatment chemicals including bromine-based biocides and

a number of other water treatment chemicals used in industrial water treatment, desalination, mining,

sugar processing, and pulp and paper manufacturing. It has regional headquarters in Atlanta, Singapore,

Tokyo, Dubai, and Manchester, England. BWA's manufacturing facilities are located in eight different

countries, with research and development centers in Atlanta and Manchester. Its global customer service

and distribution network serves more than 90 countries.[1]

BWA was a subsidiary of Chemtura until May 2006, when the British private investment group Close

Brothers Private Equity bought it for $85 million. As is common in divestitures, Chemtura allowed BWA to continue using its existing systems and IT infrastructure for a limited time while assessing BWA a hefty monthly maintenance fee. However, BWA wanted to move out of its old parent company's physical premises and build its own telephone and IT systems, including an enterprise resource planning (ERP) system, as quickly as possible. An ERP system would support multiple business units and enable sharing of data through a common, shared database of operational data.

For the chemical industry, ERP software includes modules that have the ability to provide detailed product costing and profitability analysis, forecasting and scheduling, efficient management of raw materials, improved order fulfillment and customer service, and inventory optimization.

Paul Turgeon, president and COO of BWA says that, "While we could have implemented the same systems we were familiar with from our parent company, it was critical we choose a solution... that fit our chemicals business by design and therefore could be deployed quickly and cost effectively."[2] Following his lead, BWA management conducted a thorough evaluation of leading ERP solutions to find the one that could provide the vertical functionality needed for the chemical industry while meeting several specific company requirements. In addition, BWA focused on solution providers who had proven global implementation experience because it was critical to minimize the time and effort required to complete the project.

The ERP solution chosen was Ross Enterprise, CDC Software's ERP product suite, specifically designed for chemicals manufacturers. Importantly, Ross Enterprise was compliant with global and local regulations to produce the detailed documentation needed for Material Safety Data Sheets (MSDS) documentation.[3] MSDS forms provide workers and emergency personnel with procedures for handling or working with substances in a safe manner as well as procedures for storage, disposal, and spill handling.

The system went into production in May 2007. "Rolling out an ERP system in 120 days, in two different countries, is quite an accomplishment," says Turgeon. Such rapid execution was possible because BWA took the steps to ensure that Ross Enterprise met all of the selection criteria. BWA also made sure that CDC Software understood that implementation speed was critical.[4]

BWA is very satisfied with the Ross Enterprise ERP software. In fact Turgeon adds, "Since implementing Ross Enterprise, we have exceeded our targets for working capital, inventory turns and days sales outstanding. We also now have a deeper and broader view of the business than we did with our previous systems... This out-of-the-box, cost-effective and vertical-focus application has enabled us to boost our operational effectiveness under some very challenging conditions."[5]

LEARNING OBJECTIVES

As you read this chapter, ask yourself:

- What role does management play in the selection, implementation, and operation of ERP software?
- What are the various ERP solution options available and what are their advantages and disadvantages?

This chapter will explain what an ERP system is, identify several of the benefits associated with an effective ERP system, highlight some of the potential issues associated with ERP implementation, outline a "best practices" approach to implementing an ERP system, and discuss future trends of ERP systems.

WHAT IS ERP?

An **enterprise resource planning (ERP)** system is a set of core software modules that enable organizations to share data across the entire enterprise through the use of a common database and management reporting tools. The goal is to enable easy access to business data and to create efficient, streamlined work processes. This is achieved by building one single database that is accessed by multiple software modules, which provide support for key business functions for different areas of an organization as shown in Figure 8-1.

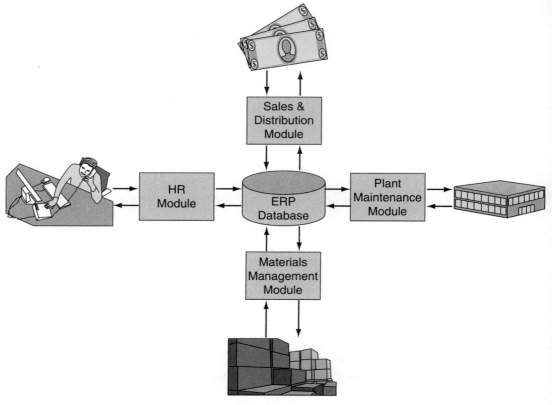

FIGURE 8-1   ERP enables sharing of data across an entire enterprise

An effective ERP system enables people in various organizational units to access and update the same information based on permission levels assigned within the system. For example, when the sales organization captures data about a new order, the information is immediately available to workers in finance, production planning, shipping, warehouse operations, and any others who need access to the records. Through the sharing of data, ERP software enables standardization and streamlining of business processes whether it is in a small, locally-based organization or in a large, multi-national organization. The leading ERP software vendors for large organizations include Infor, Microsoft, Oracle (including its two acquisitions JD Edwards and PeopleSoft), and SAP (Systems Applications and Products).

The use of a shared database and core software modules from a single software manufacturer is a much different approach than many organizations have taken in the past. Countless organizations utilize computer hardware and software products from multiple vendors implemented in their various functional units. For example, the purchasing department might have a dedicated computer running purchase order processing software, which creates a database of open purchase orders that cannot be accessed by other departments. The accounts payable organization might have its own dedicated computer running accounts payable software, which creates a separate database of purchase orders, receiving reports, and supplier invoices. In such an environment, the purchasing processes still involve conventional mail or fax delivery of purchase orders and associated documents such as supplier quotations, change orders, receiving reports, and invoices. Thus there is a high probability that the information in the purchasing department database and accounts payable department database will be inconsistent. Such lack of consistency leads to confusion and a duplication of effort, making it impossible for workers in purchasing, accounts payable, receiving, inventory control, materials management, and sourcing to operate efficiently.

Best practices represent the most efficient and effective way of accomplishing a task, based on procedures that have proven themselves repeatedly over a long period of time. Consider the procedures required to pay a supplier's invoice. For many organizations, the best practice for this process involves forming a three-way match between the supplier's invoice, the original purchase order, and the receiving report. The three are compared and if there are no significant differences between what was ordered, what was received, and what was invoiced, the supplier's invoice is scheduled to be paid as late as possible without forfeiting any supplier discount for prompt payment. ERP software is designed to support how an organization using industry "best practices" conducts business. Thus the ERP software would be programmed to follow the "three-way match" process before approving an invoice for payment.

Each industry has different business practices that make it unique. In order to address these differences, ERP vendors offer specially tailored software modules designed to meet the needs of specific industries such as consumer packaged goods manufacturing, higher education, utilities, banking, oil and gas, retail, and the public sector. Table 8-1 shows the primary software modules associated with the SAP ERP package for a manufacturing organization. Table 8-2 lists the primary software modules associated with the SAP ERP package for higher education.

TABLE 8-1 SAP ERP software modules for a manufacturing organization

Software Module	Description
Financial accounting	Records all financial transactions in the general ledger accounts and generates financial statements for external reporting.
Controlling	Supports managerial decision making by assigning manufacturing costs to products and cost centers for analysis of the organization's profitability.
Workflow	Automates the various activities in SAP ERP; performs task flow analysis and prompts employees via e-mail if they need to take action.
Plant maintenance	Manages maintenance resources and planning for preventive maintenance of plant equipment.
Materials management	Manages the acquisition of raw materials from suppliers and the subsequent handling of raw materials from storage to work-in-progress goods to the shipping of finished goods to the customer.
Sales and distribution	Maintains and allows access to customer information (pricing, shipping information, billing procedures, etc). Also records sales orders and scheduled deliveries.

TABLE 8-1 SAP ERP software modules for a manufacturing organization (continued)

Software Module	Description
Production planning	Plans and schedules production and records actual production activities.
Quality management	Plans and records quality control activities such as product inspections and material certifications.
Asset management	Manages fixed asset purchases and related depreciation.
Human resources	Aids in employee recruiting, hiring, and training. Also includes payroll and benefits tools.
Project system	Supports planning and control for new R&D, construction, and marketing projects.

Source: Ellen Monk and Bret Wagner, Concepts in Enterprise Resource Planning, 3rd edition, @ 2009 Course Technology/Cengage Learning, pages 27–29.

TABLE 8-2 SAP ERP software modules for an institution of higher education

Software Module	Description
Student lifecycle management	Supports recruiting, admissions, registration, academic advising, course management, student accounting, and academic program management.
Grants and fund management	Helps organizations compete for and manage a variety of grant programs and endowments including proposal development and submission, budgeting, award, spending and payroll, reporting, renewal, and evaluation.
Financial management, budgeting, and planning	Supports proactive financial planning, real-time budget visibility, and consolidated financial reporting. Also supports treasury management, billing, dispute resolution, collections, receivables, and payables.
Relationship management, institutional development, and enrollment management	Provides personalized multi-channel communication to internal and external constituents, such as prospective students, donors, high school guidance counselors, grant organizations, current students, and alumni.
Governance and compliance	Enables the organization to collect, document, assess, remediate, and attest to internal control processes and safeguards.
Human capital management	Supports the recruitment, training, development, and retention of employees. Also supports administration, payroll, time management, and legal reporting.
Procurement	Supports plan-driven and ad hoc purchasing, conducts accurate spend analysis and ensures compliance with procurement best practices.
Enterprise asset management	Manages the asset life cycle from business planning and procurement to deployment and reliability-centered maintenance to disposal and replacement.

TABLE 8-2 SAP ERP software modules for an institution of higher education (continued)

Software Module	Description
Business services	Streamlines administrative processes and improves efficiencies in real estate management and project portfolio management.
Performance management	Helps track, understand, and manage performance across operational areas, including student administration, student affairs, human resources, finance, and operations.

Source: SAP for Higher Education and Research Industry Overview accessed at *http://www.sap.com/industries/highered/brochures/index.epx* on August 30, 2008.

Most ERP software packages are designed so that an organization does not have to implement the entire package. Companies can pick and choose which modules to install based on business needs. Many organizations may choose to purchase one or two of the software modules and delay implementing the other software modules until the necessary resources are available.

ERP and Customer Relationship Management (CRM)

A **customer relationship management (CRM) system** is an enterprise system that supports the processes performed by all the entities involved in creating or increasing the demand for an organization's products and services. People responsible for product development, sales, marketing, and customer service are the end users of a CRM system.

A CRM module is often part of the offering from an ERP software provider. The CRM module is a tool used by customer-facing employees (front office) to increase sales and service customers. Other modules of the ERP software provide resources for employees in manufacturing, finance, human resources, and other functions (back office) to support the efforts of the front office employees.

An essential goal of a CRM system is to enable employees who interact directly with customers to provide better, more personal service thus increasing customer satisfaction and loyalty. To achieve this, the CRM system must effectively capture and present customer information so that employees can successfully use that data. It can be extremely difficult to capture useful customer information. For example, if an individual comes into a bank to open a new checking account, they will be turned off immediately if a well meaning bank employee bombards them with a series of questions not directly related to the new account—Do you own or rent? Do you own or lease a car? Do you have any children?

Vantage Credit Union is able to avoid asking members lots of unnecessary questions because it built its new CRM system using data that already resided in current systems and databases. Vantage decided to start with the data it already had and then capture new information via members' transactions and interactions with staff. The goal was to use this information to identify cross-selling opportunities and offers that would meet members' needs. Vantage integrated all of the data available from its core transaction processing system, mortgage and consumer loan processing software, and automated solutions for collections, credit card processing, and deferred compensation products. Its new CRM system includes a customer information database that stores and analyzes household and demographic data. Vantage executives have been able to enhance member relationships by maintaining a database of high quality customer information, improving the sales and service tools available to its front-line employees, and developing new methods sustaining member loyalty.

ERP and Supply Chain Management (SCM)

The **supply chain** involves the flow of materials, information, and dollars as they move from supplier to manufacturer to wholesaler to retailer to supplier. The supply chain includes the following major processes:

- **Demand planning**—Determining the demand for your products taking into account all the factors that can affect that demand—general economic conditions, actions by competitors, your own pricing, promotion and advertising activities, etc.
- **Sourcing**—Choosing the suppliers and establishing the contract terms to provide the raw materials needed to create your product and deliver them to your manufacturing locations.
- **Manufacturing**—Producing, testing, packaging, and preparing your products for delivery.
- **Logistics**—Establishing a network of warehouses for storing products, choosing carriers to deliver products to customers, and scheduling carrier pick-ups so that the product is delivered to the customers or warehouses on a timely basis. This process also includes invoicing the customer.
- **Customer Service**—Increasing customer satisfaction and improving the customer experience by, for example, dealing with problems caused by over (customer receives more of a particular item than he expected), short (customer receives less of a particular item than he expected), and damaged shipments.

Supply chain management (SCM) involves the planning, executing, monitoring, and controlling of this set of processes. The primary goal of SCM is to lower costs and inventory levels while still meeting customer requirements for timely delivery of high quality products.

Each of the major processes has dozens of activities and tasks that must be executed well for the supply chain to function effectively and efficiently. Major ERP software suppliers include software modules to handle many of these tasks, but no one supplier has a single, all encompassing software package that meets all of the SCM needs in an ideal way for every company. For example, developing a demand forecast for a beer distributor is much different than for a woman's handbag manufacturer. As a result, hundreds of software suppliers provide a myriad of software packages to support the various supply chain management tasks for companies in different industries.

Because each industry has a unique set of SCM needs, many companies elect to implement what is called "best of breed" solutions for specific tasks. For example, some beverage companies in the consumer packaged goods industry have selected Red Prairie's Warehouse Management Solution as a "best of breed" solution for providing complete raw material controls from sourcing to production and all the way through the shipping of finished products. Effective implementation of this software reduces raw material waste, enables traceability throughout the supply chain, and increases customer confidence in the quality of the end product. The Red Prairie software would not be considered as "best of breed" for companies in the oil and gas industry as it would not meet their different business needs.

SCM applications frequently draw on the data captured and stored in an ERP system—data such as orders, shipments, inventory, customers, suppliers, etc. Thus some sort of interface must be built to allow these stand-alone SCM applications to access data from the ERP system database. Also, these SCM applications may process data and then need to update data in the ERP system database. This requires another interface to be built.

For example, consider the customer shipment planning process. This process takes open (unfilled) orders and decides from which warehouse the order will be filled and when it should be shipped in order to meet the customer's desired delivery data. This process can become quite complicated as each warehouse has a maximum number of shipments it can handle each day. In addition, the warehouse must be selected to minimize the shipping cost. To perform this task, an organization may draw open order data from its ERP system, pass it to a stand-alone SCM shipment planning optimization software package that plans all the open orders, and then passes the "planned" orders back to the ERP system to update the open order data.

Ace Hardware uses an SCM application to manage its inbound shipment process to move items from its many suppliers to its more than 4000 stores in all 50 states. Suppliers provide Ace with data about upcoming shipments such as the date their shipment will be available for pick-up, the ship from location, and the contents of the shipment. Ace then enters this data into the On-Demand TMS software from LeanLogistics. The software prepares an inbound shipment plan that minimizes transportation costs by assigning the various shipments to appropriate freight carriers and scheduling the pick-ups and deliveries to avoid any out-of-stock situation. "The On-Demand TMS Supplier Inbound Module shaved about 7 percent off our freight bill," according to Brian Cronenwett, director of supply chain at Ace Hardware. "Supplier inbound improvements gave us additional benefits, including significantly lower inventory through a lead-time reduction program and greater buying leverage with our core group of carriers."[6]

BENEFITS OF IMPLEMENTING ERP

The successful implementation of an ERP system can bring many benefits to an organization including establishing standardized business processes, lowering cost of doing business, improving the overall customer experience, facilitating consolidation of financial data, supporting global expansion, and providing a compliant system. These benefits will now be discussed and several examples of companies using ERP to achieve these benefits will be presented.

Establish Standardized Business Processes

An ERP system can help an organization establish standardized streamlined business and workflow processes that eliminate redundant steps and that are based on industry-specific best business practices. Such business processes ensure that workers, even in multiple business locations, are performing their work in an efficient manner and in a way that provides a consistent interface between the organization and its customers and suppliers.

Gooch and Housego is a manufacturer of precision optical components and subsystems used in medical, research, and scientific applications. Some of its products transform lasers into industrial tools that generate high power pulses for drilling, cutting, or welding materials such as steel or diamonds. They also can be used to cauterize or cut human tissue in medical applications. The firm employs about 350 people and generates annual revenue in the neighborhood of $55 million. Its operations in the U.S., Germany, and the UK had been operating completely independently of one another before the firm commissioned Project Orion to combine the separate operations into one consolidated operation using the SYSPRO ERP system from K3 Business Technology Group. Gareth Jones, CEO and sponsor for Project Orion wanted the company to present one common, consistent face to customers, suppliers, and business partners. "Customers over the world will be able to deal with Gooch and Housego as a single operation; one common sales front and one common business style, irrespective of where the client is or where the goods come from. The ERP implementation has improved financial consolidation and control as well as provided better visibility across the group resulting in improved customer service levels and control of the supply chain."[7]

Lower Cost of Doing Business

An oft-cited benefit of ERP implementation is improved coordination and sharing of current data across functional departments leading to lower costs of doing business.

Reduced inventory costs resulting from better planning, tracking, and forecasting of customer demand and inventory levels. Gibraltar Industries is a manufacturer, processor, and distributor of products for the building, industrial, and vehicular markets. It serves customers in a variety of industries around the world, and its recent annual sales were $1.3 billion. Gibraltar employs 3700 employees and operates 70 facilities in 27 states, Canada, China, England, Germany, and Poland. The firm uses an ERP system to gain improved inventory visibility. According to John Lentz, PMG vice-president, Finance, "[with our ERP system], we now have the ability to utilize inventory fully and to move it among our facilities when needed. This gives us a lot more flexibility to meet customer requirements at the best possible cost." Matt Jacobs, manager of business processes, states that an inventory accuracy of 99.75 percent has been achieved and that "we know exactly what is available for shipment … and we have virtually eliminated shipping errors."[8]

Faster collection of receivables based on better visibility into accounts and fewer billing and delivery errors. Solectron Corporation was a global contract manufacturer for computer and electronics companies with a $12 billion annual revenue flow. (The firm became part of Flextronics in late 2007.) The company grew rapidly during the 1990s and operational efficiency suffered while management focused on meeting the increased need for its services. However, Senior Vice President Guy Rabbat recognized a need to improve the cash collection business processes and that "by predictably accelerating the velocity at which cash flows into the company, Solectron can reduce its borrowings, pay less interest, reduce its foreign currency exposure, and collect interest on the extra cash." Rabbat led the effort to use data from its ERP system to improve the efficiency of the receivables process and increase the firm's ability to control collections. As a result, Solectron was able to save $14 million per year by reducing interest payments on its working capital financing while at the same time reducing the finance department headcount which resulted in an additional savings of $1 million per year.[9]

Lower vendor costs by taking better advantage of quantity discounts and tracking vendor performance to use as leverage in negotiating prices. Montefiore Medical Center in New York City claims that successful implementation of their ERP system led to major changes in its purchasing processes and an estimated savings of $72 million over a 10 year period. The medical center uses data from its ERP system to prepare for intense negotiations with vendors. Now instead of changing suppliers every few months, the medical center establishes multi-year contracts that lower costs while improving the quality of products and services. The ERP system also enables managers to see what is being ordered and by whom, thus eliminating occasional unnecessary purchases and reducing the shrinkage of supplies including drugs and expensive medical equipment due to employee theft.[10]

Improve Overall Customer Experience

Effective use of an ERP system can improve the overall customer experience in several ways. Improved inventory management can eliminate out-of-stock situations, which drive customers to your competitors. The associated streamlined business process can dramatically shorten the lead time from receipt of order to delivery of product. More careful attention to quality control can dramatically improve overall product quality.

Toray Membrane manufactures products used by municipalities, sewage treatment facilities, and heavy industries in the water desalination and treatment process to remove contaminants from water. It is critical that its products perform at a very high level and deliver promised results. Toray implemented an ERP system to improve the firm's operations from start to finish, including quality control. Workers use the ERP system to define quality test plans and record the results for quality control. If a quality issue arises, Toray can identify the root cause of the problem and take corrective action before a minor issue raises major problems. With this tight level of quality control, the firm has improved its ability to mitigate quality problems, reduce scrap materials, and provide an improved level of customer service.[11]

Facilitate Consolidation of Financial Data

Accurate, consistent, detailed, and up-to-date financial data is of the utmost importance in today's fast moving business environment. Organizations need it in order to respond quickly to business changes and stay ahead of the competition. Operational and strategic decisions are based upon it. Precise planning depends upon it. Problem solving demands it. A well-implemented ERP system enables rapid consolidation of financial data across multiple organizational units and countries because every business unit is using the same system and same database. In addition, ERP systems are designed to deal with differences in currencies and fluctuating currency exchange rates, which can cause additional problems in consolidating financial data. Organizations in which financial data is generated by separate computer systems in accounting, purchasing, sales, and other departments can find it very difficult to obtain the financial data they need on a timely basis. They are at a distinct disadvantage.

Oxford Industries is an international apparel design, sourcing, and marketer of clothing for men, women, and children. Its brands include Tommy Bahama, Oxford Golf, and Indigo Palms among others. The firm decided to implement an ERP system. According to Tom Chubb, executive vice president, "We anticipate building an environment of robust and timely insight. To support our strategic objectives, Oxford will work with SAP to streamline global financial reporting with a planned rollout across our operations in the United States, the United Kingdom and Hong Kong." Use of the ERP system will enable Oxford to eliminate bottlenecks in data integration, simplify its reporting process, and gain the ability to view a common database easily across the organization.[12]

Support Global Expansion

U.S. firms are continuing to expand their operations overseas to find new markets, lower labor costs, and gain access to key suppliers. According to a recent survey by the Aberdeen Group, nearly 80 percent of U.S. companies view expansion into global markets as a growth opportunity.[13] ERP systems can support global expansion as they are designed to monitor supply chains thousands of miles long.

According to Alfonso Cos, vice president for global supply network solutions at Procter & Gamble, "There's a big difference between being a global company because you have operations in many countries and being global because you operate globally. A few years ago, the company moved away from country-to-country operations to really operating globally." A key to the success of the firm's global operation was a four-year project to standardize on a single ERP system to support 135 plants in 40 countries. Data from the ERP system provides excellent visibility into the supply chain so workers can see by product, the orders, production plans, and actual production at a single plant or across multiple countries. Such information allows P&G to reduce its overall inventory levels and associated costs while maintaining high customer service levels.[14]

Provide Fully Compliant Systems

Senior management, including boards of directors, of many companies have taken great comfort in the fact that one side benefit of their ERP implementations is increased compliance with many state and federal laws, such as:

- Sarbanes-Oxley Act (establishes standards for all U.S. public company boards, management, and public accounting firms)
- Health Insurance Portability and Accountability Act or HIPAA (protects the health insurance coverage for workers and their families and requires the establishment of national standards for electronic health care transactions and national identifiers for providers, health insurance plans, and employers)

Another law that requires compliance from companies in the food industry is the Public Health Security and Bioterrorism Preparedness and Response Act (Bioterrorism Act). The law was passed in 2002 to help protect the nation's food supply from a bioterrorist attack. According to the act, food processors with more than 10 employees had until June 9, 2006 to be able to provide, within 24 hours of an FDA request, the following information or be subject to civil and/or criminal penalties.

- Identify every entity in their supply chain from the grower through each intermediate link to the final wholesaler or retailer.
- Provide a lot or code number for the food product.
- Identify the specific source of each ingredient that was used to make every lot of finished product.

Benner Foods, Inc. is a cheese producer with about $50 million in annual revenue. The firm implemented an ERP system to improve its inventory control processes and was able to reduce expired inventory by over $200,000 per year. At the same time, implementation of the ERP system also enabled Benner to track and trace the ingredients it received from suppliers, as well as dairy products it ships out. The firm was then confident that it was 100 percent fully compliant with the Bioterrorism Act.[15]

ERP ISSUES

There are a number of potential issues associated with the implementation of ERP systems including post start-up problems, high costs, lengthy implementation, and organizational resistance. These issues will be discussed and examples of companies encountering these problems will be provided.

Post Start-Up Problems

A Deloitte Consulting survey of 64 Fortune 500 companies revealed that one in four confessed to an actual drop in performance for some period of time after their ERP system went live. For example, at Invacare, a leading manufacturer and distributor of medical equipment used in the home, post start-up problems with the order-to-cash process (those activities from the taking of a customer order through collection of the customer payment) caused the firm to lose $30 million in revenue. "Our systems were locking up," says Invacare Chief Financial Officer Greg Thompson. "We had a lot of hang-ups by customers when we couldn't answer the phones in a timely way, and when we did talk, we couldn't give them complete information on our stock availability and when we could ship the product."[16]

Many early ERP efforts in the 1990s and early 2000s were less than glowing successes. Indeed there are numerous examples of companies that spent tens of millions on ERP only to have problems. See Table 8-3 for a partial list of companies experiencing major ERP implementation problems.

TABLE 8-3 Organizations with major ERP start-up problems

Company	Summary of Initial ERP Implementation Results
FoxMeyer Drug Company	Attempt at ERP system implementation proved to be so disastrous, according to FoxMeyer, that it forced the company into bankruptcy and liquidation.
Nike	Botched ERP implementation cost firm over $100 million in lost sales, depressed the stock price by 20 percent, and triggered a flurry of class action lawsuits.
Cleveland State University	Filed $510 million lawsuit against ERP vendor after software failed to work as expected.
Hershey Foods	Order processing and shipping problems caused the firm to lose substantial sales during the peak Halloween and Christmas seasons.

TABLE 8-3   Organizations with major ERP start-up problems (continued)

Company	Summary of Initial ERP Implementation Results
Whirlpool	ERP system created problems with orders with quantities less than one truckload in order processing, tracking, and invoicing.
WW Grainger	Massive distribution problems due to faulty ERP implementation led to major losses in sales.
Waste Management, Inc.	Forced to terminate ERP project after incurring major implementation expenses.

High Costs

The cost of a typical ERP implementation is quite high, running from several hundred thousand dollars to hundreds of millions of dollars. Table 8-4 shows the implementation costs of ERP projects for a variety of firms categorized by their annual revenue (the number of survey respondents is shown in parenthesis). The cost of an ERP project depends on a number of factors, several of which are shown in Table 8-5.

TABLE 8-4   Average implementation costs for ERP projects including cost of internal resources

Annual Revenue	Cost to Implement Financial Modules Only	Cost to Implement All Modules
< $100 million	$.9 million (7)	$ .7 million (18)
$100 million – $499 million	$.4 million (4)	$5.4 million (25)
$500 million – $999 million	$1.3 million (2)	$8.5 million (6)
$1 billion – $5 billion	$ 60 million (1)	$30.4 million (9)
> $5 billion	$115 million (1)	$46.0 million (2)

Source: "ERP Implementations," The Controller's Report, December 2007 accessed at *www.ioma.com/fin*.

TABLE 8-5   Key cost drivers for an ERP implementation

Cost Drivers	Comment
Degree of business process change expected	The greater the degree of business process change expected, the greater the cost of training and effort required to overcome organizational resistance.
Degree of customization required	The greater the ERP software must be customized, the greater the cost.
Number of implementation locations	The greater the number of sites, the greater the cost.
Scope of business to be impacted	The more modules to be implemented, the greater the cost.
Number of people impacted	The more people impacted, the greater the cost.

TABLE 8-5   Key cost drivers for an ERP implementation (continued)

Cost Drivers	Comment
Degree to which legacy systems will be used	The more legacy interfaces, the greater the cost.
Organizational preparedness	The more employees are prepared to contribute to a successful ERP system, the lower the cost.

In developing a budget for an ERP implementation, it is best to set a realistic budget rather than an optimistic one. Has your organization successfully completed large-scale projects in the past? Has your organization worked well with outside consultants on other large projects? Do you have a high level of in-house expertise in ERP implementations and business process change? Affirmative answers to these questions provide a basis for confidence in completing the project successfully.

Recent surveys of financial executives show that 38 percent of the respondents said that their organization's ERP total project costs were 10 to 30 percent above the original budget, while 17 percent say total costs exceeded the original budget by 30 percent or more. However, in spite of the cost overruns, only 6 percent of the survey respondents considered their ERP projects to be moderately problematic, while only 2 percent said they are failures.[17]

The following kinds of costs are commonly overlooked or underestimated in setting the budget for an ERP project:

- **Hardware upgrades**—Implementation of an ERP system frequently requires a substantial upgrade to an organization's servers and personal computers.
- **Training**—Training needs are great because employees need to learn a whole new set of software and business processes for accomplishing their work. In addition, they must learn new roles and responsibilities as well as adapt to new expectations for how they are to work and interact with others.
- **Testing**—Thorough testing of new ERP software can be extremely tedious and time consuming; however, most organizations believe it is mandatory in order to avoid unexpected problems at system start-up.
- **Customization**—Any customization of the code of the standard ERP software to add new functionality or to enable interfaces to other software can require many months, and customization often fails. Such changes in the ERP software also require lots of testing and reworking of the code.
- **Data conversion**—A tremendous amount of effort is needed to extract and move data from old legacy systems into the new ERP system. Such data might include customer data, employee data, price lists, product details, manufacturing data, and supplier information. To complicate matters further, in the process of moving the data, it is often discovered that much of the existing data is inaccurate or out-of-date. Additional effort is then required to "clean up the dirty data" and replace it with current, accurate data.

233

- **Consultants**—Companies frequently fail to establish clear objectives and measures for the work to be done by ERP project consultants. This leads to a loss of accountability and can result in consulting fees spiraling out of control.

Lengthy Implementation

Large organizations view an ERP implementation project as an opportunity to improve fundamental business processes. This requires significant changes in people's roles and responsibilities. Organizational changes of that magnitude do not come easily or quickly. The time frame for full implementation can be one to four years depending on the number of ERP modules, the number of different organizational locations at which the system is implemented, and other factors. Much faster ERP implementations are possible when the scope of the effort is limited to a single business function, such as Human Resources, and a single business location.

According to a study by the Aberdeen Group, small and medium businesses (SMBs) usually achieve less than the full potential of business benefits possible from the implementation of ERP systems. SMBs look for quick, simple implementations of ERP and generally avoid making changes to fundamental business processes and people's roles and responsibilities.[18]

Difficulty in Measuring a Return on an ERP Investment

Simple return on investment is calculated by dividing the value of the net benefits directly associated with a project by the costs associated with the project. Ideally, decision makers would like to have an accurate estimate of the return on investment for an ERP project *prior* to approving it. Unfortunately, it is difficult to put an exact dollar figure on both the benefits and costs associated with an ERP project. Project costs frequently are underestimated, and project benefits often are overly optimistic.

Even after an ERP project is complete, it is difficult to measure the return on investment because the project frequently takes years to implement. Often many more years are required before substantial benefits accrue. Over this period of time, perhaps five years or more, so many other business changes are occurring that it can be difficult to isolate the benefits and costs of the ERP system.

Organizational Resistance

As discussed in the previous section, any new ERP system brings with it considerable changes to an organization's business processes and to the roles and responsibilities for employees across the organization. These changes include modification in the way employees do their work and interact with others. Furthermore, many organizations see ERP implementation as a way of cutting costs through elimination of workers and thus people fear they will lose their jobs. It is human nature to resist such major changes.

Organizational resistance manifests itself in many ways. Some valuable workers resign from the organization rather than go through the transition. Other workers, in a desperate effort to delay the oncoming changes, fail to execute the work required to transition from the old way of doing things to the new way. Still other workers avoid taking the training necessary to learn their new roles and new work processes. As a result of such organizational resistance, many ERP projects take much longer than expected and/or fail to deliver hoped for enterprise improvements.

The next section will outline a tried and proven process for successful ERP implementation.

ERP SYSTEM IMPLEMENTATION PROCESS

The major ERP vendors have all developed a recommended implementation process based on years of experience with hundreds of customers. Each vendor's process typically divides the effort up into well-defined stages with associated tasks. A representative ERP implementation process is shown in Table 8-6.

In addition to following the selected ERP software vendor's implementation process, organizations typically will try to assign employees who have previous ERP implementation experience to the project. They usually will consider hiring an experienced system integrator who is familiar with both the industry in which the organization competes and with the ERP software under consideration. These measures go a long way toward improving the probability of success of the ERP implementation project.

There are several common factors associated with failed ERP implementation projects including failure to gain senior management commitment and involvement, choosing the wrong business partners to help, not adequately assessing the level of ERP customization that may be needed, failure to contain project scope, and lack of planning for effective knowledge transfer. The following section will address how to avoid these problems.

TABLE 8-6 Representative ERP Implementation Process

Project Stage	Tasks
Initiation	Perform stage initiation tasks (at the start of each phase): • Identify, recruit, and prepare appropriate team members for this stage • Develop detailed schedule and cost estimate for this stage
	Identify desired business needs to be met through this project
	Develop business justification for project
	Determine if system integrator will be used and select one
	Perform stage closing tasks (at end of each stage): • Release team members not needed for next stage • Develop high level schedule and cost estimate for remainder of project • Review project benefits compared to costs • Make decision to continue project, re-define project, or terminate project
Requirements Analysis	Perform stage initiation tasks
	Analyze current business processes for strengths and weaknesses
	Determine business processes to be supported by ERP system
	Define mandatory business requirements

TABLE 8-6   Representative ERP Implementation Process (continued)

Project Stage	Tasks
	Define which business organizations and locations to convert to ERP system
	Perform stage closing tasks
ERP Software Selection	Perform stage initiation tasks
	Identify 2 to 4 candidate ERP software packages for in-depth evaluation
	Develop set of software package selection criteria
	Evaluate candidate ERP software packages against selection criteria
	Perform gap analysis to identify if packages fail to meet significant business requirements or are unable to support desired business processes adequately
	Assess level of customization needed to "bridge the gaps"
	Select ERP system software
	Select system support provider
	Perform stage closing tasks
Design	Perform stage initiation tasks
	Define inputs needed and sources of inputs
	Define required reports
	Define necessary ERP system interfaces
	Define other system outputs
	Perform business process re-engineering
	Define any mandatory software customization
	Perform stage closing tasks
Implementation	Perform stage initiation tasks
	Set ERP system configuration parameters
	Clean up and migrate data from old sources to ERP system
	Develop required interfaces
	Perform necessary customizations
	Implement controls and security
	Train the trainers
	Conduct end user training
	Provide training for specialists

TABLE 8-6 Representative ERP Implementation Process (continued)

Project Stage	Tasks
	Test new business processes
	Test hardware and software
	Test system interfaces
	Test interaction with system support provider
	Perform stage closing tasks
Maintenance and continuous improvement	Provide on-going technical support
	Deliver on-going training for new end users and to cover system upgrades
	Plan and implement necessary software upgrades

BEST PRACTICES TO ENSURE SUCCESSFUL ERP IMPLEMENTATION

ERP project managers must attempt to deliver a solution that meets specific scope, cost, time, and quality goals while managing the expectations of the project stakeholders. Given the broad scope, high costs, large number of project stakeholders involved, and the amount of organizational change that is required, achieving ERP project success can be a very difficult challenge. However, a set of best practices has emerged to ensure the successful implementation of an ERP System. These best practices include ensuring senior management commitment and involvement, choosing the right business partners to help, assessing the level of ERP customization that may be needed, avoiding increases in project scope, and planning for effective knowledge transfer.

Ensure Senior Management Commitment and Involvement

As with any other major organizational change project, ERP implementation requires the commitment of senior management to achieve the necessary organizational buy-in. Specifically, senior management must define a vision for the ERP system with supporting goals and visible, measurable success criteria. In addition, senior management will need to provide leadership and take action to ensure that the goals of the project are met. For example, they must be proactive in identifying and removing "roadblocks" that stand in the way of project progress. Without strong leadership and timely interventions by senior management, the likelihood of an ERP implementation failure is very high.

Choose the Right Business Partners

Choosing the right business partners to provide, implement, and support your organization's ERP project is critical. Three key business partners include the ERP system integrator, the ERP software provider, and the ERP software support providers.

An ERP system integrator provides its customers with consulting, integration, and implementation services to improve the likelihood of a successful ERP implementation.

System integrators may be very large organizations that provide services to customers in every industry, or they may be smaller firms that specialize in specific industries. Ideally, they provide in-depth knowledge of business processes, industry experience, and solution expertise. The services they provide are negotiated and agreed to prior to each engagement. The range of services can be quite broad and include such activities as helping a customer to select appropriate ERP vendors and software, deploying ERP solutions, integrating ERP software with existing legacy systems, delivering training, providing post-start up support for end users, and in general, helping the customer achieve a good return on their ERP investment.

The ERP software provider is the organization that provides the ERP software (e.g., SAP, Oracle, etc.). In many cases, the organization implementing the ERP software will also request that the ERP software provider supply many of the same services as the ERP system integrator.

The ERP software support provider is an organization that ensures that the users are able to use the software effectively once it is installed. The software support provider may provide help desk service, deliver training, and monitor and fix hardware, software, databases, and communications networks.

The chosen business partners should have a solid, verifiable track record of successful engagements with other organizations in your industry including organizations with similar operational and business issues as your own. Check references thoroughly to verify that the resources you plan to use know the software and understand your industry and business. The first step for these project partners should be to develop a solid understanding of your business needs and processes as they are now. They must be thorough in their approach, looking for fundamental, underlying issues—not just those that you and your organization tell them about.

Assess Level of Customization Needed

It is critical that the level of ERP software customization needed is understood and the option to align business processes to ERP software is agreed to before you sign a contract. A key initial step is to determine if your organization's fundamental ways of doing business can be supported by an ERP solution. If it is determined that the software under consideration does not support one or more of an organization's fundamental business processes, there are three options.

One option is to change the inconsistent business processes to accommodate the software. This means making fundamental changes in long-established ways of doing business, even though the existing processes may provide a competitive advantage. It also means changing roles and responsibilities for a lot of employees, something that makes senior managers and the affected workers extremely uncomfortable.

The second option is to modify the software to fit the process. This is a highly undesirable choice as it will slow down the project, introduce potentially dangerous bugs into the system, and make upgrading the software to the ERP vendor's next release extremely time consuming and costly because the customizations will need to be re-implemented in the new release.

The third option is to select a different ERP solution, perhaps an industry specific ERP solution. ERP software providers have long recognized that business processes that work effectively for dozens of companies in a given industry prove to be good solutions for almost all companies in that industry. For example, a customer order entry process that is highly effective for dozens of firms in the auto parts distributor industry likely will be good for almost all companies in that industry. Taking that one step further, many ERP software providers have designed and implemented ERP software tailored to support all the essential business processes for specific industries such as retail clothing, consumer packaged goods, and higher education. The availability of industry-specific ERP solutions provides an opportunity to improve the speed and likelihood of a successful ERP implementation. Organizations should be strongly encouraged to adopt such a ready-to-use template for its business processes. Organizations unwilling to use industry standard practices should question why their business processes need to be different. If they determine that their unique practices provide some sort of competitive advantage, then they must accept that any modifications to industry specific ERP solutions will raise the risk of failure and lengthen the time to execute a solution.

Avoid Increases in Project Scope

As discussed in Chapter 3, project scope management includes defining the work that must be done as part of the project and then controlling the work to stay within the scope to which the team agreed. It is very typical that as an organization implements an ERP system, it learns that there is much more that could be done than was included in the original project scope. For example, additional modules could be implemented to bring substantial new business benefits. There is a strong temptation to expand the ERP project scope to achieve these benefits. However, this will lead to an increase in cost and duration of the project and delay achievement of the benefits originally identified to justify the project. If the goal is to complete the project as quickly as possible and to minimize the risk of project failure, potential increases in scope should be rejected. Once the original project scope is complete and the organization has cut over successfully to the new ERP system, these new ideas can be further evaluated and implemented if justified.

Plan for Effective Knowledge Transfer

Obviously, employees must be prepared and trained thoroughly to use the ERP software to accomplish their work in the context of their new or revised role. But this is not enough, and this is not where the training should start. First, employees need to understand clearly the rationale of why ERP is being implemented in the first place. This will help motivate them and enable them to understand the importance of a successful ERP implementation. Employees also need to be given the big picture on the extent of change and how it will impact them personally in terms of changes in their roles, job performance expectations, and interaction with others. This will help them overcome their fear and resistance to the numerous changes they will experience.

The training should not be considered a one-time event. Over time, employees will need refresher training to eliminate any problems they are encountering in using the system and to expose them to new, better ways in response to changes in the system. Some employees will need further training to provide them with a higher level of competency so that they can help support others in their work area.

Test Thoroughly

Most organizations do everything they can to ensure a smooth start-up of their new ERP system. Key to a smooth start-up is the thorough testing of the ERP hardware and software, associated business processes, interfaces to existing systems—even interactions with the help desk that will provide post-start-up support and troubleshooting.

Key business processes must be tested from start to finish. For example, order processing would be tested with carefully selected orders that encompass all of the various types of orders the system is expected to process. Each step this process should be executed and the results checked carefully. These steps might include recording the items purchased and their quantities, verifying that sufficient inventory is available to fill the order, pricing the order, subtracting any available discounts or promotions, checking that the customer's credit limit is not exceeded, etc. The tests and test data must be set up carefully to execute a wide range of possible scenarios. Different scenarios should be tested, such as: a customer places an order for an item no longer in stock, a customer places order but has insufficient credit available, a customer places an order for an item where more than one discount or promotion applies, etc. Considerable time is required to plan and prepare the necessary test data for such thorough testing.

Plan for a High Level of Initial Support

It is wise to anticipate a heavy need for support from the systems integrator and the ERP software support provider and to contract for a heavy level of coverage during at least the first three months following implementation. This will ensure that resources are available to answer questions and address issues in the pressure packed time of initial start-up.

Interestingly, in implementing Oracle ERP, Arizona State University failed to follow these recommendations for successful ERP implementation and intentionally released only partially-tested software to users. Read further about this in the Manager Takes Charge special feature.

A MANAGER TAKES CHARGE

ERP – The Arizona State University Way

In February 2006, William Lewis, Vice Provost and Chief Information Officer of Arizona State University (ASU) requested approval for funding of the Oracle PeopleSoft Student Administration and HR/Payroll ERP system. The goal was to "improve service to the community, improve recruitment of students and staff, minimize costs, and begin to coordinate services across universities to simplify student access to university resources." The estimated cost was $23 million, including $6.7 million in staff costs over five years and $16.3 million in implementation costs over five years. The Student Information System portion of the project was targeted for completion in fall of 2006 with the HR/Payroll portion of the project to be delivered no later than the end of 2007. The project was a key component of ASU's 10-year plan to increase in size and scope while increasing academic quality. ASU is already the nation's fourth largest university and has set a goal to increase enrollment by more than 55 percent for a total enrollment of over 90,000 students. ASU's vision is to become "The New American University." [19]

ASU took an unconventional approach to installing its ERP software. They decided to follow a strategy of strict adherence to planned project milestones and to start-up various components of the software on schedule—even if it meant cutting back on planned testing and that all the glitches and issues were not been identified and resolved. Problems would be fixed on the fly as they arose. It was anticipated that there would be problems as workers and students started using a system that wasn't rock solid, but managing through the problems was part of the plan. Mr. Adrian Sannier, ASU's technology officer, calls this strategy, "'implement, adapt, grow,' since it not only relies on the IT department to fix any technical glitches, but also requires employees and students to help identify problems, as well as to adjust to working within the new system." [20]

ASU planned to implement the ERP system in a scant 18 months even though other similar sized institutions had taken over four years. Sannier was willing to spend money on additional project resources and consultants to fix problems rather than risk missing a deadline. [21]

ASU began using the new payroll system on schedule in July 2007, at the beginning of its fiscal year. Not unexpectedly, there were problems right away. For a variety of reasons, some 3000 employees were underpaid or not paid at all, while other employees were paid thousands more than they should. To compensate for these problems, the HR department was directed to write checks on the spot to any employee with an erroneous paycheck. Unfortunately, there were so many errors that the check writers could not keep up with the underpaid employees who overwhelmed the HR offices. In some cases rather than write a new check, check writers asked hundreds of employees to wait for up to a week to receive a corrected check.

Over time payroll calculation errors were corrected and timesheet data collection procedures were smoothed out. The new payroll system error rate is down to around 4 percent, which is lower than the 6 percent error rate of the old payroll system. The final cost of the project was a total of $30 million, $7 million over the Feb 2006 budget request. [22]

continued

Discussion Questions:

1. With the benefit of 20-20 hindsight, how might the problems with payroll checks been significantly reduced?

2. Do you suspect that there might have been serious problems with the Student Information System portion of this project as well? Why do you think there seems to be no documentation of problems in this area?

3. There is mixed reaction to ASU's implementation approach and its results. Imagine that you have been hired as a consultant to assess whether or not the ASU strategy was effective. Identify six people (by role or title) that you would want to interview. Identify a set of four or five questions you would ask each interview subject.

4. Would you recommend the ASU approach to other universities? Why or why not?

ERP TRENDS

There are many interesting trends in the evolution of ERP solutions including the emergence of ERP solutions targeted at SMBs, the availability of ERP as a software service, and the ready availability of open source ERP software. These trends will now be discussed.

ERP Solutions Targeted for SMBs

The ERP market continues to expand and is expected to grow to $45 billion by 2011, up from $30 billion in 2006 according to International Data Corporation. Much of this growth is expected to come as SMBs begin using ERP systems.[23] As a result, some of the large, well-established ERP vendors are creating software for this market. For example, Microsoft has integrated what were diverse application modules into its Dynamics AX 2009 ERP software to enable SMBs to build a single, integrated view of both the financial and supply chain. The Dynamics AX 2009 user interface is designed to look like Microsoft office thus reducing the time it takes users to get comfortable with the package. It also comes with management reporting tools that provide managers with key performance indicator reports and alert them to any changes in the business requiring action on their part.[24]

ERP as a Service

The high cost of ERP software licenses has traditionally kept SMBs with small IT budgets from taking advantage of these powerful applications. Besides the initial software licensing costs, a company also has to consider the huge expense of building and maintaining the IT infrastructure (hardware, data center, communication network, etc.) needed to support the application.

The emergence of the software as a service (SaaS) model for the delivery of ERP solutions offers organizations the opportunity to acquire ERP capabilities at a much lower start-up cost. Under this model, the organization pays a monthly fee for its users to access the ERP software (via the Internet) running on the service provider's hardware. This eliminates the high initial costs associated with software licensing and the building of the prerequisite hardware infrastructure. Organizations that offer ERP SaaS solutions include Aplicor, Intaact, Microsoft, NetSuite, Oracle, Plexus, SAP, and Workday.

In addition to lower start-up costs, the SaaS model is appealing to many organizations because they believe that SaaS ERP can be implemented with little or no effort. This can free up IT staff and others to work on other projects. However, it is important to recognize that there is still considerable work to be done to reap the full benefits of an ERP implementation—business processes must be redefined and streamlined, interfaces with existing systems must be designed and executed, databases must be created, users must be trained, and so forth.

A key concern with an ERP SaaS solution is the security of your firm's customer, employee, and financial data residing on another organization's computers. Another issue is the potential loss of access should some sort of disaster strike the service provider or if there is some disruption in Internet service.

Open Source ERP Software

Many SMBs elect to implement open source ERP solutions because of their lower initial acquisition cost—perhaps several hundred to tens of thousands of dollars. Another attractive feature of an open source software solution is that because the user has access to the source code, there is a wide range of resources (including the acquiring organization's own IT staff) that can make modifications to the software. Popular open source vendors include Apache, Compiere, Open for Business and Openbravo, Technology Group International, xTuple, and WebERP. Compiere claims that there have been more than 1.2 million downloads of its software.[25] Open source Web-based OpenBravo has been localized for 45 different countries and has been downloaded 600,000 times.[26]

As with any other open source software, organizations will not get the same level of support that they would receive from commercial software providers. To combat this, some open source vendors such as xTuple offer managed services in addition to software to help users better maintain and upgrade their ERP application. xTuple offers its XTN service at three levels. The base level provides ongoing maintenance and upgrades. The middle tier adds automatic nightly updates. The premium tier adds high-end database tuning and optimization.[27]

Dan Carter, Inc. Cheese Company (DCI) manufactures, distributes, and sells domestic and imported high quality specialty cheeses, as well as prepared foods. During 2006 and 2007, DCI grew rapidly through the acquisition of G & G Specialty Foods, Green Bay Cheese Company, Swissrose International, and Advantage Foods International. Recent annual sales exceeded $500 million. According to DCI's CFO Tim Preuninger, "We needed a software system that allowed us to effectively manage our multiple businesses. As we continue to grow, it's becoming more critical that we identify and use key information to help us quickly and accurate complete our financial reporting needs as well as achieve operational excellence within our supply chain. We also wanted a user-friendly software package that provided flexibility to allow us to tailor it to our needs."[28]

Preuninger and his management team selected the open source software package Enterprise 21 ERP from Technology Group International (TGI). The selection was made in large part because of TGI's experience within the cheese industry. The core software

243

already met most of DCI's basic business requirements including support for market-based pricing, lot control and reporting, production planning, and data extract capabilities. Also weighing very heavily in DCI's decision was the fact that the software is open source so that it is capable of being changed rapidly to meet changing business needs. While the implementation is not complete at all sites, DCI's Chief Financial Officer Tim Preuninger anticipates real benefits. "Once we are fully operational, we expect to achieve substantial operational efficiency improvements through better production planning, less downtime due to inefficiency and changeover, and better raw material inventory management, all of which should lower cost and improve customer service. Additionally, we expect to see our period-end closing process become more efficient, allowing us to close our financial statements faster and more accurately as well as providing our financial group more time to drive our business results."[29]

This chapter has provided information and numerous examples of how successful implementation of an ERP system can provide substantial benefits to an organization. It has also pointed out many of the potential pitfalls associated with ERP implementation and how to avoid some of the most significant problems.

The checklist in Table 8-7 provides a set of recommended actions for an organization to take to ensure the success of an ERP implementation. Use this checklist to evaluate if your organization is ready for implementation. The appropriate answer to each question is "yes."

TABLE 8-7 A manager's checklist

	Yes	No
Is senior management committed to this project and prepared to get involved to ensure its success?		
Have you chosen the right business partners to provide, implement, and support your organization's ERP software?		
Do you know the level of customization that will be needed to align business processes to the ERP software?		
Are the project and senior management teams determined to contain the scope of the ERP implementation project to complete the project as quickly as possible and minimize the risk of project failure?		
Are sufficient time and dollars budgeted to ensure effective knowledge transfer?		
Are sufficient time and people budgeted to ensure thorough testing before system cutover?		
Have you planned for a high level of support following system cutover?		

Chapter Summary

- Enterprise Resource Planning (ERP) is a set of core software modules that enable organizations to share data across the entire enterprise through the use of a common enterprise database and management reporting tools.

- In order to address differences in business processes in various industries, ERP vendors offer specially tailored software modules designed to meet the needs of specific industries.

- Organizations can pick and choose which ERP software modules to install based on business needs.

- A customer relationship management system (CRM) is an enterprise system that supports the processes performed by all the entities involved in creating or increasing the demand for an organization's products and services. It is often one of the software modules offered by an ERP software provider.

- Supply chain management (SCM) involves the planning, executing, monitoring, and controlling of the demand planning, sourcing, manufacturing, logistics, and customer service set of business processes. The primary goal of SCM is to lower costs and inventory levels while still meeting customer requirements for timely delivery of high quality products.

- An effective ERP system implementation can bring many benefits to an organization including establishing standardized business processes, lowering the cost of doing business, improving the overall customer experience, facilitating consolidation of financial data, supporting global expansion, and providing a compliant system.

- A number of potential issues are associated with the implementation of ERP systems including post start-up problems, high costs, lengthy implementation, difficulty in measuring return on investment, and organizational resistance.

- To improve the probability of success of their ERP implementation project, organizations follow their ERP vendor's recommended ERP implementation process, assign employees with previous ERP implementation experience, and frequently hire an experienced system integrator.

- Given the broad scope, high costs, large number of project stakeholders involved, and the amount of organizational change required, achieving ERP project success is a very difficult challenge.

- A set of best practices has emerged that are key to ensuring successful implementation of an ERP system. These include ensuring senior management commitment and involvement, choosing the right business partners, assessing the level of software customization needed, avoiding increases in project scope, planning for effective knowledge transfer, testing thoroughly, and providing a high level of initial support.

- Rapid growth is expected in the use of ERP systems within SMBs.

- The emergence of the software as a service (SaaS) model for the delivery of ERP solutions provides on-demand delivery of ERP capabilities at a much lower start-up cost.

However, two key concerns with an ERP SaaS solution are the security of data residing on another organization's computers and the potential loss of access should some sort of disaster strike the service provider or if there is disruption in Internet service.

- Many SMBs are electing to implement open source ERP solutions, which offer lower initial cost and ease of modification.

Discussion Questions

1. How would you define an ERP system?

2. What are best practices? Are best practices the same for all companies?

3. Imagine that you need to conduct an in-depth assessment of an ERP implementation to identify what went well and what did not go so well. Prepare a list of 10 questions that would help you gather this information. Identify the key people (by business title or organizational role) you need to interview.

4. Identify three major benefits that an institution of higher education would likely gain from the use of an effective ERP system.

5. Identify and briefly discuss the key factors that often lead to a failed ERP project.

6. How would you distinguish between ERP, CRM, and SCM? Why do some organizations elect to implement CRM and SCM software applications independent of their ERP software?

7. What are some advantages of having the ERP software provider also fulfill the role of system integrator and support provider? What are some disadvantages of this approach? In general, would you recommend using the ERP software provider to fulfill all three roles? Why or why not?

8. What options are available if the ERP software under consideration does not support important business processes of your organization? Which option do you think is best? Why?

9. Do you think that it is essential for an organization to plan to do some parts of the ERP training for its employees? If so, which parts? If not, why not?

10. Identify a few advantages associated with the use of open source ERP software. What are a few disadvantages?

11. Briefly summarize the advantages of implementing ERP in an SaaS environment over a traditional ERP implementation where the software is purchased and runs on the user organization's computers. Can you identify any disadvantages?

12. What questions would you need to have answered by an SaaS vendor in order to feel comfortable using their service?

Action Memos

1. You are the CFO of a mid-sized manufacturing firm. As you are walking out the door to go to lunch, your BlackBerry rings. It is the CEO. She informs you that the presentation to the Board of Directors on implementing a new ERP system went well; however, they did not approve the funding for the system. They insist on seeing a stronger justification for spending the $15 million on the project (this represents nearly 5 percent of the company's annual

revenue). The CEO states that the Board did not accept the "everyone else is doing it" justification she offered. She asks for your help in preparing a strong justification before the Board meets again in two weeks. In order to build a strong business case for the project, she wants you to lay out a process and identify the resources that would be required. What do you say?

2. You are the ERP integration project manager of an ERP implementation project for a small manufacturing organization. The Director of IT calls you to discuss his interest in enlarging the scope of the project from implementing two ERP software modules to four. After talking for 10 minutes about the additional benefits scope expansion will bring, he pauses and asks for your opinion. On the one hand, such an expansion will mean additional consulting fees for you and your company; on the other hand, you worry about how this might affect the ultimate success of the project. What do you say?

Web-based Case

Do a search for the article " ERP, We Did It Again," which appears in the September 11, 2008 issue of *iWeek*. Read this article and comment on Dr. James Robertson's statement about taking an engineering approach to ERP projects. How would such an approach affect how an organization takes on an ERP project? Identify two other suggestions that are made in this article to improve the odds of a successful ERP project.

Case Study

Hunter Manufacturing: Successful ERP Implementation

Hunter Manufacturing was founded in 1937 and provided tent and truck heaters for U.S. troops during World War II. Today, Hunter designs and manufactures a broad set of solutions to provide shelter, heat, power generation, and chemical, biological, radiological and nuclear (CBRN) protection for shelters and vehicles for both the military and homeland security markets. It manufactures every heater in the Army's M-151 jeep and Hummer vehicles. It also makes a heater based on thermoelectric design that provides 35,000 BTU per hour of clean heated air with no external electricity. The CBRN filters are the basis for individual and group protection systems for temporary shelters, permanent structures, military vehicles, and Navy ships. Hunter has adapted its CBRN filtration systems to meet the needs of the homeland security market for use in emergency response shelters and command centers, emergency response vehicles, and HVAC systems for buildings. In 2002, Hunter purchased the Camfire line of portable heating equipment to provide temporary heating equipment for a variety of non-military applications.

Following the events of September 11, there was a great increase in demand for the firm's products. To meet this increased demand, Hunter made two key acquisitions. First, it purchased the supplier that made the tents to house its heaters. Then it acquired the company that provided the power generation and air conditioning equipment for its shelter and tent applications. Following these acquisitions, Hunter's annual revenue increased to $170 million. Hunter now employs approximately 500 workers at two plants in Ohio and has a research & development lab in Edgewood, Maryland.[30]

Hunter's legacy information system was old and inflexible. It was also highly unreliable and crashed frequently—sometimes twice a week. Clearly this was not acceptable and the system needed to be replaced. Hunter's management team developed a set of critical requirements for a new ERP solution.

With a well-defined set of system requirements, Hunter's management team next looked at several potential ERP software solutions. Eventually SAP was selected. Hunter's CFO Steve Demko recalls, "We had the usual perceptions of SAP software: too big, too complex, and too expensive. So we looked at some smaller systems, but they just didn't stack up."[31]

The scope of business encompassed by the ERP system included purchasing, inventory, order entry, delivery processes, product configuration at the point of order entry, spare parts tracking, real-time reporting, profitability analysis, and postings for payment receipts and billing purposes.[32] However, Hunter Manufacturing was able to implement a preconfigured template of SAP that met 85 to 90 percent of its requirements in just four months.

Discussion Questions

1. Indicate "Yes" for each requirement listed in Table 8-8 that you feel was essential in the selection of an ERP system for Hunter Manufacturing. Provide a brief rationale for selecting or rejecting each requirement.

TABLE 8-8 Potential ERP system requirements for Hunter Manufacturing

Requirement	Yes/No?	Rationale
Support rapid business growth.		
Provide a common business solution across multiple and diverse product lines.		
Facilitate a rapid and smooth integration of multiple business units.		
Streamline and standardize the firm's business processes.		
Enable the consolidation of financial statements.		
Improve data and system security over existing legacy system.		
Provide a secure and fully complaint system.		
Support global expansion.		
Provide support for industry best practices.		
Avoid increases in staff even as the size of the business grows.		
Integrate operational data across all departments.		

2. Briefly outline a process that the management team could have followed to evaluate several ERP solutions objectively and overcome their original misgivings about SAP.

3. What role might industry consultants or third party ERP implementation experts have played in this successful project?

4. What role could Hunter Manufacturing management have taken to minimize organizational resistance?

Endnotes

1. "BWA Water Additives, UK Limited," *Hoovers*, accessed at http://www.hoovers.com/bwa-water-additives/--ID__113524--/free-co-profile.xhtml on August 25, 2008.

2. "BWA Water Additives Reports Improved Performance After Global Implementation of Ross Enterprise ERP Applications," *Reuters*, March 26, 2008.

3. "BWA Water Additives Reports Improved Performance After Global Implementation of Ross Enterprise ERP Applications," *Reuters*, March 26, 2008.

4. "The Benefit of Foresight," *http://www.worksmanagement.co.uk/*, May 2008.

5. "The Benefit of Foresight," *http://www.worksmanagement.co.uk/*, May 2008.

6. "LeanLogistics TMS Provides New Functionality for Inbound Management," *Food Logistics*, July 8, 2008.

7. "Gooch and Housego Standardizes Global ERP with K3 SYSPRO," *Manufacturing & Logistics IT*, June 18, 2007.

8. "ERP Paces Gibraltar's Progress," *Metal Producing & Processing*, September/October 2006.

9. "Customer Success Story (Solectron)," Emagia Web site accessed at *http://www.emagia.com/* on September 10, 2008.

10. Marianne Kolbasku McGee, "Reining in Health Care Costs Through Stricter Supply Management," *InformationWeek,* September 20, 2007.

11. Toray membrane USA Enhances Operational Visibility with Infor," Press Release from Infor Web site at *http://www.infor.com*, November 12, 2007.

12. "Oxford Industries, Inc. Selects SAP to Streamline Global Financial Reporting," *PRNewswire*, May 7, 2008.

13. "The Role of ERP in Globalization: A Low-Cost Approach to Reaching New Markets," The Aberdeen Group, 2007.

14. Gary Anthes, "Supply Chain Blind Spots," *Computerworld*, February 20, 2008.

15. George V. Hulme, "Food Chai's Fear Factor," *InformationWeek*, May 23, 2005.

16. "Difficulties in New Systems Implementation Causes Invacare Corporation to Lower Fourth Quarter Earnings Guidance," *Business Wire*, December 14, 2005.

17. "4 Surveys Analyze Impact of Implementation of ERP on Credit and Receivables Functions," *Managing Credit, Receivables & Collections*, August 2007, accessed at www.ioma.com/credit.

18. "Small Businesses Missing ERP Benefits," *BusinessWeek*, January 5, 2007.

19. Board of Regents Meeting, February 2–3, 2006, Agenda Item 23, Arizona State University.

20. Associated Press, "New Philosophy School ERP Software: Try It First, Fix It Later," *Tech Briefs*, September 25, 2007.

21. Associated Press, "New Philosophy School ERP Software: Try It First, Fix It Later," *Tech Briefs*, September 25, 2007.

22. Associated Press, "New Philosophy School ERP Software: Try It First, Fix It Later," *Tech Briefs*, September 25, 2007.

23. Bob Violino, "The Next Generation ERP," *CIO Insight,* May 2008.

24. John Pallatto, "Microsoft Finally Upgrades ERP Suite," *eWeek*, June 9, 2008.

[25] Renee Boucher Ferguson, "Open-Source Enterprise Push," *eWeek*, January 7, 2008.

[26] "Open Source EERP Gets 600,000 Downloads," *worksmanagement.co.uk*, July 2008.

[27] Renee Boucher Ferguson, "Open-Source Enterprise Push," *eWeek*, January 7, 2008.

[28] "ERP Implementation Gets A+ Progress Report," *Food Engineering*, April 2008.

[29] "Enterprise 21 ERP Software," SAP Research Library, accessed at *http://searchsap.bitpipe.com/detail/PROD/1083592072_309.html*.

[30] "Corporate Information," Hunter Manufacturing Company Web site accessed at *http://www.huntermfgco.com/* on September 8, 2008.

[31] "Hunter" accessed at *http://www.sap.com/contactsap on September 8*, 2008.

[32] "Hunter Manufacturing Company: The Quest for a Strategic Asset Solution," *SMB News*, August 24, 2005.

250

CHAPTER 9

BUSINESS INTELLIGENCE

THE VALUE OF DATA MINING

"Data mining tools are very good for classification purposes, for trying to understand why one group of people is different from another. What makes some people good credit risks or bad credit risks? What makes people Republicans or Democrats? To do that kind of task, I can't think of anything better than data mining techniques... Another question that's really important isn't which bucket people fall into, but when will things occur? How long will it be until this prospect becomes a customer? How long until this customer makes the next purchase?... Data mining is good at saying will it happen or not, but it's not particularly good at saying when things will happen."

—Peter Fader, in a conversation with Allan Alter, "Business More Intelligently", *CIO Insight*, June 2007.

PAPA GINO'S ILLUSTRATES WHY MANAGERS MUST UNDERSTAND BUSINESS INTELLIGENCE

Papa Gino's started in 1961 as a single East Boston pizza shop called Piece O' Pizza, which provided a family atmosphere and authentic Italian food. The owners, Michael and Helen Valerio, changed the name to Papa Gino's in 1968 and began expanding throughout the Boston area. In 1991, the Valerio's sold the company to a group of private investors. In 1997, Papa Gino's Holdings Corporation, the parent company of Papa Gino's, acquired D'Angelo, a chain of sandwich shops. Today you can get appetizers, subs, salads, Italian specialties, and of course, pizza at 170 corporately-owned Papa Gino's Restaurants. Sandwiches, salads, wraps, and soups are available at 200 D'Angelo sandwich shops throughout New England.[1]

Papa Gino's management understood the importance of business intelligence (BI)—a broad set of software solutions that enables an organization to gain a better understanding of its critical operations through improved analytical tools and reporting capabilities. As Papa Gino's continued to grow, it recognized an opportunity to use BI to analyze data from its point-of-sale and general ledger systems to gain an in-depth understanding of its operations and customers and to enhance management decision making across its 370 restaurants. In addition, management needed reports on key performance metrics to enable decisive action by the individuals responsible for restaurant operations. Decision makers also needed to be able to spot and analyze key trends and take advantage of them to improve customer service and increase profitability.[2]

"The success of Papa Gino's and D'Angelo hinges on our commitment to providing high-quality products, attentive service, and a premium dining experience for every guest. We needed a performance management solution that could help us build on that success—optimizing our infrastructure and supporting our future growth in current markets," according to Louis Psallidas, senior vice-president of finance and chief financial officer.[3]

The firm began evaluating business intelligence tools in 2006. In early 2007, it chose the Cognos 8 product line and began installing the software. By late 2007, Papa Gino's was using BI tools to analyze customer delivery data. This area of the business was chosen for the initial BI application because roughly one-third of Papa Gino's business comes from customers placing home-delivery orders. According to CIO and vice-president of IT, Paul Valle, "Delivery is a key piece of our business, so anything we can do to monitor and measure and improve a third of our business is a huge thing to us."[4]

At each Papa Gino's restaurant, data from a point-of-sale system records what time an individual order was received, when the customer was promised delivery, when the employee making the delivery left the store, and when he or she returned. Before BI tools became available, that information was kept in spreadsheets and was difficult to analyze. Now Papa Gino's managers use BI tools to analyze that delivery data to improve delivery-time estimates. The goal is to enable order takers to accurately set customers' delivery expectations and possibly learn how to speed up deliveries. Papa Gino's store managers are also using BI to develop more accurate customer demand forecasts so they can get a better idea of how much product they need to order and how many workers they must schedule. The results of all this BI analysis are happier customers, reduced inventories of raw materials, and overall lower costs.[5]

LEARNING OBJECTIVES

As you read this chapter, ask yourself:

- What is business intelligence and how can it be used to improve the operations and results of an organization?
- What are some of the basic business intelligence tools and how are they used?

This chapter continues with a definition of business intelligence, and then discusses how data warehouses and data marts support business intelligence. It also outlines and provides examples of the use of several business intelligence tools, and covers an increasingly important application of business intelligence called business performance management.

WHAT IS BUSINESS INTELLIGENCE?

Business intelligence (BI) includes a wide range of applications, practices, and technologies for the extraction, translation, integration, analysis, and presentation of data to support improved decision making. The data used in BI is often pulled from multiple databases and may be internally or externally generated. Many organizations use this data to build a large collection of data called a data warehouse, or data mart, for use in BI applications. Users, including employees, customers, suppliers, and business partners, can access the data and BI applications via the Web, Internet, organizational intranets—even via mobile devices such as smart phones.

Organizations often employ BI to make predictions about future conditions and then make adjustments in staffing, purchasing, financing, and other operational areas to better meet forecasted needs. For example, the Federal Transportation Security Administration (TSA) implemented a BI system it dubs Performance Information Management System (PIMS) to manage more effectively and schedule its workers. It estimates that it reduced staffing and overtime costs by $100 million over a two-year period. PIMS gathers, analyzes, and summarizes passenger and baggage screening data to report operational performance metrics such as passenger wait times at checkpoints and the types and amount of unauthorized items collected from passengers during screening. PIMS also collects and reports payroll data, TSA staff utilization, and passenger complaints and compliments. The tool can even be used to analyze detailed data related to individual checkpoints at an airport. All of this analysis enables TSA to fine tune its staffing plans to minimize payroll costs while still meeting air travelers' expectations for a smooth check-in and safe air travel.[6]

Often the data analyzed by BI software comes from the vast amount of operational information captured by a company's ERP systems. Velsicol Chemical Corp, with world headquarters in Chicago, produces and sells adhesives, paint and coating products, flexible vinyl, food additives, and plasticizer solutions used in products as diverse as children's toys and personal care products. Velsicol adopted Information Builders' WebFocus for its BI software because it needed a program that would work seamlessly with its existing SAP ERP system. WebFocus provides reporting capabilities to support decision making by workers in distribution, finance, human resources, materials management, plant maintenance, production planning, quality management, and sales.[7] "We favored WebFocus because it works well with the SAP environment and has comprehensive security, data access, and drill-down and parameterized-reporting capabilities," according to Lee Goodrich, senior business systems analyst at Velsicol.[8]

Organizations must be extremely careful to protect the data they use in their BI applications. Petrobras is a Brazilian energy company with headquarters in Rio de Janeiro. It employs over 68,000 people, produces more than two million barrels of oil per day and is a major distributor of oil products. Recent annual revenue exceeded $101 billion. In what authorities are calling a case of industrial espionage, Petrobras had four laptops and two hard drives stolen in 2008. They contained secret and important information about a huge new oil ocean reservoir that in the next few years could produce up to eight billion barrels of oil.[9] (Organizations increasingly are prohibiting or at least limiting the amount and kinds of critical data that is stored on laptops to prevent just such a problem.)

The most widely used BI software comes from SAP, IBM, Oracle, and Microsoft. In 2007, three of these vendors each acquired a major BI player. Oracle acquired Hyperion; SAP acquired Business Objects; and IBM acquired Cognos. Vendors such as JasperSoft and Pentaho also provide open source business intelligence software, which is appealing to some organizations. Delta Dental, Lifetime Networks, Monsanto, Orbitz, and Sun Microsystems all have adopted Pentaho's open source software because they think it will help them to achieve their business goals more quickly and at a lower cost.[10]

BI tools frequently operate on data stored in a data warehouse or data mart. The next section will provide an overview of the concept of a data warehouse/data mart.

Data Warehouse/Data Marts

A **data warehouse** is a database that stores large amounts of historical data in a form that readily supports analysis and management decision making. Data warehouses frequently hold a huge amount of data—often containing five years or more of data. Wal-Mart built a data warehouse that contains some 2.5 petabytes (2.5 million gigabytes) worth of sales data generated from 800 million business transactions created each day. Wal-Mart uses the data warehouse and BI to determine the ideal mix of products to stock in each store, figure out how to stock product in shelves so as to maximize sales, and perform profit analysis relating to markdowns.[11] Other organizations with extremely large data warehouses include eBay (5 petabytes), Bank of America (1.5 petabytes), and Dell Inc. (1 petabyte).[12]

The data in a data warehouse typically comes from numerous operational systems and external data sources. An **extract-transform-load (ETL)** process is used to pull data from these disparate data sources to populate and maintain the data warehouse (Figure 9-1). An effective ETL process is essential to ensure data warehouse success.

The extract step in the ETL process is designed to access the various sources of data and pull from each source the data desired to update the data warehouse. For example, the extract process may be designed to pull only a certain subset of orders from the Orders database—say for orders that were shipped only after a certain date. During the extract step, the data is also screened for unwanted or erroneous values; data that fails to pass the edits is rejected. For example, the extract process may be designed to reject all shipped orders that are under a certain dollar value or that are shipped to certain geographical locations.

In the transform step in the ETL process, the data that will be used to update the data warehouse is edited and, if necessary, converted to a compatible format. For example, the store identifier present in a detailed transaction record (e.g., Home Depot on Glenway Avenue, Cincinnati, Ohio) may be converted to a less specific identifier that enables a useful aggregation of the data (e.g., Home Depot, Midwest Sales Region). Because the data comes from many sources (e.g., Access databases, Oracle databases, Excel spreadsheets, etc.), it often must be transformed into a format that can be handled easily in the load step.

The load step in the ETL process updates the existing data warehouse with the data that have passed through the extract and transform steps. This creates a new, updated version of the data warehouse.

The ETL process is run as frequently as necessary to meet the needs of the decision makers who use the data warehouse. Every organization must balance the cost and time required to update the data warehouse with the need for current data. Many companies update their data warehouse on a monthly or weekly basis; some execute the ETL process daily.

The Internal Revenue Service built a 150 terabyte Compliance Data Warehouse (CDW) that stores all tax returns and related information from the past 10 years. The IRS has about 500 researchers who use it to identify trends, flag those groups of taxpayers most likely to fall behind on their payments, and conduct simulations to analyze proposed tax changes. The CDW has even enabled IRS investigators to identify areas where tax cheating is prevalent such as tax shelters for small businesses and the Earned Income Tax Credit. When the CDW was first created in 1998, it required up to eight weeks to load a single year's worth of tax returns. Today, thanks to vast improvements in computer hardware and software, it takes only about four hours.[13]

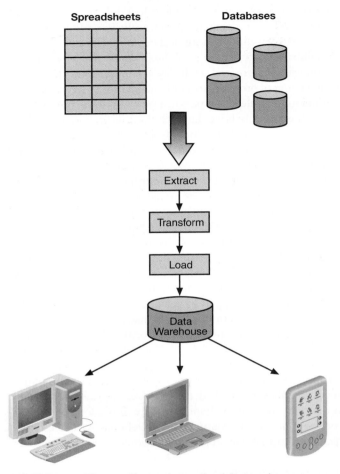

FIGURE 9-1 The creation and use of a data warehouse

A **data mart** is a smaller version of a data warehouse—scaled down to meet the specific needs of a business unit. Some organizations have multiple data marts, each designed to meet the needs of a different organizational business unit. Data marts are sometimes designed from scratch as a complete, individual, miniature data warehouse. Sometimes the data mart is simply created by extracting, transforming, and loading a portion of the data in a data warehouse.

Business Intelligence Tools

This section will introduce and provide examples of many BI tools including spreadsheets, reporting and querying tools, online analytical processing, drill-down analysis, data mining, and reality mining.

Spreadsheets

Business managers often import data into a spreadsheet program, which then can perform operations on the data based on formulas created by the end user. Spreadsheets are also used to create useful reports and graphs based on that data. End users can even employ tools such as the Excel scenario manager to perform "what if" analysis to evaluate various alternatives or Excel Solver to find the optimal solution to a problem with multiple constraints (e.g., determine a production plan that will maximize profit subject to certain limitations on raw materials).

Indiana Botanic Gardens is the world's largest mail-order seller of herbs. The firm uses software from Taurus Software to extract business data and translate it into a form that can be exported into Excel to create charts and reports to aid decision making. The Fulfillment/Inventory Manager was able to utilize the software to pull data into an Excel spreadsheet, which she used to compare various shipping methods (e.g., overnight, regular, air, and truck) to determine the impact of a shipping rate increase. Without the tools, she would have had to make a formal request for assistance from the IT department and wait perhaps weeks to get the results.[14]

Reporting and Querying Tools

Most organizations have invested in some reporting tools to help their employees get the data they need to solve a problem or identify an opportunity. Reporting and querying tools can present that data in an easy to understand fashion—via formatted data, graphs, and charts. Many of the reporting and querying tools enable end users to make their own data requests and format the results without the need for additional help from the IT organization.

PepsiCo is a global leader in convenience foods and beverages with recent annual revenue in excess of $25 billion and more than 142,000 employees. PepsiCo's Frito-Lay division implemented reporting and querying tools from Business Objects and deployed them to some 3000 users. They use these tools to analyze spending patterns and identify cost-savings opportunities, such as reducing the number of suppliers for basic raw materials and office supplies. The division has also used the tools to "provide vendors with an itemized statement that details each bill of lading, invoice number, the amount of each check, and grand total, via an extranet."[15]

Online Analytical Processing (OLAP)

Online analytical processing (OLAP) is a method to analyze multidimensional data from many different perspectives. It enables users to identify issues and opportunities as well as perform trend analysis. Databases built to support OLAP processing consist of **data cubes** that contain numeric facts called measures, which are categorized by dimensions such as time and geography. A simple example would be a data cube that contains the unit sales of a specific product as a measure. This value would be displayed along the metric dimension axis shown in Figure 9-2. The time dimension might be a specific day (e.g., September 30, 2012) while the geography dimension might define a specific store (e.g., Krogers in the Cincinnati, Ohio community of Hyde Park). Figure 9-2 depicts a simple three-dimensional data cube.

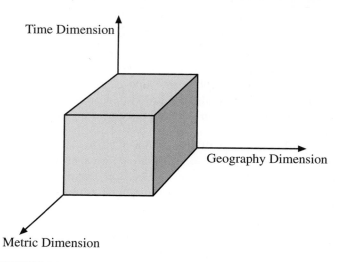

Time Dimension

Geography Dimension

Metric Dimension

FIGURE 9-2 A simple three-dimensional data cube

The key to the quick responsiveness of OLAP processing is the pre-aggregation of detailed data into useful summaries of data in anticipation of questions that might be raised. For example, data cubes can be built to summarize unit sales of a specific item on a specific day for a specific store. In addition, the detailed store level data may be summarized to create data cubes that show unit sales for a specific item, on a specific day for all stores within each major market (e.g., Boston, New York, Phoenix, etc.), for all stores within the United States, or for all stores within North America. In a similar fashion, data cubes can be built in anticipation of queries seeking information on unit sales on a given day, week, month, or fiscal quarter.

It is important to note that if the data within a data cube has been summarized at a given level, for example, unit sales by day by store, it is not possible to use that data cube to answer questions at a more detailed level, such as what were the unit sales of this item by hour on a given day.

Data cubes need not be restricted to just three dimensions. Indeed, most OLAP systems can build data cubes with many more dimensions. In the business world, we construct data cubes with many dimensions but usually look at just three at a time. For example, a consumer packaged goods manufacturer might build a multidimensional data cube with information about unit sales, shelf space, unit price, promotion price, level of newspaper advertising—all for a specific product, on a specific date, in a specific store.

In the retail industry, OLAP is used to help firms to predict better customer demand and maximize sales. For example, each week over 13 million customers visit one of Lowe's 14,000 stores, each of which carries more than 40,000 items. Point-of-sale scanners collect tens of millions of customer sales transactions each week.[16] Information about each item purchased is summarized into a data cube that depicts the sales of a specific item, on a specific day, for a specific store. Lowe's uses OLAP to track sales in order to forecast the right level of inventory to meet future customer demand. The amount of data and transaction processing power required is so great that Lowe's OLAP operation runs on some 3000 servers.[17]

Drill-Down Analysis

The small things in plans and schemes that don't go as expected can frequently cause serious problems later on—the devil is in the details! Drill-down analysis is a powerful tool that enables decision makers to gain insight into the details of business data to better understand why something happened.

Drill-down analysis involves the interactive examination of high level, summary data in increasing detail to gain insight into certain elements—sort of like slowly peeling off the layers of an onion to reach the core. For example, in reviewing the worldwide sales for the past quarter, the sales vice-president might want to drill down to view the sales for each country. Further drilling could be done to view the sales for a specific country (say Germany) for the quarter. A third level of drill-down could be done to see the sales for a specific country for a specific month of the quarter (e.g., Germany for the month of September). A fourth level of drill-down could be accomplished by looking at the sales by product line for a particular country by month (e.g., each product line sold in Germany for the month of September).

The Federal Election Commission (FEC) is responsible for administering U.S. election laws regarding campaign funding. The laws cover set limits and define rules regarding donations by individuals, political action committees, and campaign groups. The FEC developed a set of database tools that enables it to click on a map of the United States to drill down into campaign contributions by state, zip code, or candidate name. The tools are also available for use by the public so that any citizen can learn who is funding presidential campaigns.[18]

Texas Assessment of Knowledge and Skills (TAKS) exams are taken each year by Texas public school students in grades 3 through 11. These tests assess the level of improvement for each school district by grade and subject. The test result data is combined with neighborhood and school district community demographic data. Specially designed drill down analysis tools are used to examine the data at a region, school district, or individual school level. The overall goal is to evaluate and plan instruction for the upcoming school years. Often the data is used to answer specific questions such as:

- Which districts in the state (or a region) have performance levels above the state average for a specific subject?
- How do the median household income levels and performance results within one district compare to another district?
- Which districts or schools are exhibiting steady improvement? Which are declining?

Data Mining

Data mining is a BI tool used to explore large amounts of data for hidden patterns to predict future trends and behaviors for use in decision making. Used appropriately, data mining tools enable organizations to make predictions about what will happen so that managers can be proactive in capitalizing on opportunities or avoiding potential problems.

Among the three most commonly used data mining techniques are association analysis (a specialized set of algorithms sorts through data and forms statistical rules about relationships among the items), neural computing (historical data is examined for patterns that are then used to make predictions), and case-based reasoning (historical if-then-else cases are used to recognize patterns).

The process of data mining consists of three primary processes: data repository creation, pattern recognition, and deployment. Data repository creation begins with an ETL process to create an appropriate set of data to support the data mining technique to be used. Pattern recognition involves trying various models and selecting the best one based on its ability to explain the variability in the existing data. Deployment of the model involves use of the model selected to generate estimates of the outcome based on new data.

Here are a few common examples of data mining:

- Based on past responses to promotional mailings, identify those consumers most likely to take advantage of future mailings.
- Examine retail sales data to identify seemingly unrelated products that are frequently purchased together.
- Monitor credit card transactions to identify likely fraudulent requests for authorization.
- Use hotel booking data to adjust room rates so as to maximize revenue.
- Analyze demographic data and behavior data about potential customers to identify those who would be the most profitable customers to recruit.
- Study demographic data and the characteristics of an organization's most valuable employees to help focus future recruiting efforts.
- Recognize how changes in an individual's DNA sequence affect the risk of developing common diseases such as Alzheimer's or cancer.

The doctors with HMO Sentara Health System suspected they had a fundamental health care delivery problem—nearly 12 percent of their pneumonia patients died and many pneumonia patients required a hospital stay of two weeks or longer. The HMO's quality improvement team used data mining techniques to analyze claims data in a data mart. The analysis revealed that, surprisingly, doctors were ordering multiple sputum cultures for patients. Upon further investigation, it was uncovered that there was such a significant delay in getting test results back from the lab that doctors were ordering second and third cultures because they were concerned that the first culture had been lost. The health of the pneumonia patients deteriorated while doctors waited for the lab results. The quality improvement team implemented a streamlined process that dropped the lab turnaround time to just two hours. The result was that the mortality rate fell to 9 percent (significantly lower than the national average of 14 percent) and the average hospital stay was reduced to one week.[19]

Cablecom, a Swiss telecom company uses data mining and online surveys to identify those customers at high risk of moving to a competing firm. The data mining software analyzes customer problem ticket data such as the average duration of each customer's tickets and number of tickets the customer generated in the last month. In addition, Cablecom offers customers a satisfaction survey in the seventh month of service (previous research shows customer dissatisfaction often begins around the ninth month of service). Combining the results of its data mining efforts with the results of the surveys provides Cablecom with an accurate picture of customers at risk. The firm can then elect to take a number of various actions in an attempt to retain the customer.[20]

There are numerous privacy concerns associated with data mining, especially concerning the source of the data and the manner in which the data is gathered. The National Security Agency (NSA) initiated a massive data mining program following the 9/11 attacks. The White House asked the major telephone carriers for their assistance in providing call detail records on U.S. phone numbers found in captured al Qaeda laptops and cell phones. It was hoped that useful patterns would emerge and future terrorist plots could be thwarted. The government bypassed established legal procedures to collect this data for the sake of speeding up the process. Many of the carriers went along with the expedited process. As time went by, the request for call records expanded to two or three calls removed from the original requests with the number of records involved in the millions. Eventually, the NSA also gained access to wire transfers, bank transactions, and other personal financial data.[21] There are many disturbing and, as yet, unanswered questions about the means used by the government to gather this data and the volume of data gathered.

Reality Mining

Reality mining is the study of human interaction based on data gathered from mobile phones and other portable communicating devices.[22] For example, each time we use our cell phone, we leave behind bits of information—our location, number called, and the length of call.

Inrix is a small company that uses GPS-enabled mobile phones and tracking devices installed on nearly one million commercial vehicles in 129 cities to gather real-time traffic congestion data. The data is used to provide live traffic information to vehicle navigation devices made by Garmin and TomTom.

Researchers are exploring the use of reality mining for many other applications:[23]

- By analyzing where and with whom people spend their time, scientists can improve computational models of how communicable diseases spread.
- Speech analysis software running on the cell phone can look for changes in the user's speech pattern that could be an early indicator of a health problem. For example, depressed people frequently speak more slowly.
- All cell phones have built-in microphones that can be used to analyze your tone of voice, how long you talk, and how often you interrupt people. These patterns can tell what roles people play in groups, for example, if they are a leader or a follower.

BUSINESS PERFORMANCE MANAGEMENT

Business performance management (BPM) is an application of BI that enables the continuous and real-time analysis of operational data to measure actual performance and forecast future performance. BPM creates improved feedback loops for the critical processes of the organization so that problems can be identified and eliminated before they become serious. BPM also can be used to model work processes and predict future performance under various what-if scenarios. This enables business managers and analysts to modify existing work processes to better achieve organizational goals.

This section will discuss and provide examples of the use of the Balanced Scorecard and the dashboard—two commonly used BPM tools. Next, various types of BPM software will be discussed including monitoring software, workflow designer software, and reporting and insight software. Also some of the BPM software vendors will be identified. Finally, a general process to apply BPM to improve a business process will be discussed.

Balanced Scorecard

The Balanced Scorecard is a performance management tool used to track performance over time, communicate and drive organizational strategy, identify strategic initiatives, and conduct periodic performance reviews to assess if goals are being met. It was introduced by Robert Kaplan and David Norton in the early 1990s as a means of translating an organization's vision and strategy into implementation activities working from four perspectives—financial, customer, business process, and learning and growth.[24]

- **Financial**—This perspective defines the long-term strategic objectives of the organization in traditional financial terms using measures such as revenue growth, costs, profit margins, cash flow, and net operating income.
- **Customer**—This perspective identifies what must be done from the customer's perspective in order to meet organizational objectives using measures such as percent of on-time deliveries; value of customer claims for over, short, and damaged shipments; and customer retention rate.
- **Business Process**—This perspective defines those internal business processes that indicate how well the business is running and whether its products and services are meeting customer requirements. These metrics must be designed carefully by those that know and are most involved in these processes. They might include such measures as number of process bottlenecks and degree of process automation.
- **Learning and Growth**—This perspective describes the employee training and corporate cultural attitudes required for both individual and corporate self-improvement. Measures might include number of training opportunities taken by employees, number of mentors and tutors available, and employee turnover.

For each perspective on the balanced scorecard, four things are defined and monitored: objectives, measures, targets, and initiatives undertaken to meet the object. As discussed in Chapter 2, measures are metrics that track progress in executing chosen strategies to attain organizational objectives and goals. These metrics are also called **key performance indicators (KPIs)** and consist of a direction, measure, target, and time frame. To enable comparisons over different time periods it is also important to define the KPIs and to use the same definition from year to year. Over time, some existing KPIs may be dropped and new ones added as the organization changes its objectives and goals. Obviously, just as different organizations have different goals, various organizations will have different KPIs. Here are a few examples of well defined KPIs.

- For a university—increase (direction) the five year graduation rate for incoming freshman (measure) to at least 80 percent (target) starting with the graduating class of 2014 (time frame).
- For a customer service department—increase (direction) the number of customer phone calls answered within the first four rings (measure) to at least 90 percent (target) within the next three months (timeframe).

- For an HR organization—reduce (direction) the number of voluntary resignations and terminations for performance (measure) to six percent or less (target) for the 2011 fiscal year and subsequent years (time frame).

Effective use of a Balanced Scorecard can provide the following benefits: greater customer satisfaction, improved financial results, more effective information systems and business processes, and more motivated and better educated employees.

The mission of the Illinois Department of Transportation (IDOT) is to provide safe, cost-effective transportation in ways that enhance quality of life, promote economic prosperity, and demonstrate respect for our environment. A KPI linked to the public safety portion of its mission relates to the number of motor vehicle fatalities per 100 million vehicles driven in the state. This number has decreased from a high of 2.4 in 1987 to a low of 1.18 in 2006—below the national average of 1.46.[25]

A key idea behind the Balanced Scorecard is that financial measures tend to be lag measures, which are difficult to influence directly without addressing issues in the other three perspectives (indicated by the arrows between the various perspectives depicted in Figure 9-3). Customer, business, and learning and growth measures are considered leading indicators of future financial results. If customers are not satisfied, they eventually will find other suppliers that will meet their needs resulting in an eventual negative financial impact on the firm. If business processes are not effective and efficient, it will put the organization at a distinct disadvantage relative to competitors. If learning and growth measures are not met, the employees will become less skilled and dissatisfied.

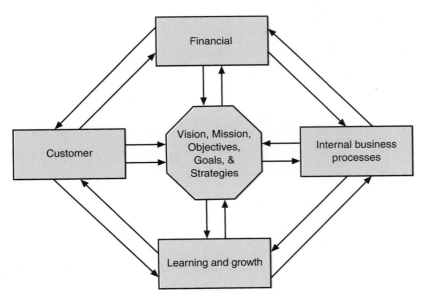

FIGURE 9-3 Balanced Scorecard template

Dashboards

Digital dashboards present a set of key performance indicators about the state of a process at a specific point in time. The ability to have rapid access to information, in an easy-to-interpret and concise manner, helps organizations run more effectively and efficiently.

A wide range of options for displaying results in a dashboard include maps, gauges, bar charts, trend lines, scatter diagrams, and other representations as shown in Figures 9-4 and 9-5. Often items are color coded (e.g., red = problem; yellow = warning; and green = OK) so that users can see at a glance where attention is needed. Many dashboards are designed in such a manner that users can click on a section of the chart displaying data in one format and drill down into the data to gain insight into more specific areas. For example, Figure 9-5 represents the results of drilling down on the sales region of Figure 9-4.

Dashboards provide users at every level of the organization the information they need to make improved decisions. Operational dashboards can be designed to draw data in real time from various sources including corporate databases and spreadsheets, so decision makers can make use of up-to-the-minute data.

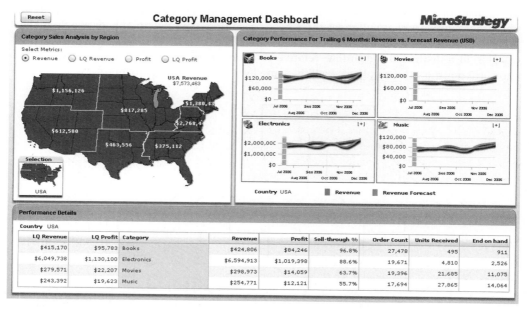

FIGURE 9-4 Sample summary dashboard

Welch's is a marketer of over 400 grape-based consumer products including grape juice, jams, jelly, preserves, and other fruit-based products. These products are broadly distributed throughout the U.S. and more than 50 countries worldwide. The firm contracts with many different freight carriers to make some 50,000 customer shipments each year. Welch's logistics managers implemented a BI system to better manage its multi-million dollar transportation expenses. Each night logistics data is loaded into the BI system from the firm's ERP system and freight audit and payment system. The data includes customer

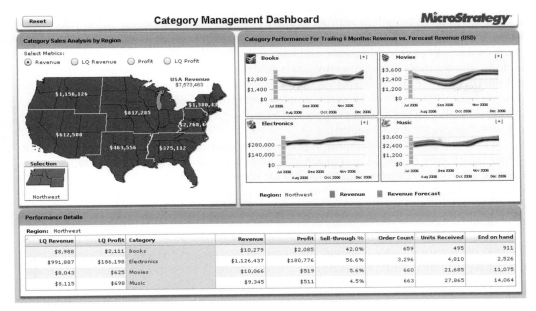

FIGURE 9-5   Sample drill-down results for one sales region

orders and freight carriers' bills of lading that provide details for each shipment including origin, destination, product weight, and product shipping cost. Overnight the data is processed, reports are produced, and dashboards refreshed with the latest data. One dashboard presents summary data in the form of a map that shows freight movements and associated costs. Managers can study the map to identify potential cost savings opportunities and make changes where warranted.[26]

BPM Software

There is a wide range of software available to support organizations in their BPM efforts including BPM efficiency monitoring BPM software, workflow designer software, and reporting and insight BPM software.

Process efficiency monitoring BPM software provides built-in application programming interfaces to connect with each of the systems that a company uses to support a particular process and then monitors the process to identify bottlenecks and inefficiencies. Once problems are uncovered, it is up to the organization to implement the necessary process changes.

AIC is Canada's largest privately held mutual fund company with roughly $12 billion U.S. in assets under management. The firm used BPM monitoring software to identify the need to speed up its process for updating clients' accounts whenever a transaction occurs. Now instead of waiting for an overnight batch processing job to update clients' accounts, a Web service updates the accounts in real time so that current information is always available.[27]

Workflow designer BPM software enables business managers and analysts to design a business process complete with all of the associated forms, business rules, role definitions, and integration to other systems involved in the process. For example, such software could help people design an efficient and effective invoice-payment process that includes the business rule that no invoice is paid without a visual comparison to the original purchase order and receiving report in order to detect possible discrepancies. Workflow designer BPM software typically employs an intuitive, easy-to-use graphical user interface for the business users to apply to design the process. Such software facilitates rapid process changes for greater agility.

Wyeth is a global manufacturer of prescription pharmaceuticals, non-prescription consumer health care products, and pharmaceuticals for animal health. Its products are sold in more than 145 countries, and its worldwide resources include 47,500 employees with manufacturing facilities on four continents.[28] For a company like Wyeth, research and development to find new drugs that will lead to patents and profits is critical to its future growth and success. Wyeth is using BPM software on a number of projects in the research and development arena to define and improve business processes with the goal of improving the collaboration and innovation among its various geographic and organizational units. One of Wyeth's BPM initiatives is directed at improving the process for creating the medical labeling documents that the FDA requires to be placed inside each bottle or package. The document includes a molecular diagram that shows the composition of the medicine, explains any restrictions on its use, and outlines all known drug interactions. [29]

Reporting and insight BPM software gather data from a business process and provide reports and dashboards to create actionable information. This software provides a wide range of BI capabilities such as trend analysis and drill-down analysis for users to gain complete visibility across all of an organization's business processes. Real-time event management and Business Activity Monitoring (BAM) capabilities ensure that users can identify and manage the highest priority items quickly.

Pitt Ohio Express, a Pittsburgh-based freight carrier with 21 freight terminals serving the Midwest and Mid-Atlantic region, delivers over 10,000 shipments daily with an admirable 97 percent on-time delivery rate. Gross revenue has exceeded $200 million since 2003.[30] The firm uses BPM reporting and insight software to provide a series of reports and charts to gain full visibility into its business operations. Users can request a report for any customer depicting monthly sales totals, year-over-year revenues, tonnage, shipments, and on-time percentage. The software can also produce "trigger reports" that show a sales representative's complete book of business and are color coded to show at a glance which customer performance parameters are above, within, or below a target value.[31]

Some of the leading BPM software vendors include BEA Systems (acquired by Oracle), Cognos (acquired by IBM), Metasystems, Pegasystems, Software AG, and Tibco. Most of these vendors provide all the types of BPM software mentioned previously.

Employing the BPM Process

BI is often used to augment the traditional four-step Plan-Do-Check-Act problem solving process as outlined below.

- **Plan**—In the Plan step, the problem-solving team selects the problem to be analyzed, clearly defines the problem, sets a measurable goal for the problem-solving effort, and gathers further data related to the problem. BI can be used to gather basic data about the operations of the organization from various sources. BI applications access this data to provide standard operational and managerial "report cards" on the current state of the business. End users of the report cards identify exceptions that represent unusual performance situations that need attention.

- **Do**—In the Do step, the problem-solving team establishes criteria for selecting a solution, generates solution alternatives, selects a solution, and plans how to implement the solution. BI tools can analyze the data to identify the root causes behind the identified exceptions. BI can also be used to develop a model to simulate the impact of different solution alternatives.

- **Check**—In the Check step, data is gathered to see if the implemented solution has solved the problem and achieved the desired goal set in the Plan step. If the solution is not acceptable, then the team goes back to the Plan step. If the solution was successful, then the team proceeds to the Act step. BI can be used to gather additional data to evaluate the effectiveness of the recommended actions.

- **Act**—In the Act step, the problem-solving team identifies systemic changes and training needs to ensure a full, successful implementation of the solution. The solution continues to be monitored to ensure that it remains effective. Just as in the Plan step, BI can be used to gather basic data about the operations of the organization from various sources, provide reports, and enable users to identify exceptions that represent unusual performance situations that need attention.

Lowe's is a big believer in the use of BI, and it used the process outlined above to improve its cash flow by more than $30 million per year. Lowe's generates some 170,000 reports per week for its employees and managers at roughly 1000 of its suppliers. Lowe's uses operational data and BI tools to identify improvement opportunities at each of its 1500 stores—from failing to collect delivery fees to analyzing the effectiveness of the more than 4000 quantity discount programs it has in place at any one time. Lowe's follows the P-D-C-A cycle to implement the necessary changes to address the issues it uncovers.[32]

An effective process for applying BPM involves the following steps depicted in Figure 9-6. First, work with business process owners, customers, business partners, and other appropriate stakeholders to identify a specific, high priority process that is sorely in need of improvement. Second, study the existing process to learn how it works. You may elect to use process efficiency BPM software in your analysis. It is important to determine the KPIs for this process in the evaluation. Third, the business process improvement team, process owner, and other stakeholders need to figure out how to change the process and redesign it accordingly. Workflow designer BPM software can come in handy here to evaluate various change alternatives. Fourth, implement the new process and monitor the new process to see if it is performing as expected and yielding the desired results. Advanced reporting and insight BPM software can be used to monitor the new process. If the new process is not working well, the business process improvement team goes back to step two and repeats the process.[33]

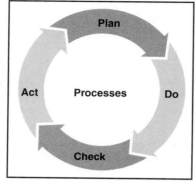

Gather basic data
about the process

Gather/analyze data
to monitor process

Analyze data to identify
root cause of problem

Gather/analyze data to determine
effectiveness of change

FIGURE 9-6   Using BPM software to augment the classic P-D-C-A problem-solving approach

Read the special feature to see a successful example of the effective application of the BPM process.

A MANAGER TAKES CHARGE

Qwest Uses BPM Software to Improve Operations

Qwest Communications International, Inc., (Qwest) is a provider of voice, data, Internet, and video services. The firm has 35,000 employees with recent annual operating revenue of nearly $14 billion. Its headquarters is in Denver, CO. Most of its business is generated from within its local service area that consists of 14 western U.S. states.[34]

Provisioning operations is a critical activity for all telecommunications providers and includes the following business processes: customer requirements capture, order data collection, price and offer management, and work distribution. Qwest's provisioning operations were highly fragmented—so much so that the various business processes were supported by "small islands of automation that had to be bridged by manual processes. [This resulted in] many people doing a lot of untracked, unmeasured, manual work that led to inefficiencies and a sore lack of process standardization."[35] As a result, Qwest was experiencing higher than necessary operating costs, rework, and missed business opportunities.

To address these problems, Qwest established a centralized BPM team to "set and document standards, facilitate change control, and manage the company's business process repositories."[36] The BPM team collaborated with already established business process improvement teams in each line of business to identify two business processes needing immediate attention—order data collection and price and offer management. Together they also identified two KPIs to measure the improvement results—held order rates (work delayed due to insufficient or incomplete information) and cycle time (the total time to complete a business process).

continued

The team recognized that BPM software could help in their analysis of the ailing business processes. They evaluated several BPM software packages to "help optimize, standardize, and improve its business processes." [37] The BPM vendor was selected only after demonstrating a system prototype that met the firm's needs.

The BPM team collected existing data from its provisioning operations business processes for input to the BPM software's business process modeling and simulation modules. The BPM team was then able to identify and evaluate possible changes in the processes. Wisely, the BPM team distributed business-process modeling and simulation tools broadly and encouraged those actually doing the work to look for business process improvement opportunities as well. By taking this approach, the BPM team was able to build trust and buy-in for business process change and build support for its recommendations.

Upon implementation of the recommended changes, Qwest quickly experienced measurable improvements in both the order data collection and price and offer management processes. Held orders dropped by 10 percent and there was a 20 percent reduction in the price and offer management process cycle time. Being able to show measurable benefits created a positive environment and the willingness of employees to participate in further business process improvement efforts. [38]

Discussion Questions:

1. Outline a reasonable process that would enable the BPM team to single out the order data collection and price and offer management business processes as the place to start. Who should be involved in this process?

2. Discuss the importance of knowing the KPIs for a business process when trying to implement improved business processes.

3. In trying to improve these two processes at Qwest, what sort of data would prove most useful for input to the BPM software's business process modeling and simulation modules? What if such data were not available at the start of the study?

This chapter has outlined a number of different business intelligence tools and discussed how they can be combined with the traditional P-D-C-A business problem solving approach. Table 9-1 recommends a set of actions that managers can take to be effective problem solvers and ensure that BI tools are used appropriately. The appropriate answer to each question is "yes."

TABLE 9-1 A manager's checklist

Recommended Management Actions	Yes	No
Do you take the time to ensure that you fully understand the problem before recommending a solution process and associated problem-solving tools?		
Do you use a general problem-solving approach such as P-D-C-A?		
Do you include the problem stakeholders in the solution process?		
Do you consider the use of information from a data warehouse in your analysis?		
Do you consider the use of BI tools in your analysis?		

Chapter Summary

- Business intelligence includes a wide range of applications, practices, and technologies for the extraction, translation, integration, analysis, and presentation of data to support improved decision making.

- An extract-transform-load (ETL) process is often employed to gather data from multiple sources to create data warehouses for use with BI tools.

- Spreadsheets, report and query tools, online analytical processing, drill-down analysis, data mining, and reality mining are examples of commonly used BI tools.

- Business performance management is an increasingly important application of BI used to measure the actual performance or forecast the future performance of critical operations of an organization.

- The balanced scorecard is a performance management tool that tracks performance over a period of time from four perspectives—financial, customer, business process, and learning and growth.

- A dashboard presents a set of key performance indicators about the state of a process at a specific point in time.

- Business performance management is often used in conjunction with the traditional P-D-C-A business problem solving process.

Discussion Questions

1. How would you define business intelligence? Do research on the Web to identify two recent real-world applications of BI that are of interest to you.

2. What is a data mart? In what ways is it different from a data warehouse? Imagine that you are a member of the customer service organization for a large retail business that sells products over the Internet and through a nationwide network of stores. How might you employ a data mart and BI tools to improve the performance of your organization?

3. Briefly describe the ETL process for building a data warehouse. Provide two examples of the data transform step.

4. As discussed in this chapter, the IRS maintains a large data warehouse of 10 years of tax return data. What other types of data warehouses do you think the federal or state government might have pertaining to U.S. residents? What purposes might these serve?

5. Define the term reality mining. Would you permit your primary care physician to install speech analysis software on your cell phone to monitor your speech pattern for early indications of diseases, such as depression or Parkinson's? Why or why not?

6. What is the difference between OLAP analysis and drill-down analysis? Provide an example of the effective use of each technique.

7. Identify at least four key performance indicators that could be used to define the current state of operations of a fast food restaurant. Sketch what a dashboard displaying those KPIs might look like.

8. Outline the P-D-C-A process for business problem solving. Identify someone in your class who has employed this technique. Ask them to describe the problem that was addressed

and to provide an assessment of what worked well and what did not work so well in using this process.

9. What do you think are some of the strengths and weaknesses of combining BPM software and the P-D-C-A process to solve a business problem?

10. Your non-profit organization wishes to increase the efficiency of its fund drive efforts. How might data mining help to increase the amount of donations with a decrease in volunteer effort?

Action Memos

1. You are the sales manager of a software firm that provides BI software for reporting, query, and business data analysis via OLAP and data mining. Your sales reps have asked you to prepare a paragraph they can use when calling on potential customers to help them understand the business benefits of BI.

2. You are the new operations manager of a large customer call center for a multinational retailer. The call center has been in operation for several years, but has failed to meet both the customers' and senior management's expectations. You were hired three months ago and challenged to "turn the situation around." As you are sitting at your desk one day, you get a phone call from your boss asking that you lead a pilot project to implement the use of dashboards in the call center. The goal is to demonstrate the value of dashboards to help monitor and improve the operations in many of the firm's business units. You, along with over 20 middle managers, just completed a two-day in-house training course of the use of dashboards. How do you respond to your boss's request?

Web-Based Case

Go online to read the current annual report of a company in which you are interested. Identify the organization's objectives, measures, targets, and key initiatives. Develop a balanced scorecard for that company.

Case Study

Blue Mountain Resorts Makes Effective Use of BI

Intrawest ULC develops and manages experiential destination resorts in some of North America's most popular vacation destinations and ski resorts (see Table 9-2). Its headquarters is in Vancouver, British Columbia, and it employs 22,000 employees. In 1999, Intrawest bought a 50 percent interest in Blue Mountain, Ontario's largest mountain resort. Since then Intrawest has financed several development projects including construction of Blue Mountain Village built at the base of the Silver Bullet chairlift, the addition of four high speed six-passenger chairlifts, and the creation of a new conference center, as well as a large number of condominium and hotel units.

TABLE 9-2 A partial list of Intrawest ULC North American properties

Venue	Comment
Canadian Mountain Holidays	The world leader in heli-skiing with remote mountain lodges accessible only by helicopter.
Compagnie des Alpes, France	Intrawest owns 16.5 percent of Compagnie des Alpes, the largest ski company in the world in terms of skier visits.
Intrawest Golf, Arizona	Intrawest Golf owns or manages more than 20 golf courses in the United States and Canada.
Mammoth, CA	One of CA's favorite mountain playgrounds with 36 lifts and over 30 feet of snow each winter. Summer activities include an 18-hole championship golf course, mountain biking, and fishing.
Mountain Creek, NJ	A four-season resort with a base-of-the-mountain village, 44 trails, state-of-the-art snowmaking system, and 11 lifts. In the summer and fall, it transforms into a mecca of extreme sports and water park rides.
Panorama Mountain Village, British Columbia	Boasts the highest vertical in the Canadian Rockies region with four distinct neighborhoods linked by pathways and a village gondola, a 6,000-sq. ft. heated slope-side water park and Greywolf, an 18-hole championship golf course awarded Best New Golf Course in Canada.
Snowshoe, WV	Nearly a mile high in the Allegheny Mountains of West Virginia with superior regional conditions, and challenging terrain.
Steamboat, CO	Also known as Ski Town USA, Steamboat is an internationally known winter resort destination.

Blue Mountain Resorts has 34 ski/snowboarding trails and a snow tubing park served by 12 ski lifts. The resort is a two hour drive from Toronto, Canada, and it is the third busiest ski resort in Canada with some one million guests each winter. It employs 320 full time, year-round workers with an additional 1150 winter and 150 summer seasonal workers. During the warm months, Blue Mountain Resorts guests can enjoy caving, rock climbing, sailing, fly-fishing, mountain biking, eco-adventures, golf, tennis, canoeing, and even off-road Hummer adventures that take them into the backcountry.[39]

Before the acquisition by Intrawest, Blue Mountain Resorts relied on a simple spreadsheet-based system and an IT staff of just three people to manage its operations. The system was not automated and required a lot of manual effort to gather data from the resort's 13 different business operations including call centers, conference rooms, a golf course, lodging, restaurants, skiing, and tennis. Reports were often delayed due to difficulties gathering and processing the disparate data.[40]

At the time of the Blue Mountain acquisition, Intrawest was a publicly held company and needed an easier, more efficient way to draw on the various data sources to gain a current and accurate picture of costs and revenue. (Intrawest is now privately held.) It was clear that the old methods and system needed to be replaced.

John Gowers, Director of Information Technology and David House, Revenue Manager, were the primary drivers behind the project to find a comprehensive software package that could do traditional budgeting, consolidation, reporting, and forecasting as well as performance analysis on revenue trends, especially daily sales. Together they defined a set of requirements for the solution.

Because Blue Mountain had only a three person IT organization, it was imperative that the software be easy to install, require minimum customization and maintenance, and make it possible for end users to generate their own reports without extensive IT support. The ideal solution would include software modules specifically designed for not only the hospitality industry, but also for the different business operations (call center, conferences, caterings, etc.).[41]

Gowers and House performed a thorough review of many software package options. The review included software demonstrations and even on-site pilot projects for the leading contenders. Following this evaluation, Applix's performance management solution TM1 was selected. This software streamlines the performance management process and includes planning, budgeting, forecasting, reporting, and analysis.[42] TM1 operates using OLAP technology and enables analysts, managers, and executives to view data across multiple dimensions and drill down into the specific data underlying those summaries. Shortly after Blue Mountain Resort chose the Applix Business Intelligence application TM1, Applix was acquired by the Canadian-based Cognos software firm. Then IBM acquired Cognos a few months after that. As a result, the full name of the software is now IBM Cognos TM1.

The TM1 implementation team consisted of two people from Blue Mountain and an IBM Cognos TM1 consultant. Their approach to implementation was to tackle each line of business one at a time. It took five months to deploy the solution first in the lodging side of the business and then across the skiing, rental, golf, and conference room businesses. [43] Lodging was the logical place to begin given the huge number of guests that visit Blue Mountain each year and the fact that lodging is one of Blue Mountain's main lines of business. In lodging, the challenge is to price rooms at a level that maximizes capacity utilization and growth margin.

For each business area, data cubes of key performance indicators were created so that managers could combine historical data, current weather, booking information, and employee schedules. The data then could be manipulated to perform detailed analyses and determine ways to minimize costs while optimizing customer service. As a side benefit of building the data cubes across each line of business, the TM1 implementation team was able to standardize how basic operational data is captured. This simplified the data capture process and made it possible for managers to compare results across various departments. During the deployment, 10 Blue Mountain employees received formal IBM Cognos TM1 training.[44]

With TM1, all of the budgeting information is prepared by line of business and resides in a single database. As the daily operating data is captured, it is possible to see how much actually is being spent for each business area. This provides management with a clear picture of each department's performance. Furthermore, because the TM1 software updates revenue in real time and can slice and dice data in any number of ways, executives at Blue Mountain can compare year-to-date actual values to projected revenues and to the previous year's totals every day. This daily snapshot and compilation of trends enables senior managers to intervene quickly should corrective action be needed in a particular department. [45]

As with any resort, staffing is a major component of Blue Mountain's operating budget. Effectively managing this line item is critical to the profitability of the resort. While approximate staffing forecasts are created during the budgeting process, actual and forecasted staffing levels must be adjusted daily depending on the weather, number of pre-sold tickets, hotel arrivals and departures, major conferences booked for the resort, and historic business patterns. The TM1 system is used to build six-day forecasts of the number of guests and guest activity across the resort. This enables line-of-business managers to adjust their worker schedules quickly to meet anticipated demand and move workers from one department to another. It is estimated that this capability has reduced Blue Mountain staffing costs by $2.5 million per year. [46]

TM1 is also used to track the age and turnover rate of all items carried in Blue Mountain's retail stores. Slow moving items are marked down for quick sales so that the company is able to reduce retail inventory levels and make room for faster moving, more profitable items. Blue Mountain also uses TM1 to analyze the use of the various ski boot sizes that are available for rental. This enables the resort to correctly order more of the frequently used sizes and less of the infrequently used sizes. This again reduces overall inventory levels and ensures that visitors will be able to rent the boots they need to enjoy their skiing. [47]

For most guests, their first interaction with Blue Mountain Resort is through the call center. If potential guests must wait on hold too long before they are helped, they will try elsewhere to book a family vacation or business conference. The call center managers use TM1 to estimate inbound call volumes based on time of year, time of day, proximity to a given holiday, and current resort promotions. Call center staffing levels can be set so that there is a sufficient number of call center associates to ensure no bookings are lost due to poor customer service. [48]

Table 9-3 summarizes the costs and benefits associated with the implementation of the TM1 BI software at Blue Mountain. (All numbers in thousands of dollars; assumes 50 percent tax rate.)

TABLE 9-3   Financial analysis for Blue Mountain Resort's TM1 project

Expensed Costs	Pre-start	Year 1	Year 2	Year 3
Direct	$0	$2500	$2500	$2500
Indirect	0	130	130	130
Depreciated Assets	**Pre-start**	**Year 1**	**Year 2**	**Year 3**
Software	$40	0	0	0
Hardware	0	0	0	0
Depreciation Schedule	**Pre-start**	**Year 1**	**Year 2**	**Year 3**
Software	$0	$8	$8	$8
Hardware	0	0	0	0
Expensed Costs	**Pre-start**	**Year 1**	**Year 2**	**Year 3**
Software	$0	$8	$8	$8
Hardware	0	0	0	0

TABLE 9-3 Financial analysis for Blue Mountain Resort's TM1 project (continued)

Expensed Costs	Pre-start	Year 1	Year 2	Year 3
Consulting	2	0	0	0
Personnel	41.5	0	0	0
Training	20.8	0	0	0
Other	0	0	0	0
Financial Analysis	**Pre-start**	**Year 1**	**Year 2**	**Year 3**
Net cash flow before taxes	($104.3)	$2622	$2622	$2622
Net cash flow after taxes	(72.2)	1315.0	1315.0	1315.0

Source: ROI Case Study IBM Cognos TM1 Blue Mountain Resorts, *Nucleus Research*, May 2008.

Discussion Questions

1. Intrawest is a partial owner of several other resorts. What issues might arise if the other resorts have different tools and processes for providing budget and operational performance data? Do you think Intrawest should attempt to have all the resorts standardize the use of the TM1 software?

2. Imagine that you are the manager of the golf line of business for Blue Mountain Resort. What sort of data and reports could you use from the TM1 system to help you better manage your business?

3. Due to a lack of snow and mild temperatures, Blue Mountain Resort was forced to close during the normally highly busy Christmas period in December 2006. What sort of data about the weather do you think is stored in the TM1 database? What additional data not directly related to the resort would be useful in forecasting the number of guests?

4. Blue Mountain Resort has not implemented an ERP system. Do you think there is sufficient business justification to do so? Why or why not?

Endnotes

[1] "Who We Are," Papa Gino's Web site accessed at *www.papaginos.com/corporate/who_we_are_pg.html* on September 21, 2008.

[2] "PapaGino's, D'Angelo Select Cognos as Their Enterprise Performance Management Solution," accessed at Cognos Web site at *www.cognos.com/news/releases/2007/0507.html* on September 21, 2008.

[3] "PapaGino's, D'Angelo Select Cognos as Their Enterprise Performance Management Solution," accessed at Cognos Web site at *www.cognos.com/news/releases/2007/0507.html* on September 21, 2008.

4   Patrick Thibodeau, "BI Makes Pizza Delivery More Efficient," *Computerworld*, February 18, 2008.

5   Patrick Thibodeau, "BI Makes Pizza Delivery More Efficient," *Computerworld*, February 18, 2008.

6   Heather Havenstein, "TSA Leans on BI to Save $100 Million," *ComputerWorld*, July 23, 2008.

7   "Intelligent Exploitation of Existing ERP," *www.worksmanagement.co.uk/*, February 2008.

8   "Velsicol Chemical to Transform Operations on Company-Wide Business Intelligence," *Manufacturing Computer Solutions*, December 13, 2006.

9   Kim S. Nash, "Gas Prices: How Oil Companies Use Business Intelligence to Maximize Profits," *CIO*, June 9, 2008.

10 Mary Hayes Weier, "Delta Dental Signs Up for Open Source Business Intelligence Software," *Information Week*, June 5, 2008.

11 Mary Hayes Weier, "Hewlett-Packard Data Warehouse Lands in Wal-Mart's Shopping Cart," *Information Week*, August 4, 2007.

12 Eric Lai, "Teradata Creates Elite Club for Petabyte-Plus Data Warehouse Customers," *ComputerWorld*, October 14, 2008.

13 Eric Lai, "Been Audited Lately? Blame the IRS's Massive, Superfast Data Warehouse," *ComputerWorld*, March 22, 2008.

14 "Indiana Botanic Relieves Reporting Bottlenecks with Taurus Software Business Intelligence Solutions," Taurus Software Press Release, July 8, 2008 accessed at *www.taurus.com/* on November 6, 2008.

15 Tien Nguyen, "Customers in the Spotlight – PepsiCo," Business Objects Web site at *www.businessobjects.com/* accessed on November 4, 2008.

16 John Caulfield, "Lowe's Looks to Exploit Opportunities in Downturn," *Builder*, September 26, 2007.

17 Heather Havenstein, "Lowe's Builds Up Infrastructure to Support BI," *NetworkWorld*, January 24, 2007.

18 Larianne McLaughlin, "How One CIO Performed Database Magic in 6 Weeks," *CIO*, June 12, 2007.

19 Jennifer Bresnahan, "A Delicate Operation," *CIO*, October 10, 2007.

20 Mary Hayes Weier, "Keeping Customers from Walking Out the Door," *Information Week*, June 16, 2007.

21 Michael Isikoff, "Uncle Sam is Still Watching You," *Newsweek*, July 21, 2008.

22 Arik Hesseldahl, "There's Gold in 'Reality Mining'," *Business Week Online*, March 25, 2008.

23 Kate Greene, "TR10: Reality Mining," *Technology Review*, March/April 2008.

24 Robert Kaplan and David Norton, "The Balanced Scorecard – Measures that Drive Performance," *Harvard Business Review*, January–February 1992.

25 Mark Kinkade, "Produce Review, Microsoft Business Intelligence," *DM Review*, July 2008.

26 Jean Thilmany, "No Wasted Movement: Business Analytics Tool Slashes Welch's Transportation Costs," *Manufacturing Business Technology*, March 16, 2008.

27 Ben Worthen, "Business Process Management: A New Glue or the Old Soft Shoe?" *CIO,*
August 27, 2007.

28 "Wyeth at a Glance," Weyth Web site at *www.wyeth.com/aboutwyeth/whoweare/wyethglance*
accessed on September 26, 2008.

29 David F. Carr, "Wyeth's prescription for Business Process Management Success," *CIO,*
May 30, 2008.

30 "About Pitt Ohio," accessed at the Pitt Ohio Web site *http://works.pittohio.com/* on
September 26, 2008.

31 Jim Ericson, "Handled with Care," *DM Review,* September 2008.

32 Heather Havenstein, "IT Struggles to Show BI Value," *Computerworld*, January 29, 2007.

33 Ben Worthen, "Business Process Management: A New Glue or the Old Soft Shoe?" *CIO,*
August 27, 2007.

34 2007 Qwest Annual Report.

35 "Case Study: Qwest Uses Process Simulation to Move at the Speed of Business Change,"
Forrester Research, April 23, 2008.

36 "Case Study: Qwest Uses Process Simulation to Move at the Speed of Business Change,"
Forrester Research, April 23, 2008.

37 "Case Study: Qwest Uses Process Simulation to Move at the Speed of Business Change,"
Forrester Research, April 23, 2008.

38 "Case Study: Qwest Uses Process Simulation to Move at the Speed of Business Change,"
Forrester Research, April 23, 2008.

39 Blue Mountain Resort Web site accessed at *www.bluemountain.ca/* on August 29, 2008.

40 "Ski Resort Gets a Lift from Business Intelligence," *Baseline*, August, 2008.

41 Gowers, John, "Blue Mountain Resort Scales Large Amounts of Data for Better Customer
Service to Resort Guests," *DM Review*, July 2007.

42 "ROI Case Study, IBM Cognos TM1 Blue Mountain Resorts," *Nucleus Research*, May 2008.

43 "Ski Resort Gets a Lift from Business Intelligence," *Baseline*, August, 2008.

44 "ROI Case Study, IBM Cognos TM1 Blue Mountain Resorts," *Nucleus Research*, May 2008.

45 "Blue Mountain," Customer Success Stories accessed at the Cognos Web site
www.cognos.com/ on September 1, 2008.

46 "Ski Resort Gets a Lift from Business Intelligence," *Baseline*, August, 2008.

47 "Ski Resort Gets a Lift from Business Intelligence," *Baseline*, August, 2008.

48 "Blue Mountain," Customer Success Stories accessed at the Cognos Web site
www.cognos.com/ on September 1, 2008.

CHAPTER 10

KNOWLEDGE MANAGEMENT

HOW KNOWLEDGE MANAGEMENT TOOLS CAN AFFECT YOUR ORGANIZATION

"When a company 'gets it' with how social media works, it changes the way they use e-mail. They begin to use e-mail for communications that are one-to-one, one-to-few, or transient messages that have little or no value in being retained (e.g. 'Can you make the meeting tomorrow?'). Content that has persistent value is best conveyed in a community where it can be cataloged, searched, and retained for future employees. And that kind of content is best entered and shared via a Web 2.0 social media community."[1]

— Eric Schurr, vice-president of marketing and direct sales, Awareness, Inc.

GOODWIN PROCTER ILLUSTRATES WHY MANAGERS MUST UNDERSTAND KNOWLEDGE MANAGEMENT

Goodwin Procter LLP is a major law firm with more than 900 attorneys serving clients from its offices in Silicon Valley, San Francisco, San Diego, New York, Los Angeles, Boston, Washington, D.C., and London. The firm's stated mission is to help its clients achieve success by developing and delivering innovative solutions to complex legal problems.[2]

Goodwin Procter has more than 60,000 active cases with more than 10 million associated documents stored in several different systems including the firm's document management system and CRM system. Additional case documents originate in the Nexis system, a searchable archive of U.S. statutes and laws, and published case opinions.

In the past, when the firm's attorneys needed to assemble all the documents pertaining to a specific case, they had to log in to various software applications and pull the necessary information from each application. The process could take hours and cause a delay in responding to a client's questions or preparing to try a case.[3]

Goodwin Procter needed a way to reduce the time attorneys and their assistants spent gathering and summarizing all of this data. Over the course of a year, the firm developed a knowledge management system called Matter Pages—a Web-based system that extracts and integrates documents from various sources into an easily readable and searchable format. "The Matter Pages system places client information at their fingertips, which means attorneys spend less time compiling information and more time focusing on their legal practice," says Peter F. Lane, chief information officer at Goodwin Procter.[4]

All documents in the Matter Pages system are identified by the client number and the matter number. (The term *matter* refers to all the aspects of an individual case). The numbers provide the key to integrating the data through a Microsoft software program called SharePoint. SharePoint builds a set of Web pages within the Goodwin Procter intranet based on the selected matter number. Once a user selects a matter, the pages with the relevant documents are generated dynamically and can be accessed via a tabbed menu.

Use of the Matter Pages system has reduced the document-gathering process from hours to minutes per case, saving thousands of hours for the firm and enabling it to be much more responsive to its clients' needs.

LEARNING OBJECTIVES

As you read this chapter, ask yourself:

- What is knowledge management, and what organizational benefits can it deliver?
- How can you help sell and successfully implement a knowledge management project?

This chapter will identify the challenges associated with knowledge management, provide guidance to overcome these challenges, present best practices for selling and implementing a successful knowledge management project, and outline various technologies that support knowledge management. We begin with a definition of knowledge management and identify several knowledge management applications and their associated benefits.

WHAT IS KNOWLEDGE MANAGEMENT?

Knowledge management (KM) "is a practice concerned with increasing awareness, fostering learning, speeding collaboration and innovation, and exchanging insights."[5] Much of KM involves creating value from an organization's intellectual assets through codifying what employees, suppliers, business partners, and customers know, and then sharing that information with employees and even with other companies to devise best practices.[6] The expansion of the services sector, globalization, and the emergence of new information technologies have caused many organizations to establish KM programs in their Information Technology or Human Resource Management departments. The goal is to improve the creation, retention, sharing, and reuse of knowledge.

An organization's knowledge assets often are classified as either explicit or tacit (see Table 10-1). Explicit knowledge is knowledge that is documented, stored, and codified—such as standard procedures, product formulas, customer contact lists, market research results, and patents. Tacit knowledge is "personal knowledge embedded in individual experience and involves intangible factors, such as personal beliefs, perspective, and the value system. Tacit knowledge is hard to articulate with formal language (hard, but not impossible). It contains subjective insights, intuitions, and hunches."[7] Tacit knowledge is not documented and encompasses the things we do when we don't have a formal checklist or written procedures to follow. Examples include the process used by an experienced coach to make adjustments when his team is down at halftime of a big game, a physician's technique for diagnosing a patient's rare illness and prescribing a course of treatment, and an engineer's approach to cutting costs for a project that is over budget. This knowledge cannot be documented easily; it is the "knowhow" people have in their heads.

TABLE 10-1 Explicit and tacit knowledge

Asset Type	Description	Examples
Explicit knowledge	Knowledge that is documented, stored, and codified	Customer lists, product data, price lists, a database for telemarketing and direct mail, patents, standard procedures, and market research results
Tacit knowledge	Personal knowledge not documented but embedded in individual experience	Expertise and skills unique to individual employees

Much of the tacit knowledge that people carry with them is extremely useful but cannot be shared with others easily. This means that new employees might spend weeks, months, or even years learning things on their own that more experienced coworkers might have been able to convey to them. In some cases, these nuggets of valuable knowledge are lost forever, and others never learn them.

A major goal of knowledge management is to somehow capture and document the valuable work-related tacit knowledge of others and to turn it into explicit knowledge that can be shared with others. This is much easier said than done, however. Over time, experts develop their own processes for their areas of expertise. Their processes become second nature and are so internalized that they are sometimes unable to write down step-by-step instructions to document the processes.

Two frequently used processes exist for capturing explicit knowledge—shadowing and joint-problem solving. Shadowing involves a novice observing an expert executing his job to learn how he performs. This technique often is used in the medical field to help young interns learn from experienced physicians. With joint problem solving, the novice and the expert work side-by-side to solve a problem so that the expert's approach is slowly revealed to the observant novice. Thus a plumber trainee will work with a master plumber to learn the trade. Shadowing is a more passive learning technique while joint-problem solving is more active.[8] The next section will discuss how KM is used in organizations and will illustrate how these applications lead to real business benefits.

Knowledge Management Applications and Associated Benefits

Organizations employ KM to deliver real benefits by fostering innovation, leveraging the expertise of people across the organization, and capturing the expertise of key individuals before they retire. Examples of knowledge management efforts that led to these results and their associated benefits will now be discussed.

Foster Innovation by Encouraging the Free Flow of Ideas

Only the fittest survive. Organizations must continuously innovate to evolve, grow, prosper, and stay fit. Organizations that fail to innovate will soon fall behind their competition. Many organizations implement knowledge management projects to foster innovation by encouraging the free flow of ideas among employees, contractors, suppliers, and other business partners. Such collaboration can lead to the discovery of a wealth of new opportunities to be evaluated and tested. Some of the opportunities can lead to an increase in revenue, a decrease in costs, or creation of new products and services.

Giant Eagle Inc. provides an excellent example of the successful use of knowledge management to foster innovation. As you read this example, note the role that senior managers played in ensuring the success of this knowledge management initiative. First, they ensured that others within the organization became aware of good ideas that were generated. Second, they publicly recognized those who provided good ideas to reward them for their efforts and to further motivate others to contribute.

Giant Eagle is a grocery retailer and distributor with 223 locations mainly in western Pennsylvania and Ohio. It is one of the largest, privately owned, and family-operated companies in the nation, with recent annual sales of more than $7 billion and roughly 36,000 employees.[9] The firm began a knowledge management effort under a set of conditions that would seem to guarantee project failure. Giant Eagle's managers were constantly competing against each other to deliver the highest sales per store, the lowest amount of shoplifting, and the most contented employees. Such competitiveness did not motivate managers to work together or share knowledge that might provide a competitive advantage. In addition, few Giant Eagle employees had ever used a computer in their work. They had to learn how to log in to the knowledge management system and take the time to read messages from other employees on proven practices. Even more difficult, employees had to overcome the long standing competitive culture and become comfortable with sharing their own ideas with others.[10]

Previously, "there was no tradition of sharing ideas in the store environment," says Jack Flanagan, executive vice president of Giant Eagle business systems. One success story helped to change that culture, however. A Giant Eagle deli manager discovered an especially effective way to display shrimp platters that boosted his weekly sales by $200. Taking a risk, he posted his idea in the knowledge management system. Another deli manager read his posting, tried his idea, and generated a similar increase in shrimp sales. The total payoff from this idea at just the two stores was more than $20,000 in annual sales. If that idea had been implemented successfully at all Giant Eagle deli departments, the total sales increase during that holiday period would have been over $300,000.[11] Senior managers made sure that other Giant Eagle managers became aware of this example of sharing knowledge and motivated them to overcome their reluctance to use the knowledge management system by continuing to recognize other successes. These actions had the desired effect—managers are now competing to come up with the best suggestions and ideas. "They're competing in the marketplace of ideas," says Russ Ross, Giant Eagle senior vice president of IS and CIO. Giant Eagle anticipates more than $100,000 in additional annual revenue by the sharing of ideas via the knowledge management system.[12]

Leverage the Expertise of People Across the Organization

It is critical that an organization enable its employees to share and build on one another's experience and expertise. In this manner, new employees or employees moving into new positions are able to get up to speed more quickly. Workers can share thoughts and experiences about what works well and what does not, thus preventing new employees from repeating many of the mistakes of others. Employees facing new (to them) challenges can get help from coworkers in other parts of the organization whom they have never even met to avoid a costly and time-consuming "reinvention of the wheel." All of this enables employees to deliver valuable results more quickly, improve their productivity, and get products and new ideas to market faster.

iCrossing, Inc., provides an excellent example of how leveraging the expertise of people across the organization can provide a tremendous productivity boost. One key to the success of knowledge management at iCrossing is that a senior manager took charge of the project and drove it to conclusion. A second key to success was that the technology employed was simple to use and one with which people were already familiar.

iCrossing is a digital marketing firm that employs tools and tactics such as Web analytics and social media analyses to gain insight into what drives user attention to, and engagement with, brands online. Its clients use this information to launch products, increase visibility, manage leads, acquire customers, sell products, and manage their reputations.[13] The firm employs more than 620 professionals in 15 offices in the United States and Europe; and it works with more than 40 Fortune 500 companies. The firm is growing rapidly through acquisition. "We're adding not only products, but we were growing in people and the knowledge they bring. We needed a way to put all this knowledge in one location," states Matthew Schultz, the firm's VP of technology.

At the time Schultz joined iCrossing, the firm had a basic company intranet that provided access to the company phone directory and some corporate documents. There was no way to update the intranet without getting the corporate IT group involved, and they were busy with other priorities. Schultz decided to build a corporate wiki and make it a repository for iCrossing's knowledge. (As discussed in Chapter 6, a wiki is a collaborative Web site that allows users to create and edit Web page content freely using any Web browser.) He purchased the necessary software and support from a firm called Socialtext. Schultz chose this firm because its software was simple to use for end users who had no knowledge of programming or HTML. Also the Socialtext business model generated revenue based on licensing fees thus eliminating the distraction of ads associated with firms whose business model relied on ad revenue.

Managers and "thought leaders" were the first ones given access to the wiki. These early users provided basic content about the company and industry articles on topics such as search engine optimization and other firms in the industry. Within weeks, the wiki was opened to all employees who add information about their projects and other industry news. Because the wiki is a repository of all organizational information on a topic, it has become a useful source of current information used by new and experienced employees alike. Employees can access data in the wiki much more quickly and easily than by going through e-mails. In addition, with the use of wikis, organizations can help ensure that employees do not miss on out on any relevant information simply because they were not included on the distribution list for an individual e-mail.[14]

Capture the Expertise of Key Individuals as They Retire

In the United States, 3 to 4 million employees will retire each year for the next 20 years or so. Add to that a five to seven percent employee turnover as workers move to different companies, and it is clear that organizations are facing a tremendous challenge in trying to avoid the loss of valuable experience and expertise. "Not only is intellectual property (such as software) and expertise (such as services) increasingly the product, the value of intellectual capital behind physical goods routinely outweighs that of factories and infrastructure (which can be outsourced)."[15] Many organizations are using knowledge management to capture this valuable expertise before it simply walks out the door and is lost forever. The permanent loss of expertise related to the core operations of an organization can result in a significant loss of productivity or a decrease in the quality of service.

Consolidated Edison Company of New York (Con Edison) provides an excellent example of using knowledge management to capture and make available to others the tacit knowledge of experienced workers. Con Edison managers recognized the need to seek out the help of experienced industry resources to ensure the success of their effort.

Con Edison provides electric service to 3.2 million customers and gas service to approximately 1.1 million customers in New York City and Westchester County, a service area that covers 660 square miles. To serve these customers, Con Edison operates one of the most complex power distribution systems in the world, including six steam generating facilities, 125,000 miles of electric cable and wires, and 4000 miles of gas mains.[16]

Key to the safe, reliable, and efficient operation of Con Edison's power distribution system was the chief district operator, Bob Blick. Over the years, the rules, protocols, and procedures that govern the operation of the power delivery system had become second nature to him. Blick had also developed a unique but highly effective approach to recognizing and solving power system switching and protection problems. As Blick contemplated retirement, Con Edison knew that it must take action to make his mission-critical expertise, judgment, and approach to problem solving available to his successor and other personnel.[17]

Con Edison called on the Electric Power Research Institute (ERPI) for help. This nonprofit organization conducts research on topics of interest to the U.S. electric power industry. ERPI has developed tools and an effective process for capturing and transferring expert knowledge that is unique to the industry. ERPI employed its knowledge-capture process, using a series of detailed interviews with Blick to capture his expertise and gain insights into his problem-solving approach. ERPI then transformed this into a roadmap that represented Blick's thought processes for addressing a variety of specific situations. The result was a detailed model to help Blick's replacement acquire his knowledge and understand his unique approach for identifying, diagnosing, and solving problems. The result was a smooth turnover, and Con Edison was able to maintain its high levels of system reliability and personnel safety.[18]

Best Practices for Selling and Implementing a KM Project

Many challenges exist in trying to establish a successful KM program. Most of these challenges have nothing to do with the technologies or vendors employed. Instead they are challenges associated with human nature and the manner in which people are accustomed to working together. A set of best practices for selling and implementing a KM project will now be presented.

Connect the KM Effort to Organizational Goals and Objectives

When starting a KM effort, just as with any other project, you must clearly define how that effort will support specific organizational goals and objectives like increasing revenue, reducing costs, improving customer service, or speeding up the time to bring a product to market. This will help you sell the project to others and elicit their support and enthusiasm. This will also determine if the project is worthwhile before the organization commits resources to it. While many people may intuitively believe that sharing knowledge and best practices is a worthy idea, there must be an underlying business reason to do so. "Without a solid business case, KM is a futile exercise."[19] Once it was shown that KM could help Giant Eagle increase revenue, the program gained broad support and users overcame their reluctance to use the system.

Identify Valuable Tacit Knowledge

It is important to recognize that not all tacit knowledge is equally valuable and that priorities must be set in terms of what to go after. "Quantity rarely equals quality, and KM is no exception. Indeed, the point of a KM program is to identify and disseminate knowledge gems from a sea of information."[20] Con Edison recognized that they were about to lose invaluable tacit knowledge about how to run their operation, which would have a negative impact on the quality of customer service. They made capture and documentation of this knowledge a high priority.

Start with a Small Pilot Involving Enthusiasts

Containing the scope of a project to impact only a small part of the organization and a few employees is definitely less risky than trying to take on a project very large in scope. With a small scale project, you have more control over the outcome, and if the outcome is not successful, the organization is not seriously impacted. Indeed, the failure can be considered a learning experience on which to build future KM efforts. In addition, obtaining the resources (people, dollars, etc.) for a series of small, successful projects is much easier than getting large amounts of resources for a major organization-wide project.[21]

Furthermore, defining a pilot project to address the business needs of a group of people who are somewhat informed about KM and are enthusiastic about its potential can improve greatly the odds of success. Targeting such a group of users reduces the problem of trying to overcome skepticism and unwillingness to change, which have doomed many a project. Also, such a group of users, once the pilot has demonstrated some degree of success, can serve as strong advocates who go out and communicate the positive business benefits of KM to others. When Shell Exploration & Production (Shell EP), a division of Royal Dutch/Shell Group, began piloting KM, perhaps 20 percent of the people could be classified as enthusiastic and willing to try KM. Seven years later, thanks to a series of small, successful projects, the number of enthusiastic users has grown to roughly 55 percent (about 16,000 out of a total workforce of 30,000).[22]

Get Employees to Buy In

Managers must create a work culture that places a high value on tacit knowledge and that strongly encourages people to share it. It can be especially difficult to get workers to surrender their knowledge and experience in a highly competitive work environment as these traits make them more valuable as individual contributors. For example, it would be extremely difficult to get a highly successful mutual fund manager to share his stock-picking technique with other fund managers. Such sharing of information would tend to put all fund managers on a similar level of performance and also tend to level the amount of their annual compensation.

Some organizations believe that the most powerful incentive for experts to share their knowledge is to receive public recognition from senior managers and their peers. For example, both Shell EP and Giant Eagle provide recognition by mentioning the accomplishments of contributors in a company e-mail or newsletter, or during a meeting. Other companies identify knowledge sharing as a key expectation for all employees and even build this expectation into the employees' formal job performance reviews. Many organizations provide incentives in a combination of ways—linking KM directly to job performance, creating a work environment where sharing knowledge seems like a safe and natural thing to do, and recognizing people who contribute.[23]

Technologies That Support KM

We are living in a period of unprecedented change where knowledge is expanding rapidly. As a result, there is an increasing need for knowledge to be quality filtered and distributed to people in a more specific task relevant and timely manner. Technology is needed to acquire, produce, store, distribute, integrate, and manage this knowledge. Those organizations interested in piloting KM need to be aware of the wide range of technologies that can support KM efforts. These include communities of practice, social network analysis, a variety of Web 2.0 technologies, business rules management systems, and enterprise search tools. These technologies will now be discussed.

Communities of Practice

A **community of practice (CoP)** is a group whose members share a common set of goals and interests and regularly engage in sharing and learning as they strive to meet those goals. A community of practice develops around topics that are important to its members. Over time, a CoP typically develops resources such as models, tools, documents, processes, and terminology that represent the accumulated knowledge of the community. It is not uncommon for a CoP to include members from many different organizations. CoP has become associated with knowledge management because participation in a CoP is one means of developing new knowledge, stimulating innovation, or sharing existing tacit knowledge within an organization.

The origins and structures of CoPs vary widely. Some may start up and organize of their own accord; in other cases, there may be some sort of organizational stimulus that leads to their creation. Members of an informal CoP typically meet with little advanced planning or formality to discuss problems of interest, share ideas, and provide advice and counsel to one another. Members of a more formal CoP meet on a regularly scheduled basis with a planned agenda and identified speakers.

Software from Socialcast supports collaboration and knowledge sharing among members of a CoP. The software enables employees to create their own fully customized community in which they can create, expand, and exchange knowledge and expertise across the enterprise. The National Aeronautics and Space Administration (NASA) is piloting use of the software to share knowledge as it begins its Constellation Program for developing new spacecraft to replace the Space Shuttle. NASA is concerned that many experienced employees are approaching retirement, and it needs a way to enable those workers to share what they know with newer employees. NASA has many large space centers located around the country and runs multiple decades-long space projects. This tends to create "silos of expertise" and limit collaboration and sharing of knowledge. The goal is to break down barriers and ensure that NASA's institutional memory carries forward.[24]

Social Network Analysis (SNA)

Social network analysis (SNA) is a technique to document and measure flows of information between individuals, workgroups, organizations, computers, Web sites, and other information sources (see Figure 10-1). Each node in the diagram represents a knowledge source; each link represents a flow of information between two nodes. Many software tools support social network analysis including NetMiner, UCINET, and NetDraw.

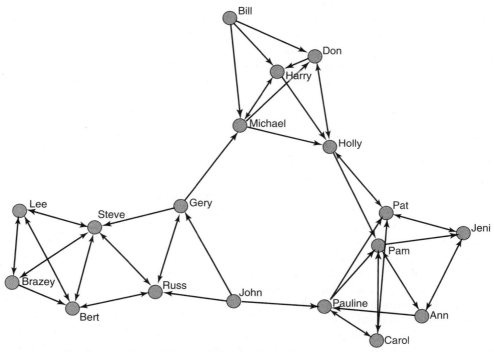

Source: www.analytictech.com/Netdraw/NetdrawGuide.doc

FIGURE 10-1

SNA has many knowledge management applications, ranging from mapping knowledge flows and identifying knowledge gaps within organizations to helping establish collaborative networks. SNA provides a clear picture of how geographically dispersed employees and organizational units collaborate (or don't collaborate). Organizations frequently employ SNA to identify subject experts and then set up mechanisms (e.g., communities of practice) to facilitate the passing of knowledge from those experts to colleagues. Software programs that track e-mail and other kinds of electronic communications may be used to identify in-house experts. For example, Mars, maker of a variety of food products that are consumed in over 100 countries worldwide, successfully employed SNA to identify how knowledge flows through its various organizations, which employees hold influence, which employees provide the best advice, and how employees share information.[25] It then established both formal and informal communities of practice to facilitate the sharing of knowledge among people with similar interests.

Web 2.0 Technologies

As discussed in Chapter 7, Web 2.0 is a term describing changes in technology and Web site design to enhance information sharing, collaboration, and functionality on the Web. Major corporations such as McDonald's, Kodak, The New York Times Company, Northwestern Mutual, and Procter & Gamble have integrated Web 2.0 technologies such as blogs, forums, mashups, podcasts, RSS newsfeeds, and wikis to support knowledge management to improve collaboration, encourage knowledge-sharing, and build a corporate memory. For example, many organizations are using Web 2.0 technologies such as podcasts and wikis to capture the knowledge of longtime employees, provide answers to cover frequently asked questions, and save time and effort in training new hires.[26] Read the following feature, "A Manager Takes Charge" to learn how one company was able to leverage the use of Web 2.0 technologies to transfer knowledge to its employees.

A MANAGER TAKES CHARGE

JetBlue Pilots Web 2.0 Technologies

New York-based JetBlue Airways is known for its ticketless travel with all seats assigned and low one-way fares with no overnight stay required. JetBlue was the first U.S. airline to offer its own Customer Bill of Rights, with meaningful and specific compensation for customers inconvenienced by service disruptions within JetBlue's control. The airline serves over 50 cities with more than 550 flights each day.

JetBlue University is a corporate university responsible for the orientation and training of all the airline's employees. In January 2008, Murry Christensen, director of Learning Technologies at JetBlue University, initiated a pilot project to evaluate a social network portal for the 200 employees who work as faculty members at the university's three locations in Orlando, New York, and Salt Lake City. The portal enables them to use Web 2.0 technologies such as wikis and blogs to share best practices on how to train employees. Indeed, JetBlue is creating an entirely new medium for capturing and sharing important company knowledge and information, as well as enhancing collaboration on projects between employees in disparate programs and locations.[27]

continued

One of the driving factors behind the move to the social network portal was the recognition that many communication "misses" occur when employees are left off the distribution list for e-mails. "E-mail is unstructured and ephemeral. With blogs and wikis, you can capture process improvements more visibly," says Christensen. For example, if the reservations clerk faculty in Salt Lake City try a new training technique that does not work well, the flight crew faculty in Orlando won't know it flopped unless it is copied in an e-mail message. "We need to turn that implicit knowledge into explicit knowledge," Christensen says.[28]

If this pilot project proves successful, JetBlue plans to expand the use of the portal across the entire enterprise so that all of the airline's 12,500 employees can use it to communicate and collaborate on key projects and improve operations.[29] Christensen adds that the faculty makes a perfect test group because of the nature of its work. Instructors are very open to learning new technologies so that they can keep their training techniques up-to-date. Choosing this test group will come in handy when it comes time to expand because the training faculty can become an advocate for the technology. During a pilot project "you want to get a sense of how well it works, but you also want to do it to a relatively receptive audience," according to Bob Koplowitz, a Forrester research analyst.[30]

The software behind the portal comes from Awareness, Inc., and costs $50,000 per year. The costs will increase if the pilot project expands or if additional capabilities are added. Christensen has involved the IT organization to ensure that the portal is fully compliant and is capable of providing an appropriate level of security and privacy.

Discussion Questions:

1. What other Web 2.0 technologies should JetBlue consider investigating for potential use in JetBlue University?
2. How could JetBlue determine if the pilot project was successful? Is there any data that could be captured to demonstrate measurable improvements in training due to the use of Web 2.0 technologies?
3. If the pilot project proves successful, how might JetBlue continue the expansion of the use of Web 2.0 technologies to other areas of the business? What area of the business do you think would be a good candidate for further expansion?

Business Rules Management Systems

Change is occurring all the time and at a faster and faster pace—changes in economic conditions, new government and industry rules and regulations, new competitors, product improvements, new pricing and promotion strategies, and on and on. Organizations must be able to react to these changes quickly to remain competitive. The decision logic of the operational systems that support the organization—systems such as order processing, pricing, inventory control, and customer relationship management—must continually be modified to reflect these business changes. Decision logic, also called business rules, includes policies, requirements, and conditional statements that govern how the systems work.

The traditional method of modifying the decision logic of information systems involves heavy interaction between business users and IT analysts. They work together over a period of weeks, or even months, to define new systems requirements and then to design, implement, and test the new decision logic. Unfortunately, this approach to handling system changes has proven too slow, and in some cases, results in incorrect system changes.

A **business rule management system (BRMS)** is software used to define, execute, monitor, and maintain the decision logic that is used by the operational systems to run the organization.[31] If the business logic of an application can be separated from its data validation logic and overall program flow logic, a BRMS enables business users to make changes and updates to the decision logic without requiring involvement from IT resources. This process eliminates lengthy delays in implementing changes and improves the accuracy of the changes.

BRMS components include a business rule engine that determines which rules need to be executed and in what order. Other BRMS components include an enterprise rules repository for storing all rules, software to manage the various versions of rules as they are modified, and additional software for reporting and multi-platform deployment. Thus, a BRMS can become a repository of important knowledge and decision-making processes that includes the learnings and experiences of experts in the field. The creation and maintenance of a BRMS can become an important part of an organization's knowledge management program.

BRMS is increasingly used to manage the changes in decision logic in applications that support loan applications, underwriting, complex order processing, and complex scheduling. The use of BRMS leads to faster and more accurate implementation of necessary system changes.

Samsung Life Insurance is a South Korean insurer with over $23 billion in annual sales and some 10 million policyholders. The firm relied on a manual claims fraud detection process but recognized a need for improvement when the level of detected insurance fraud rose by 46 percent from 2005 to 2006. The firm drew on knowledge gleaned from past insurance fraud management work to define an extensive set of business rules, which are executed by a new automated BRMS system. The system analyzes claims based on 800 different factors including claimants' demographics, claim amount, and previous claims' history. The system has greatly reduced the instances of fraudulent claims and cut the inspection time for processing 10,000 claims from 2 weeks to 1 day. As new rules are developed that help identify fraudulent claims, they can be added easily to the rules repository by the end users.[32]

Enterprise Search Software

It is estimated that unstructured data, mostly in the form of text, accounts for about 85 percent of an organization's knowledge.[33] Unfortunately, such data is not easy to locate, access, or analyze. **Enterprise search** is the application of search technology to find information within an organization.

Enterprise search software indexes documents from a variety of sources such as corporate databases, departmental files, e-mail, corporate wikis, and document repositories. When executed, the search software uses the index to present a list of relevance-ranked documents from these various sources. The software must be capable of implementing access controls so users are restricted to viewing only documents to which they are granted access. Enterprise search software may also allow employees to move selected information to a new storage repository and apply controls to ensure that the files cannot be changed or deleted.

Autonomy, Endeca, Google, IBM, Kazeon, Microsoft, Oracle, Recommind, and StoredIQ are among the software vendors that offer competing enterprise search tools. There are two main types of enterprise search software—compliance and business search software.

Compliance enterprise search software is used by members of the IT and Human Resources organizations to enforce corporate guidelines on the storage of confidential data on laptops that leave the office; by legal counsel to gather up all e-mails and documents related to upcoming litigation; and by governance officials to ensure that all guidelines for the storage of information are being followed.

Electronic discovery is an important application of compliance enterprise search software. "Electronic discovery (e-discovery) refers to any process in which electronic data is sought, located, secured and searched with the intent of using it as evidence in a civil or criminal legal case."[34] The Federal Rules of Civil Procedures governs the processes and requirements of parties in federal civil suits and sets the rules regarding e-discovery. These rules were significantly strengthened in 2006 to expand the breadth of data that organizations are expected to find and produce in litigation. These rules compel civil litigants to both preserve and produce electronic documents and data related to a case. This includes e-mail, voice mail, instant messages, graphics, photographs, contents of databases, spreadsheets, Web pages, etc. "We can't find it" is no longer an acceptable excuse for not producing information relevant to a lawsuit. As a result of the more stringent set of rules, Qualcomm Incorporated was hit with an $8.5 million penalty because it mishandled the e-discovery process and failed to produce e-mail relevant to a lawsuit with Broadcom Corporation.[35]

Effective e-discovery software solutions preserve and destroy data based on approved organizational policies through processes that cannot be altered by unauthorized users. To be useful, this software must also allow users to locate all of the information pertinent to a lawsuit quickly, with a minimum amount of manual effort. Furthermore, the solution must work for all data types across dissimilar data sources and systems and operate at a reasonable cost. The legal departments of many organizations are collaborating with their IT organization and technology vendors to identify and implement a solution that meets these e-discovery requirements.

Business search software can be used by employees to find information in various repositories or to find mislaid documents. Unilever is a manufacturer of consumer goods in the food, beverage, and home and personal care categories. It operates a consumer call center that supports millions of consumers in the U.S. and Canada who use any of its more than 90 brands. Consumer call center agents must be prepared to answer a wide range of questions about the many products—how best to use the product, what are the ingredients, which product meets a specific need, etc. Unilever implemented a business search system from Astute Solutions to enable call center agents to access information quickly about its products to answer consumers' questions and, at the same, capture information during the conversation to better understand the consumer's needs. The search system is state-of-the-art; it allows users to input a question using natural language and then delivers a precise answer.[36] The system has enabled Unilever to provide consistent, accurate answers to consumers in less time, thus improving consumer service and reducing the number of call center agents needed.

This chapter has defined knowledge management and identified both the challenges of implementing a KM program and approaches for overcoming these challenges. It has also covered a number of the more commonly used technologies in a KM program. Table 10-2 recommends a set of actions an organization can take to implement a successful KM program. The appropriate answer to each question is "yes."

TABLE 10-2   A manager's checklist

Recommended Management Actions	Yes	No
Does your organization have information systems and face-to-face communication vehicles that enable people to learn from past innovation successes and failures?		
Does your organizational culture and reward system encourage the sharing of explicit and tacit knowledge?		
Has your organization carefully considered the use of a business rules management system to maintain the decision logic of operational systems?		
Are any Web 2.0 technologies being used within your organization to improve collaboration and share tacit knowledge?		
Is the organization engaged in any KM pilot projects?		

Chapter Summary

- KM is a practice concerned with increasing awareness, fostering learning, speeding collaboration and innovation, and exchanging insights.

- Knowledge is often classified as either explicit or tacit. Explicit knowledge is knowledge that can be documented, stored, and codified easily. Tacit knowledge is not documented and is subconscious and internalized; an individual with important tacit knowledge may not even be aware how he accomplishes certain results.

- Shadowing and joint-problem solving are two frequently used processes for capturing explicit knowledge.

- KM is used to encourage the free flow of ideas, leverage the expertise of people across the organization, and capture the expertise of key individuals before they retire.

- There are several recommendations to help sell and implement a KM project—connect the KM effort to organizational goals and objectives, identify the valuable tacit knowledge worth capturing, start with a small pilot with enthusiastic participants, and get employees to buy in.

- The technologies that support knowledge management include communities of practice, social network analysis, the whole range of Web 2.0 technologies, business rules management systems, and enterprise search tools.

Discussion Questions

1. In what ways are data management and knowledge management the same? How are they different?

2. Provide three examples of tacit knowledge. Provide three examples of explicit knowledge.

3. Can you identify a subject area in which you possess tacit knowledge that would be valuable to others? Would you readily share this knowledge with others? Why or why not? If you were so inclined, how would you go about sharing this tacit knowledge with others?

4. What are the primary organizational benefits that can be gained through a successful knowledge management program? How might you attempt to justify investment in a knowledge management project?

5. Identify one community of practice you are willing to help form and contribute to. How might you go about finding others who are willing to join and participate? What would you hope to gain from your participation in this community of practice?

6. Perform a social network analysis to identify your primary sources and sinks of knowledge including people, organizations Web sites, and information systems. (You may wish to limit this exercise to just your school or work-related activities.) What insights can your draw from this exercise?

7. Identify an example you have observed of applying Web 2.0 technologies to support knowledge management.

8. Develop a set of rules that capture your thought process in completing a frequently performed task—choosing which clothes to wear to school or work, deciding what route to take

to school or work, etc. Test the accuracy and completeness of your rule set by having a class-mate follow your rules to complete the task under a varying set of conditions.

9. Imagine that you are a senior executive in the human resources group of a large organization faced with an alarming number of retirements of critical employees over the next three years. How might you deal with this situation to avoid losing valuable expertise needed for the organization's continued growth and success? What reasonable kinds of incentives/motivation might you offer reluctant employees to share their highly valuable tacit knowledge before retirement?

10. Imagine that you are the CEO of a large organization, and you strongly support the need for a greater level of collaboration in most areas of the organization. Discuss how you might be able to stimulate the formation and growth of communities of practice.

Action Memos

1. You are a talent scout for a professional sports team. Over the years, the players you have recommended to be selected in the draft have had an outstanding performance record for your team. Indeed, although you are only in your late-thirties, you are frequently cited as the top talent recruiter in the entire league. You read and re-read the study guide on knowledge management your general manager provided you two weeks ago. In addition to some basic definitions and discussion of KM, it includes several examples of successful applications of KM to the selection of top recruits for academic and athletic scholarships. Now you are sitting in your hotel room staring at the e-mail from the general manager. He wants you to become the subject of a KM experiment for the team. He plans to assign an expert in KM to study and document your approach to identifying top talent. The goal is to train the other three talent scouts for the team in your approach. He asks if you will participate in the experiment. How do you respond to this e-mail?

2. You are the CIO of a company facing a potential class action lawsuit over damages caused by one of its products. You are shocked when you receive an unsigned e-mail message sent to your personal e-mail account stating: "Destroy the contents of the e-mail back-up server. This is not a joke; your position with the firm is at stake." What do you do?

Web-Based Case

Visit the Web sites of three enterprise search software firms that provide e-discovery capabilities. Write a brief report that compares the strengths, weaknesses, and capabilities of the three software providers. Which of the three do you think offers the best solution? Why did you choose this software provider?

Case Study

Defense Threat Reduction Agency Implements KM

The Defense Threat Reduction Agency (DTRA) is an agency of the U.S. Department of Defense. It was established in 1998 with headquarters in Ft. Belvoir, Va., and is comprised of some 2000 civilian and military personnel scattered around the world in 14 locations. These people are dedicated to "providing capabilities to reduce, eliminate, and counter the threat of WMD (weapons of mass destruction including chemical, biological, radiological, nuclear, and high explosive weapons) and mitigate its effects."[37] The DTRA has three major areas of responsibility:[38]

1. Prepare U.S. and allied combatants for problems that can occur in fighting an enemy with WMD capability

2. Aid in consequence management to define what is needed for the nation to respond to a WMD attack

3. Work with Russia to secure and dismantle weapons of mass destruction in former Soviet Union states (e.g., Belarus, Kazakhstan, and Ukraine) to make sure they do not fall into the wrong hands

In meeting its challenging mission, the procurement professionals at DTRA award large and complex contracts to organizations to provide worldwide support and services; it also makes many multi-million dollar purchases. Here are a few examples of its recent contracts:

- SRA International, Inc. was awarded a $10 million, two-year contract to provide the logistics, network support, software engineering, and Web services for the research and development operations at DTRA.[39]

- Black & Veatch Holding Company (a global engineering, consulting, and construction company) was awarded a contract to provide services to Ukraine in the area of defenses against bioterrorism and bio-weapons proliferation worth $175 million, this is part of a much larger contract worth up to $4 billion.[40]

- Lockheed Martin Corporation was awarded a $45 million, 5-year contract to modernize the IT systems at DTRA.[41]

- Defense Threat Reduction Agency (DTRA) is awarding several contracts to find new anti-viral compounds that are effective in combating hemorrhagic fever viruses, a class of deadly viruses that includes Ebola.[42]

DTRA is the merged product of five different defense agencies and programs. Because of this, functional and process information about purchasing and contracting was scattered throughout a myriad of DTRA Web sites, shared and private computer disk drives, and on paper documents in file cabinets. Much of the valuable information was tacit in nature and simply was not documented in any form. Finding needed information in a timely and effective manner simply was not possible. Such complications hindered DTRA in fulfilling its mission in a cost-effective manner. Acquisition professionals, program managers, and others needed a faster means of getting to the how-to and reference information needed to fulfill their responsibilities.

DTRA initiated a knowledge management effort to codify, standardize, and streamline its many and sometimes conflicting acquisition guidelines, processes, and procedures. As part of this effort, it developed a Web-based system called Acquisition ToolBook to provide valuable "how to" and reference information to aid DTRA acquisition professionals. The system provides a top-down, comprehensive, well-organized view of the entire acquisition process. ToolBook users log in to the agency's main Web portal to see an overview of the entire acquisition process summarized into 24 activity boxes of related acquisition information and tasks. Clicking any activity box provides further details, guidelines, forms, and procedures associated with completing that acquisition activity. The information is presented in a multi-level format to better meet the needs of the users. For example, level one provides basic training on concepts, requirements, and responsibilities for completing the chosen activity. Level two provides more detailed information such as guides, tools, manuals, and standard operating procedures. Level three provides actual examples, checklists, tools, and templates to complete a specific activity. It took 12 months to design and implement the Acquisition ToolBook. Much of this effort was directed at defining a common set of practices to be followed and then documenting these practices in a simple, easy to follow manner.

The valuable acquisition information provided by ToolBook is tailored to meet the information needs of program and project managers. It also benefits others associated with the acquisition process including those people involved in negotiating, reviewing, analyzing, documenting, and monitoring the execution of contracts by assisting them in the performance of their specific acquisition and procurement functions. The ToolBook helps users obtain key information quickly when they need it, reduces the time required to get a new acquisition professional up to speed, and ensures that users follow "best practices" for completing their work. All this reduces the time required to obtain valuable goods and services needed for DTRA to fulfill its critical mission.

Discussion Questions

1. The DTRA ToolBook has now been in operation for a few years. Imagine that you are assigned to head a special project to identify further improvements to provide procurement professionals with the knowledge they need to be even more effective. The scope of the project includes evaluating the usefulness and completeness of the knowledge that the ToolBook provides to its users. It also has been suggested that more extensive use of Web 2.0 technologies be considered to enhance the system usefulness. How would you evaluate the effectiveness of the current system and identify potential opportunities for improvement? Identify three potential ideas you feel are worthy of further investigation.

2. Find the online article "A Different Kind of Web-Based Knowledge Management" by Dr. Joseph P. Avery in the May–June 2008 issue of *Knowledge Management*. Read and comment on the Eight Key Principles of Successful KM-Based Systems defined in this article.

3. DTRA strongly advocates relationships with small and minority-owned businesses. Imagine that you are in charge of developing a knowledge management system to provide guidance for such organizations interested in doing business with DTRA. Who would you work with to define the contents of such a system? How might you deliver the knowledge from this system to the end users?

4. Identify three specific acquisition tasks for which there is a high need to document the tacit information of experienced DTRA acquisition professionals.

5. Comment on how well you think other government agencies might be able to reapply the Acquisition ToolBook to support their acquisition activities. What barriers might exist that would make reapplication difficult?

Endnotes

[1] Eric Schurr, "Awareness Powers a Community for JetBlue," accessed at *http://ericschurr. awarenessnetworks.com* on November 10, 2008.

[2] "Our Firm" accessed at *www.goodwinprocter.com/OurFirm.aspx* on October 25, 2008.

[3] Jarina D'Auria, "Goodwin Procter Makes Strong Case for Knowledge Management," *CIO,* August 1, 2008.

4 "Goodwin Procter a Recipient of the CIO 100 Award for the Second Consecutive Year," News & Events, June 1, 2008 accessed at *www.goodwinprocter.com*.

5 "What is KM?" accessed at *kmwiki.wikispaces.com* on October 27, 2008.

6 Meridith Levinson, "ABC: An Introduction to Knowledge Management (KM)," *CIO*, March 7, 2007.

7 D.R. Clark, "Knowledge," May 10, 2004 accessed at *www.skagitwatershed.org/~donclark/knowledge/knowledge.html*.

8 Meredith Levinson, "ABC: An Introduction to Knowledge Management (KM)," *CIO*, March 7, 2007.

9 "About Us – Corporate Overview – Fast Facts," accessed at *www.gianteagle.com* on October 25, 2008.

10 Lauren Gibbons Paul, "How to Create a Know-It-All Company," *CIO*, June 13, 2007.

11 Lauren Gibbons Paul, "How to Create a Know-It-All Company," *CIO*, June 13, 2007.

12 Lauren Gibbons Paul, "How to Create a Know-It-All Company," *CIO*, June 13, 2007.

13 Gavin O'Mallery, "Search Centric iCrossing Enlists Socially Minded Pluck," Media Post's Media Daily, June 26, 2008.

14 C.G. Lynch, "Building a Better (and Useful) Corporate Intranet Starts with a Wiki," CIO, October 1, 2008.

15 Doug Henschen, David Stodder, Penny Crosman, Neal Mcwhorter, and David Patterson, "Seven Trends for 2007," *Intelligent Enterprise*, January 2007.

16 "About Us," Consolidated Edison Web site at *www.coned.com/aboutus* accessed October 27, 2008.

17 "Consolidated Edison Captures Expertise of Retiring Chief District Officer to Preserve Safety and Reliability," Electric Power Research Institute, October 2007.

18 "Consolidated Edison Captures Expertise of Retiring Chief District Officer to Preserve Safety and Reliability," Electric Power Research Institute, October 2007.

19 Meridith Levinson, "ABC: An Introduction to Knowledge Management (KM)," *CIO*, March 7, 2007.

20 Meridith Levinson, "ABC: An Introduction to Knowledge Management (KM)," *CIO*, March 7, 2007.

21 Meridith Levinson, "ABC: An Introduction to Knowledge Management (KM)," *CIO*, March 7, 2007.

22 Lauren Gibbons Paul, "How to Create a Know-It-All Company," *CIO*, June 13, 2007.

23 Lauren Gibbons Paul, "How to Create a Know-It-All Company," *CIO*, June 13, 2007.

24 John Foley, "One Small Step for Socialcast, One Giant leap For Enterprise Social Networking," *Information Week*, May 9, 2008.

25 Meridith Levinson, "ABC: An Introduction to Knowledge Management (KM)," *CIO*, March 7, 2007.

26 Michael Laff, "Knowledge Walks Out the Door," *T+D*, January 2008.

27 "Awareness, Inc. Powers Online Web 2.0 Community for JetBlue," Awareness Press Release, April 29, 2008 accessed at *www.awarenessnetworks.com/* on October 24, 2008.

28 C.G. Lynch, "JetBlue to Pilot the Use of Internal Wikis and Blogs," *CIO*, December 14, 2007.

29 C.G. Lynch, "JetBlue to Pilot the Use of Internal Wikis and Blogs," *CIO*, December 14, 2007.

30 C.G. Lynch, "JetBlue to Pilot the Use of Internal Wikis and Blogs," *CIO*, December 14, 2007.

31 James Owen, "Business Rules Management Systems," Infoworld, June 25, 2004.

32 "Samsung Life Insurance Selects ILOG BRMS for its Advanced Insurance Fraud Detection System," Press Release, November 6, 2007 at *www.ilog.com*.

33 Drew Robb, "Text Mining Tools Take on Unstructured Data," *ComputerWorld*, June 21, 2004.

34 "Electronic Discovery," accessed at *http://searchfinancialsecurity.techtarget.com/* on November 13, 2008.

35 Andrew Conry-Murray, "Enterprise Search: Microsoft, Google, Specialized Players Vie for Supremacy," *InformationWeek*, September 27, 2008.

36 "Real Dialog," accessed at *www.astutesolutions.com/* on November 15, 2008.

37 "About DTRA," accessed at *www.dtra.mil/* on November 11, 2008.

38 Threat Reduction Agency Marks 10 Years of Operations," FDCH Regulatory Intelligence Database, October 7, 2008.

39 David Hubler, "SRA to Continue Assisting DTRA," *Washington Technology*, May 16, 2008.

40 "U.S. Awards Threat Reduction Contract for Former Soviet States," *Defense Procurement News*, October 14, 2008.

41 "DTRA Awards IT Modernization Contract," *Defense Procurement News*, September 23, 2008.

42 "DTRA Researching Hemorrhagic Fever Anti-Viral Compounds," *Defense Industry Daily*, July 1, 2008.

CHAPTER 11

ENTERPRISE ARCHITECTURE

WHAT ROLE DOES ENTERPRISE ARCHITECTURE PLAY IN BUILDING A SUCCESSFUL BUSINESS?

Sarah Winchester, heir to the Winchester Rifle fortune, built the Mystery House over nearly 40 years, adding on bit by bit. The house had three elevators, 47 fireplaces, rooms built around rooms, stairways leading to nowhere, and doors that open into blank walls. The house ended up being not very functional. It wasn't for lack of money or highly skilled workers; it was lacking an architectural plan. The enterprise architecture in some organizations is like the Winchester Mystery House.[1]

ENTERPRISE ARCHITECTURE GIVES GOOGLE A COMPETITIVE EDGE

Google overcame established competition to become one of the world's leading innovators. Its mission is to organize the world's information and make it universally accessible and useful.[2] Google is now the primary gateway to the world's largest digital network of publicly available information and knowledge. The company is so dominant that its name is now a verb, as in "Google it."

Google has implemented an enterprise architecture that can handle more than one billion searches a day. It uses information gleaned from Web searches to improve its search engine continually, enabling the company to implement enhancements and keep it ahead of rivals Yahoo! and Microsoft.

Most commercial advertising is not customer-specific, which makes it inefficient. Consider an advertisement on TV that is targeted to a specific drug used to treat Alzheimer's disease. The advertisement is broadcast to a much wider audience than is necessary. By contrast, Google uses an advertising model based on combining a Web page ranking mechanism and targeted advertising. Customers conduct free searches with Google, but vendors pay Google to match consumers with their relevant products and services, resulting in more effective targeting.

The power behind Google's Web platform is a well-designed enterprise architecture, which consists of a vast array of interconnected computers and software systems hosted by a large number of regional data centers.[3] The enterprise architecture enables Google to run its core business processes and manage huge amounts of data in a specialized database. The database includes information about customer searches and the content of Google e-mails. The enterprise architecture is designed to access new Web content continuously, index the content, and manage the advertising business, thus freeing up Google employees to perform high-order thinking and pursue other innovations.

Google's high technology is balanced with a high-touch, "people-centric" culture that has a strong influence on helping the company provide a sustainable competitive advantage through its selective recruitment and retention of top talent. Google places a high priority on innovation and expects its workers to devote 70 percent of their time to the core business process of search and advertising, 20 percent to other business functions, and 10 percent to experimenting with new ideas.

Google's combination of enterprise architecture and innovation-oriented culture enables the company to thrive. The firm has achieved innovation through in-house efforts such as Google Earth, a service that combines the power of Google Search with satellite imagery, maps, terrain, and three-dimensional buildings to put the world's geographic information at the user's fingertips. Google also has achieved innovation through strategic acquisition. For example, it acquired YouTube, the popular Web site used to share video clips. YouTube has helped to change the way people are entertained and the way they communicate.

WHAT IS ENTERPRISE ARCHITECTURE?

Enterprise architecture is a set of models that describe the technical implementation of an organization's business strategy and business processes.[4] Figure 11-1 is a representation of the Google enterprise architecture, which includes an estimated half-million servers.[5] The tenet "form ever follows function"[6] describes the relationship between an organization's form (enterprise architecture) and the customers' need for functionality (purpose, utility, and desired value). The organization's business processes must be able to provide the desired functionality. The business processes determine the form. This chapter will discuss why enterprise architecture is important, indentify some architecture styles, and outline a process for developing an enterprise architecture. The key role of managers in ensuring that the proper architecture is designed and built will be emphasized throughout.

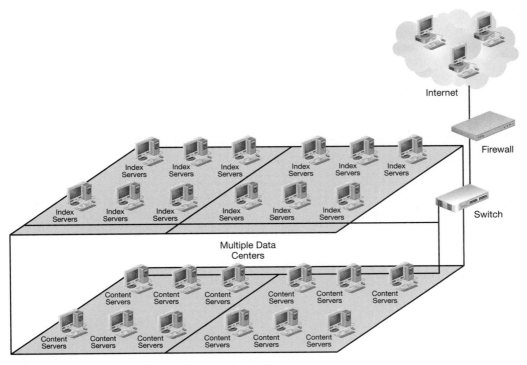

FIGURE 11-1 Google's high-level enterprise architecture

Why Is Enterprise Architecture Important?

Enterprise architecture provides the overall foundation for achieving an organization's strategic vision. Enterprise architecture can enable organizations to facilitate the delivery of new products and services, to be the catalyst for change, to be more agile, and to provide meaningful value propositions from their strategic initiatives, all at the lowest possible total cost of ownership. To meet changing business requirements, enterprise architecture must be in a state of constant evolution. Ineffective enterprise architecture can jeopardize the success of an organization. If enterprise architecture is not in place, is mismanaged, or is built without flexibility, it can cause service outages, problems with products, and failure.

The example of JetBlue Airways illustrates the consequences of not implementing an adequate enterprise architecture.[7, 8, 9] JetBlue was forced to cancel a large number of flights in February 2007 due to severe ice and snow storms. JetBlue was unable to give its passengers and many of its employees sufficient information about the canceled flights. As a result, many passengers had to wait in grounded airplanes for as long as 11 hours during the storms.[10] The airline's enterprise architecture was incapable of performing all of the required passenger and crew rescheduling tasks. In addition, no system was in place to keep track of off-duty crews of pilots and flight attendants. JetBlue's reservation system was overloaded, customers could not reschedule flights themselves, and not enough employees were trained to use the reservation system. To compound matters, JetBlue did not have a baggage tracking system that was capable of dealing with the snafu. After JetBlue's major service disruptions, staffers at the company posted then-CEO David Neeleman's apology to customers on YouTube. By comparison, other airlines were able to cancel their flights much earlier than JetBlue, so their passengers were able to avoid much of the airport congestion. In addition, other airlines use baggage tracking systems to track luggage throughout the route.

Today a company can do all the right things but still fail unless it has viable enterprise architecture to guide its strategic direction. The enterprise architecture enables managers to do the following:

- Increase employees' effectiveness by enabling high-order thinking.
- Develop new value propositions of interest to customers.

Enabling High-Order Thinking

The new global economy is becoming increasingly complex and competitive. Organiza-tions need to look beyond product differentiation and cost reduction to provide customer value. An organization's employees must be able to transition from routine execution of daily tasks to high-order thinking: understanding, forming perspectives, and thinking critically about new ideas or ways of doing things differently. High-order thinking is vital for adapting to the new global economy, and a major type of high-order thinking is the ability to innovate. According to Clayton Christensen, there are two broad categories of innovation:

- Radical innovation (disruptive technology), which creates such dramatic change that it transforms existing industries or creates new ones. Such innovation generally accomplishes one or more of the following: creates an entirely new set of performance features, improves performance by a factor of five or more, or reduces costs by 30 percent or more.[11]
- Incremental improvement (sustaining technology), a process of implementing continual small enhancements to a process, resulting in slow but steady improvement.

If an organization "stays too close to customers," bringing out only products and changes requested by the customer, the result may be less emphasis on more viable, disruptive technologies.[12] As Christensen wrote, leading firms cannot allow themselves to be "held captive by their customers, enabling attacking entrant firms to topple the incumbent industry leaders each time a disruptive technology emerges."[13]

The problem with staying too close to customers is that sometimes they cannot tell you what they really want. For example, in 1995, Ford Motor Company asked its customers if they wanted a second sliding door on the Windstar minivan. When research showed that customers were not interested, Ford nixed the idea. However, Chrysler correctly anticipated that such a feature would be highly desirable. Ford was wrong, Chrysler was right, and Ford paid more than $500 million to correct its mistake. The Sony Walkman is another example of a product for which no customer asked, yet it was a great success.

The new global economy will require more innovation, which in turn will require more intellectual capital. Intellectual capital is the knowledge of the workforce.[14] It is the aggregate of an organization's capacity for deep thinking, domain knowledge, problem solving, and creative skills. Intellectual capital enables an organization to provide value to customers and to create innovations.

Figure 11-2 represents the three-layer approach to thinking about innovation. Forward-thinking organizations will digitize their stable basic processes and incrementally improve them, liberating the managers to work on strategic thinking and innovation.

A research study of 147 companies from 1998 to 2002 found that 24 percent of a company's sales were from new products (level 2 and level 3) introduced in the prior three years. In that time period, one-third of the companies were able to achieve 50 percent of total sales from new products. The 50 percent group had a higher percentage of their basic business functions digitized (level 1).[15] This lends credence to the theory that automation of standard business processes liberates the high-order thinking required for innovation.

As we saw with Google in the opening vignette, an effective enterprise architecture must support the business strategy and enable an organization to increase its competitiveness. If the enterprise architecture is implemented thoughtfully, then routine business tasks are handled smoothly, efficiently, and reliably. Employees do not have to devote as much effort to ensuring the ongoing operation of fundamental business processes. This enables managers to spend more time on high-order thinking and innovation and less time on routine tasks such as reordering spare parts.

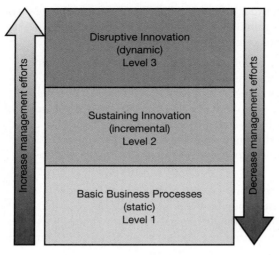

FIGURE 11-2   Three levels of innovation

Developing New Value Propositions

A well-designed enterprise architecture creates a foundation of common business processes and shared data that ultimately allow for the development of new value propositions. **Value propositions** provide a clear statement of the tangible benefits that a customer obtains from using a company's products or services.

Managers need to participate as stakeholders in the development and implementation of enterprise architecture so they can convey the type of value propositions being conceived. Effective organizations use technology to provide value propositions to external and internal customers. Internal customers are interested in the value that the enterprise architecture provides, including what capabilities it has and what problems it solves, rather than the underlying technology. External customers will attempt to simplify their efforts by searching for products and services that are:[16, 17]

- The most innovative (Apple, Google, Toyota, General Electric, Microsoft)
- The least expensive (Wal-Mart, Southwest Airlines, JetBlue)
- The best quality for their needs (Procter & Gamble, UPS)
- The most distinctive and familiar brand (McDonald's, Coca Cola, Harley Davidson, BMW)

The result of enterprise architecture is to provide a solid foundation for both internal and external customers: "companies with a solid foundation had higher profitability, faster time to market, and lower IT costs."[18]

For example, the GM OnStar System is a byproduct of the GM enterprise architecture, and it illustrates a unique value proposition. GM's OnStar is a telematics subscription service for in-vehicle safety, diagnostics monitoring, security, wireless communications, location tracking, auto navigation, and Web access. Telematics is the transmission of data communications between systems and devices. Drivers can communicate with advisors at the OnStar Center any time. OnStar consists of four different types of technology that work together as subsystems: hands-free cellular service, a virtual advisor that can use voice recognition to search the Web for information, a global positioning system (GPS) that uses satellites for location identification, and vehicle telemetry for diagnostic and emergency information.[19]

Software Architecture Styles

A city has many different types of housing styles, such as colonial, Cape Cod, art deco, contemporary, Georgian, French provincial, Queen Anne, Tudor, and ranch. Similarly, software architecture includes multiple styles of computing, which are separated broadly into two categories: centralized and distributed. Almost all new software applications are built using the distributed model because it provides for lower costs and overall higher value. New solutions are likely to be added to older solutions, resulting in a mixture of different architectures. Packaged solutions also come with their own software architecture styles that ideally complement the existing enterprise architecture. Figure 11-3 describes the lineage of the principal software architectural styles.

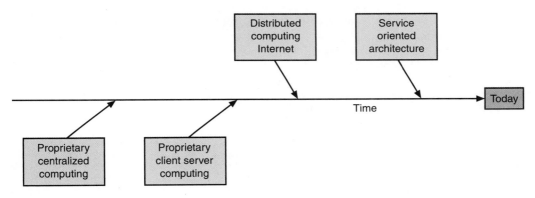

FIGURE 11-3 History of computing styles

Centralized Architecture

Centralized architecture is based on the use of a mainframe computer that supports a variety of local and remote devices, such as printers, terminals, and workstations. The mainframe computer maintains tight control over the software applications that run on it as well as the associated data that it manages. On the other hand, it is difficult to add incremental amounts of mainframe computing capacity to handle increased demands for additional processing, because mainframes are not easily scalable. Centralized architecture and mainframe computing are used frequently to process high volumes of transactions, such as credit card transactions, customer billing, and automated teller machine transactions. IBM has developed a large selection of technology to support centralized computing, including mainframe operating systems, transaction systems such as the Customer Information Control System (CICS) Transaction Server, middleware such as WebSphere, and databases such as the DB2 database and IMS (Information Management System). An estimated 95 percent of the world's banks use IBM mainframe computers.[20]

Distributed Architecture

John Gage of Sun Microsystems created the phrase "The Network is the Computer" to describe distributed computing.[21] In a distributed computing model, processing functions and data can reside anywhere on the network or the commercial Internet.

Distributed applications share the processing, formatting, presentation, and storage functions across clients and servers. There is less reliance on proprietary software and more emphasis on open standards. With the distributed model, processing capacity is much more scalable by adding more servers or upgrading to faster servers. Such upgrades are much simpler and less costly than upgrading to a faster mainframe computer. Almost all new development is moving toward the distributed model built using Web-oriented tool sets.

Client/Server Architecture

Client/server, a type of distributed architecture, is a general-purpose model of network computing with the following parts (see Figure 11-4):

- The client requests services and resources over the network.
- The server provides resources and services over the network.
- The network provides the mechanism for the client and server to communicate.
- The database provides the functionality to create, read, update, and delete stored values.

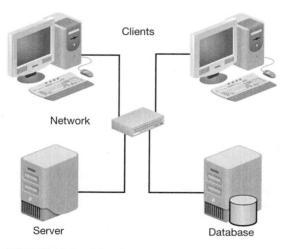

FIGURE 11-4   Client/server processing

In a client/server system, a client requests information from a server and the server performs a database request to the database server. Client/server architecture provides for a separation of responsibilities and enables the application to be organized in layers:

- The Presentation layer manages the customer experience and user interfaces.
- The Application layer allows for programming and codifying of business rules.
- The Database layer stores and accesses data values.

Figure 11-5 is an example of the separation of layers using client/server architecture.

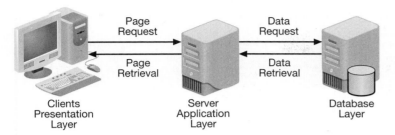

FIGURE 11-5   Separations of layers

Tiers are physical units such as servers or mainframes. Client/server layers can be deployed on separate physical tiers. For example, each of the three layers can be deployed onto one to three physical tiers.

Service-Oriented Architecture

Service-oriented architecture (SOA) is a software application development approach based on building user applications out of software services. A software service is a unit of work developed by a service provider (a piece of software) to achieve desired results for a service consumer (another piece of software with which the end user interacts). Services could include such activities as completing an online credit card application, booking a reservation online, or requesting an online mortgage rate quote. A well-defined set of rules or protocols describe how one or more services can "talk" to each other. As a result of the standards and protocols associated with SOA, an SOA-based application structure works similarly to the way your Web browser accesses information and services on the Internet. Regardless of what browser you use or what computer you have, you can still access and interact with any Web services.

To use SOA to build a comprehensive set of services that support a business initiative—for example, a system to support the online customers of a stock brokerage firm—business managers and IT people must define the services to be offered and design how to link and sequence the necessary services. This process is sometimes called orchestration. For example, Fidelity Investments uses the SOA architecture to provide services to its online investors (see Figures 11-6 and 11-7). The services must be well thought out and designed in advance. The services are linked so that after logging in, the investor can display the current value, original value, and the gain/loss for each investment in the portfolio. The investor also can buy and sell stocks, bonds, or mutual funds; perform research on various investments; and administer the account (for example, change the password or specify that dividends automatically be reinvested). Each of these actions is handled by a separate service; the orchestration of the entire set of services enables Fidelity to provide excellent business support for its customers.

FIGURE 11-6 Fidelity Investments Web services support online investors

The ability to respond to unanticipated changes in the business environment is a key advantage of SOA. This flexibility is achieved by establishing a *loosely coupled* relationship between services so little or no dependency exists between the services. Thus, a software module can be modified to meet new business needs with little or no impact on other services. Or, new services can be added easily without requiring a change to an existing service.

Another advantage of SOA architecture is that services can be implemented and made available gradually. Over time, the collection of services can provide a comprehensive set of capabilities for end users. Such an approach avoids the cost, delay, and risk associated with a large-scale, all-at-once implementation of all services.

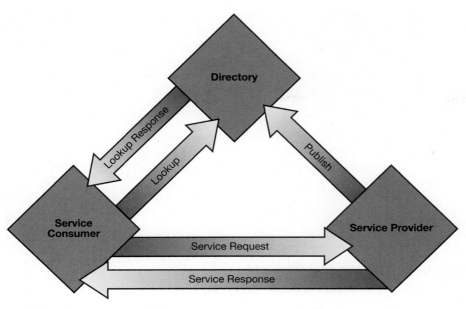

FIGURE 11-7   Web Services support service-oriented architecture

A MANAGER TAKES CHARGE

American Modern Converts to Service-Oriented Architecture

American Modern Insurance Group is the specialty insurance subsidiary of the Midland Company, with more than 40 years of experience in manufactured housing insurance. American Modern provides specialized products and services for owners of motor homes, travel trailers, boats, personal watercraft, classic cars, motorcycles, and snowmobiles.

American Modern decided that its aging enterprise architecture was no longer capable of keeping up with the changing needs of the business. Vice President of Infrastructure Patrick Law led a $62 million project to replace the insurer's 30-year old casualty policy administration system. The scope of the project included converting from a proprietary, centralized architecture to a service-oriented architecture. To accomplish this goal, American Modern had to revamp its existing infrastructure of mainframe computers, databases, and core business applications. The firm replaced its two Unisys mainframe computers with a single IBM zSeries mainframe. It is migrating from Unisys and Oracle databases to IBM's DB2. The firm's internally developed casualty policy system is being replaced with an IBM application.[22]

continued

American Modern has been concentrating on choosing the right technology components required to make everything work and then training employees to use these new tools. As the firm shifts to SOA, it is converting its original COBOL modules into hundreds of smaller component modules that will provide all the capability of the original application and provide much-needed agility to add new features as business needs change.

Discussion Questions:

1. What was the business case for American Modern to spend $62 million to move to a service-oriented architecture?
2. What role would business managers play in justifying and implementing this project?
3. What advantages does American Modern gain from using an SOA approach? Are there any drawbacks to using this approach?

Developing an Enterprise Architecture

Developing an enterprise architecture is like planning a city. The vision of a city is to provide the overall future layout for streets, schools, businesses, retail areas, parks, and infrastructure. The city provides basic common services such as fire and police protection. The plan for the city can include sketches of each property's layout, blueprints of important buildings, and three-dimensional models.

Enterprise architecture also is planned using models. The **Unified Modeling Language (UML)** is a language for specifying, constructing, visualizing, and documenting the artifacts of a software-intensive system. Like musical notation, which specifies and documents music to enable the conductor and musicians to perform harmoniously, the UML specifies and documents software systems to enable system builders and users to work well together. With the knowledge of customer needs and trends in place, the process of developing enterprise architecture can begin. Each organization needs to develop its own approach to realizing enterprise architecture. The key objective of enterprise architecture is to build a foundation that will enable change and meet the next generation of needs.

The Boeing Story

The Wright Brothers invented the first powered fixed-wing airplane and were the first to fly it. Their innovation was to design the capability to navigate an airplane using three controls simultaneously. The three controls were the ability to roll the wings to the right or left, the ability to raise and lower the nose, and the ability to turn the nose from side to side. These capabilities are the basis for modern aircraft, submarines, and spacecraft.[23] The Wright brothers built their airplanes by hand from spruce, a strong lightweight wood. After a number of years of experiments and refinements, they created the Wright Flyer in 1905, the first practical airplane.

While the basic design of the plane remained the same for several decades, the methods used to build it changed dramatically with the introduction of standardized parts and the assembly line. The airplane evolved into an increasingly complex set of subsystems that must work together. During World War II, the aircraft industry was required to develop war planes very quickly. Their approach was to develop the subsystems separately without necessarily considering the whole. For example, the engines, weapon system, and airframe were developed independently. When the World War II fighter airplanes were assembled, it was amazing that they worked effectively because they were not designed in an integrated manner.[24]

Today we can design complex aircraft holistically and expect the separately designed subsystems to function together as a whole. For example, Boeing plans to deliver the first 787 Dreamliner, a high-performance and low-emissions airplane, in the third quarter of 2009.[25] The Dreamliner will use 20 percent less fuel than similar planes. More than half its components will be built out of lightweight composite materials instead of heavy metal. For example, the fuselage of the 787 is a one-piece part made of a lightweight composite material. This one-piece part eliminates 1,500 aluminum sheets and 40,000 to 50,000 fasteners, simplifying the assembly.[26] The fuselage performs a single, well-defined cohesive function and minimizes the coupling of aluminum sheets and connectors that was formerly required.

The plane will use two manufacturers' engines interchangeably (General Electric and Rolls-Royce). It will be built in pieces from plants in Japan and Italy and then assembled in the state of Washington.[27] A plane can be assembled at the rate of one every three days. The Boeing story is representative of both risk-taking and innovation in the new global economy. The Boeing 787 Dreamliner is an innovative and complex product that required a tremendous amount of planning to design and build.

In information technology, organizations need to develop the same level of sophistication as Boeing's to build the enterprise architecture and enable the assembly of large-scale reusable components that can sustain an enterprise's business needs. In the context of hardware and software, **cohesion** is a measure of how strongly related and focused the various responsibilities of a software or hardware component are. **Coupling** is a measure of the degree to which each software and hardware component relies on other modules to perform its function. In software and hardware, the ideal component is one that is highly cohesive and has low coupling, just like the Boeing fuselage. This simplifies component design and makes it easier to modify components in the future without affecting other components.

The notion of standard parts in software was envisioned many years ago. But the industry struggled with the concept as it continued to develop software in an ad hoc manner, often building it from scratch without much consideration for building reusable components like standard parts.

An example of a complex, large, high-risk computer system is the one that operates the International Space Station. Its computer system has five major subsystems and 100 computers. The computers' subsystems are the U.S. Command and Data Handling (CDH) System, the Russian Onboard Complex Control System (OCCS), the Canadian Computer System, the Japanese Data Management System, and the European Data Management System. However, even with the extensive organization of the International Space Station into subsystems and components, computers and related software are difficult to diagnose and correct. When something goes wrong, as it did with the International Space Station in June 2007, it is much easier to localize the problem when the system consists of discrete subsystems and components that perform well-defined functions.[28, 29]

Business Processes

Enterprise architecture enables the implementation of a set of digitized business processes. The commonality of business processes across business units determines a set of potential repetitive patterns that add value if digitized. An example of a business process is the highly secret computer programs used to power the Google searches. The most useful business processes provide value propositions that deliver substantial customer benefits, that are difficult for competitors to imitate, and that can be used for innovative products and services. It takes the unique knowledge and skills of the organization's workers—the intellectual capital of the organization—to develop these business processes effectively.

Cross-business unit common processes are illustrated in Figure 11-8. Examples of these processes include activities required to capture a customer order, plan its shipment to meet the customer's desired delivery date, build and ship the order, bill for the order, and provide post-sales customer service. These processes all require the ability to access, update, and share data about orders, shipments and customers.

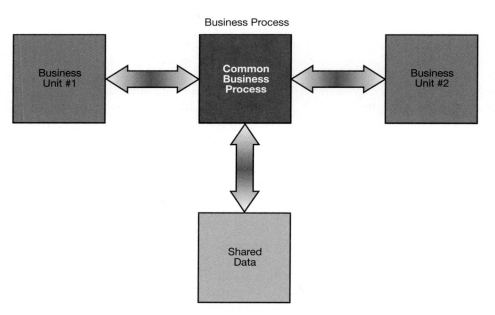

FIGURE 11-8 Cross-business unit common processes

The pharmaceutical industry provides an example of how organizations must design an enterprise architecture that not only meets business needs but allows data and information to be used to save lives. Roche is an innovation-driven healthcare organization that provides products to the global diagnostics market and supplies pharmaceuticals for cancer treatment. In Roche's vision of the future, therapies will be available for many of today's untreatable diseases, drug efficacy and safety will be optimized, and effective strategies will be in place for preventing disease. Early diagnosis and improved new treatments will significantly reduce the need for expensive surgeries and long hospital stays. To this end, Roche is developing a wide range of products and services to determine disease predisposition, provide health information to help prevent or delay the onset of illness, diagnose disease, treat numerous diseases and conditions, and monitor the progress of therapy.

Roche employs a global pharmaceutical R&D network that includes more than 5,000 scientists at four research centers, which are dedicated to providing clinically differentiated drugs to address unmet medical needs. Roche couples strong in-house R&D capabilities and alliances with numerous partners around the world, including Genentech and Chugai. In China, Roche is collaborating with the Chinese National Genome Centres on genetic epidemiology studies that identify genetic predispositions to conditions such as diabetes and Alzheimer's disease.

Roche had to take all these needs into consideration in designing its enterprise architecture. The architecture consists of a secure, reliable global network; computers and software applications that can be used by employees and contractors; and databases of clinical and trial data required for FDA approval of new products.

Process for Developing an Enterprise Architecture

Many organizations have tried and failed to define an effective enterprise architecture. In 1996, Congress passed the Clinger/Cohen act, which gave the Office of Management and Budget the authority to dictate standards for "developing, maintaining, and facilitating the implementation" of an enterprise architecture. Since then, several organizations within the U.S. government have been audited by the General Accounting Office and found to be sorely lacking in terms of a well-developed enterprise architecture—the IRS, Department of Homeland Security, the FBI, FEMA, Census Bureau, Federal Aviation Agency, National Air and Space Administration, HUD, Health and Human Services, Medicare, and Medicaid.[30] The private sector has had problems as well, although it has not been as widely publicized. McDonald's, Ford, and Kmart all had major IT failures that cost more than $100 million and were attributed to poor enterprise architecture.

Numerous approaches exist for creating and documenting an enterprise architecture. One popular approach is The Open Group Architecture Framework (TOGAF), an industry standard architecture framework that has been evolving since the mid-1990s. TOGAF is fully documented and may be used freely by any organization that wants to develop an IT architecture.[31]

TOGAF divides enterprise architecture into four components:

1. Business architecture that describes the processes the business uses to meet its goals
2. Application architecture that describes how specific applications are designed and how they interact with each other
3. Data architecture that describes how the enterprise data is organized and accessed
4. Technical architecture that describes the hardware and software infrastructure that supports applications and their interactions

The TOGAF process involves nine steps, as outlined in Table 11-1. It is typically applied within a given business area (such as human resources) or a couple of related business areas rather than across the entire enterprise at one time. This constraint keeps the process more manageable and ensures that useful results are delivered within a reasonable time. Once an architecture is defined for one area, the process can move on to other business areas.

TABLE 11-1 The TOGAF approach to generate an enterprise architecture

Phase	Objectives
1. Framework and principles	1. Ensure that everyone understands the process and is comfortable with it 2. Modify the process as necessary to fit the organization and its needs 3. Set up a governance process that will oversee future architectural work
2. Architecture vision	1. Ensure that the necessary support exists within the organization for the enterprise architecture project 2. Define the scope of the project 3. Identify project constraints such as cost and/or time 4. Document business requirements that must be supported by the architecture 5. Establish first-cut, high-level definitions of both the existing and desired future architecture, in terms of the business, application, data, and technical architecture
3. Business architecture	1. Create a more detailed description of the current and desired future business architecture 2. Clearly delineate the gaps between the existing and desired future business architecture
4. Information systems architecture	1. Develop a detailed description of the existing data architecture, including logical data models and relationship models that relate business functions to Create, Read, Update, and Delete data operations 2. Define requirements for performance, reliability, security, and integrity 3. Identify gaps between the existing state and desired future architecture
5. Technology architecture	Define technology choices appropriate to support the proposed new architecture
6. Opportunities and solutions	Identify and evaluate various implementation approaches and projects required
7. Migration planning	Work with the established governance body to prioritize and sequence projects based on costs, benefits, and risks
8. Implementation governance	Create an architectural specification for each project to be implemented
9. Architecture change management	Update the existing architectural plan with information gained from the latest projects

No matter what process is used, developing an enterprise architecture requires defining a future structure for an organization's processes, information systems, personnel, and organization subunits, so that they align with the organization's core goals and strategic direction.[32] While developing the enterprise architecture, an organization must answer the following questions:

- What objectives, goals, strategies, and measures are important to the organization?
- What new products or services does the business want to deliver?
- How is the business organized into autonomous business processes?
- How are those business processes related to each other?
- Which business processes are most in need of improvement?
- What is the plan for making those improvements?

The process of defining the enterprise architecture is never really finished. Instead, the enterprise architecture is a continually evolving set of documents that guides the use of technology.

This chapter has explained what is meant by the term enterprise architecture, discussed why enterprise architecture is important to the future of the organization, identified some architecture styles, and outlined a process for developing an enterprise architecture. It is critical that managers recognize that they have an important role in helping the technical people define the desired enterprise architecture. They must convey the value propositions being considered so that the enterprise architecture is designed and built to support these future needs.

Chapter Summary

- Enterprise architecture provides a foundation for achieving an organization's strategic vision and delivering new products and services.

- The value of enterprise architecture is the timely availability of information both to internal and external customers that enables both incremental and radical innovation.

- The enterprise architecture enables routine business tasks to be handled smoothly, efficiently, and reliably, which frees up managers to perform high-order thinking.

- Software architecture styles represent how processes and information can be organized to achieve the strategic direction of an organization.

- The service-oriented architectural style consists of a set of standard reusable and extensible building blocks known as services, which enable managers to spend more time focusing on high-order thinking instead of routine managerial functions.

- Service-oriented architecture provides the loose coupling of application building blocks for reuse that supports a multivendor and multiplatform technology infrastructure.

- If the enterprise architecture is built thoughtfully, an organization can deal with inevitable changes by reassembling the building blocks and adapting to increased competition in a more agile way.

- No matter what process is used, developing an enterprise architecture requires defining a future structure for an organization's processes, information systems, personnel, and organization subunits, so that they align with the organization's core goals and strategic direction.

Discussion Questions

1. What is enterprise architecture?
2. Why is it important for managers to understand enterprise architecture?
3. How does the tenet "form ever follows function" apply to the design of an enterprise architecture?
4. What value proposition does enterprise architecture provide?
5. What is an enterprise architecture style?
6. What are the two major enterprise architecture styles?
7. What is service-oriented architecture? What are its advantages?
8. How does enterprise architecture allow for the development of new value propositions?
9. Identify and briefly discuss the three levels of innovation.
10. Can an organization be successful without enterprise architecture? Discuss fully and identify an example to support your position.
11. What is meant by high-order thinking? How does enterprise architecture enable high-order thinking?
12. Describe a centralized architecture. Give an example of such an architecture.

Action Memos

1. You are the manager of human resources for a midsized manufacturing organization. Your team is about to embark on a six-month project to develop an intranet-based, self-service Web site for the Human Resources (HR) Department. The consultants leading this effort have asked you to send a brief e-mail to HR members, in which you request their cooperation and participation in a one-day session to define the services to be provided. Draft an e-mail that would encourage participation and make clear why it is important to the project's success.

2. You are a new CIO of a large financial organization. Your department has a history of not meeting customer needs, and it has spent too much on proprietary technology. The department's schedules have not been met, and the perception around the company is that the department is not responsive to change. As you sit in a staff meeting, you are surprised to hear the CEO propose that the organization develop a major new product line that will require a substantial investment in Web technologies. The CEO has asked for a brief response from each of her direct reports. How would you respond?

Web-Based Case

Use the Web to research an organization that employed UML to define its enterprise architecture. Discuss the process that the organization followed. What were the benefits of defining their enterprise architecture?

Case Study

Healthcare: A Model Needing Transformation

In his book "Crossing the Chasm,"[33] Geoffrey Moore identified a technology adoption life cycle that included five segments: innovators, early adopters, early majority, late majority, and laggards (see Figure 11-9). Healthcare can be considered part of the laggards segment.

Why is healthcare in the laggards segment? As Maggie Mahar wrote in 2007, "The U.S. spends more on healthcare than any other nation. Does that money buy what it should? Not according to decades of Dartmouth research on regional variations in spending and outcomes."[34]

There is nothing more fundamental than healthcare. In the United States, however, it is one of the most neglected of disciplines.[35] According to *Consumer Reports* magazine, "The U.S. spends an average of $7,000 per capita on healthcare. According to a 2007 analysis by McKinsey Global Institute, that's 28 percent more than any other industrialized country even after adjusting for its relative wealth."[36]

Consider the following:

- There is a gap between the knowledge and practice of medicine, with poor service and a 500 percent variation in rates of some surgical procedures from city to city. This gap is being addressed by evidence-based medicine, an approach that adopts standardized procedures for the treatment of diseases.[37]

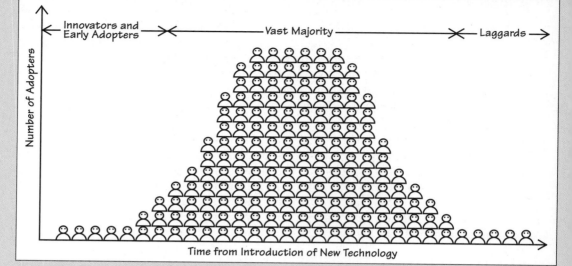

FIGURE 11-9   Adoption of New Technology

- In 2004, nearly 46 million people in the United States had no medical insurance.[38] The cost of healthcare in the United States is 40 percent higher than that of the next most expensive nation. According to Mahar, "Chronically ill patients who receive the most intensive, aggressive, and expensive treatments fare no better than those who receive more conservative care. In fact, their outcomes are often worse."[39] According to authors Anne Gauthier and Michelle Serber, "High spending has not translated into better health: Americans do not live as long as citizens of several other industrialized countries, and disparities are pervasive, with widespread differences in access to care based on insurance status, income, race, and ethnicity."[40]

- Adverse events harm patients far too often: "There are also significant issues with the safety and quality of care. As many as 98,000 deaths result annually from medical errors, and U.S. adults receive only 55 percent of recommended care. Inefficiencies, such as duplication and use of unnecessary services, are costly and compromise the quality of care. High administrative costs in health insurance and healthcare delivery are also problems."[41]

- Patients with chronic diseases, who account for 75 percent of all healthcare expenditures, are most vulnerable.

- A large majority of hospitals do not use an electronic health record (EHR) system. According to Marianne Kolbasuk McGee, "There's a troubling lack of urgency in much of the industry toward EHRs and data sharing, despite the lives being lost to mistakes that IT-enabled healthcare might help prevent and the potential for cost savings. We all have a stake, as users of the healthcare system. Companies want more progress because they're feeling the pain of rising healthcare costs for employees, and they believe in IT's role in lowering that—just as they've applied tech to improve processes at their own companies. Anyone who's pushed tech-driven transformation can understand why it's difficult."[42]

Electronic Health Records (EHRs)

Healthcare involves a complex set of interrelated processes of vital importance, but healthcare organizations are lagging in the implementation of information technology solutions. Gartner Research indicates that less than 10 percent of healthcare organizations have fully implemented EHR systems.[43]

Healthcare organizations use a large number of disparate systems that were implemented at different times in different divisions and for different purposes, resulting in a lack of integration. Most healthcare organizations use paper-based manual procedures for their medical records, which cause inefficiencies in the practice of healthcare. EHR systems could alleviate a number of current problems with healthcare organizations.

McGee wrote, "Despite several years of concerted national effort, including President Bush's rallying cry in 2004 to get most Americans on e-health records by 2014, the use of digital records is in a precarious place. Just 10 percent of doctors' offices use them. And while hospitals are expanding their use, the most difficult work—the exchange of data among healthcare providers, especially with rivals—has barely begun."[44]

Veterans Health Administration (VHA)

The Veterans Health Administration is the largest integrated healthcare system in the United States, with 1,400 hospitals, clinics, and nursing homes. According to the American Customer Satisfaction Index, patients served by the VHA scored their care 10 percent higher than patients at private hospitals.[45]

The VHA is a leader in the implementation of EHRs. The VHA does better than private hospitals in customer satisfaction and costs. The VHA's cost per patient has remained steady at $5,000 for the past 10 years. Meanwhile, the consumer price index for medical care (what families pay in the private sector for care) has increased about 40 percent.[46]

Douglas Waller wrote, "With 51 percent of its patients 65 or older, the VA has pioneered research in geriatric care. In 2006 the journal *Medical Care* reported that Boston University and the VA reviewed 1 million records from 1999 to 2004 and found that males 65-plus who received VA care had about a 40 percent decreased risk of death compared with those enrolled in Medicare Advantage's private health plans or HMOs."[47]

The Veterans Health Information Systems and Technology Architecture (VistA) is an integrated system of software applications that directly supports patient care at VHA healthcare facilities. VistA connects VHA facilities' workstations and PCs with nationally mandated and locally adapted software applications that are accessed by users through a graphical user interface known as the Computerized Patient Record System (CPRS).[48] The VHA has electronic health records for all of its patients. Regardless of where a patient is treated, the physician has access to the patient's complete medical history. At the VA, a bar code on the patient's wrist is scanned, and then the prescription bar code is scanned to make sure that it matches the physician's prescription.

In addition, VHA prescription drug costs are relatively low. The VHA is able to negotiate with the pharmaceutical industry to achieve better pricing. In contrast, the industry successfully prevented Medicare from negotiating the prices it pays for drugs.[49]

Discussion Questions:

1. Discuss why it is important for healthcare organizations in general and the VHA in particular to develop a robust IT enterprise architecture.

2. Identify issues or problems that make it difficult for the VHA to develop an enterprise architecture.

3. Outline a process for developing an enterprise architecture and identify some of the people who should be involved in each step of the process.

4. Do research on the Internet to find at least three documents related to the VHA enterprise architecture. Briefly summarize each document in a paragraph that describes the specific project or business area addressed.

Endnotes

1 Sarah Winchester House (accessed at *www.winchestermysteryhouse.com/*, 18 June 2007).

2 "Company Overview" (accessed at *www.google.com/corporate/index.html*, 29 August 2007).

3 Jena McGregor, "The 50 Most Innovative Companies," *BusinessWeek*, 4 May 2007.

4 Bill Barr, *opensourceto.blogspot.com*, 29 November 2006.

5 Google platform (accessed at *http://en.wikipedia.org/wiki/Google_platform*, 30 June 2007).

6 Louis Henry Sullivan, "The Tall Office Building Artistically Considered," *Lippincott's Magazine*, March 1896.

7 Jeff Bailey, "JetBlue's C.E.O. Is 'Mortified' After Fliers Are Stranded," *New York Times*, 19 February 2007.

8 Doug Bartholomew and Mel Duvall, "What Really Happened At JetBlue," *CIO*, 5 April 2007.

9 Susan Carey and Darren Everson, "Lessons on the Fly: JetBlue's New Tactics," *Wall Street Journal*, 27 February 2007.

10 CBS News, "JetBlue Attempts To Calm Passenger Furor" (accessed at *http://cbs2.com/national/topstories_story_046063757.html*, 6 July 2007).

11 Richard Leifer, Lois Peters, and Gina O'Connor, *Radical Innovation*, Harvard Business School Press, 2000.

12 Clayton Christensen, "The Innovators Dilemma" (accessed at *www.businessweek.com/chapter/christensen.htm*, 21 April 2007).

13 Clayton Christensen, *The Innovator's Dilemma*, Harvard Business School Press, 1997.

14 Boeing Web site, "International Industrialization," Boeing Frontiers (accessed at *www.boeing.com/news/frontiers/archive/2006/july/*, 31 May 2007).

15 Jeanne W. Ross, Peter Weill, and David Robertson, *Enterprise Architecture As Strategy: Creating a Foundation for Business Execution*, Harvard Business School Press, June 2006.

16 Ravi Kalakota and Marcia Robinson, *E-Business 2.0: Roadmap for Success*, Addison-Wesley, Reading, MA, 2002.

17 Efraim Turban, D. King, J. Lee, and D. Viehland, *Electronic Commerce: A Managerial Perspective*, Prentice Hall, Inc., Upper Saddle River, NJ, 2006.

18 Jeanne W. Ross, Peter Weill, and David Robertson, *Enterprise Architecture As Strategy: Creating a Foundation for Business Execution,* Harvard Business School Press, June 2006.

19 GM's OnStar Web site (accessed at *www.onstar.com/us_english/jsp/index.jsp*, 15 June 2007).

20 "IBM Eyes Mainframe Security," *CIO*, 30 March 2006.

21 John Gage (accessed at *http://en.wikipedia.org/wiki/John_Gage*, 17 June 2007).

22 Steve Ulfelder, "American Modern Pioneers SOA in Infrastructure Overhaul," *Computerworld*, 13 March 2006.

23 The Wright Story (accessed at *www.first-to-fly.com/ History/Wright%20Story/wright%20story. htm*, 4 July 2007).

24 Wright Brothers (accessed at *http://en.wikipedia.org/wiki/Wright_brothers*, 19 June 2007).

25 J. Lynn Lunsford and Daniel Michaels, "Airbus Meets Pressure to Deliver on A350," *Wall Street Journal*, *http://online.wsj.com/article_print/SB117859020100095343.html*, 8 May 2007.

26 Boeing Web site (accessed at *www.boeing.com/news/feature/sevenseries/787.html*, 6 July 2007).

27 Boeing Web site, "International Industrialization," Boeing Frontiers (accessed at *www.boeing. com/news/frontiers/archive/2006/july*, 31 May 2007).

28 "Computer crash hits space station," *http://news.bbc.co.uk/2/hi/science/nature/6752459.stm*, 14 June 2007.

29 Mission Operations Directorate, Space Flight Training Division, "International Space Station Familiarization" (accessed at *www.vision-play.com/products/game1/ISS_Manual.pdf*, 15 June 2007).

30 Roger Sessions, *A Better Path to Enterprise Architectures*, Microsoft Corporation, April 2006.

31 The Open Group Web site (accessed at *opengroup.org/togaf/*, 25 October 2007).

32 Wikipedia.

33 Geoffrey Moore, *Crossing the Chasm*, Harper Business, 1991.

34 Maggie Mahar, "The State of the Nation's Health," *http://dartmed.dartmouth.edu/spring07/html/ atlas.php*, Spring 2007.

35 Institute for Health Improvement Web site (accessed at *www.ihi.org/IHI/Topics/Improvement/ caseforimprovement.htm, 21 July 2007*).

36 "Are You Really Covered?" *Consumer Reports*, page 19, September 2007.

37 Institute for Health Improvement Web site (accessed at *www.ihi.org/IHI/Topics/Improvement/ caseforimprovement.htm, 21 July 2007*).

38 Julia King, "The Grill: Dealing with Darwin Author Geoffrey A. Moore on the Hot Seat," *Computerworld*, 16 July 2007.

39 Maggie Mahar, "The State of the Nation's Health," *http://dartmed.dartmouth.edu/spring07/html/ atlas.php*, Spring 2007.

40 Anne Gauthier and Michelle Serber, "A Need to Transform the U.S. Health Care System: Improving Access, Quality, and Efficiency," The Commonwealth Fund, *www.cmwf.org/ publications/publications_show.htm?doc_id =302833&#areaCitation*, October 2005.

41 Ibid.

[42] Marianne Kolbasuk McGee, "Why Progress Toward Electronic Health Records Is Worse Than You Think," *www.informationweek.com/news/showArticle.jhtml?articleID=199702199*, 26 May 2007.

[43] Gartner Research, "Magic Quadrant for North American Enterprise CPR," 31 March 2006.

[44] Marianne Kolbasuk McGee, "Why Progress Toward Electronic Health Records Is Worse Than You Think," *www.informationweek.com/news/showArticle.jhtml?articleID=199702199*, 26 May 2007.

[45] Government Satisfaction Scores, American Customer Satisfaction Scores (accessed at *www.theacsi.org/index.php?option=com_content&task=view&id=162&Itemid=62*, 10 May 2007).

[46] Douglas Waller, "Vetting the VA," *www.aarp.org/bulletin/medicare/bulletin/yourhealth/vetting_va.html*, May 2007.

[47] Ibid.

[48] Veterans Health Information Systems and Technology Architecture (accessed at *www.virec.research.va.gov/DataSourcesName/VISTA/VISTA.htm*, 8 May 2007).

[49] Jerome Groopman, *How Doctors Think*, Houghton Mifflin, 2007.

324

Additional Bibliography

Chris Zook, "Googling Growth," *Wall Street Journal*, page A-12, 9 April 2007.

Deborah Perelman, "Google Aims to Extend Data Mantra into Health Care," *eWeek, www.eweek.com/article2/0,1895,2138333,00.asp?kc=EWKNLDAT053107STR1*, 29 May 2007.

Sergey Brin and Lawrence Page, "The Anatomy of a Large-Scale Hypertextual Web Search Engine," Computer Science Department, Stanford University (accessed at *http://infolab.stanford.edu/~backrub/google.html*, 30 June 2007).

"How Google Grows...and Grows...and Grows" (accessed at *www.fastcompany.com/magazine/69/google.html*, 30 June 2007).

Susan Kuchinskas, Internet News, "Peeking Into Google," *www.internetnews.com/bus-news/print.php/3487041*, 2 March 2005.

Paul Sloane, *The Leader's Guide to Lateral Thinking Skills*, Kogan Page, pages 1–2, 2006.

George Polya, *How to Solve It, 2nd ed.,* Princeton University Press, 1957.

Richard E. Mayer, *Learning and Instruction*, Prentice–Hall, Pearson, 2003.

John W. Satzinger, Robert B. Jackson, and Stephen D. Burd, *Object-Oriented Analysis and Design with the Unified Process*, Thomson, 2005.

Amrit Tiwana and Mark Keil, "The one-minute risk assessment tool," Communications of the ACM, Volume 47, no. 11, pages 73–77, 2004.

Ivar Jacobson, Grady Booch, and James Rumbaugh, *The Unified Software Development Process*, Addison Wesley, 1998.

Ivar Jacobson, Grady Booch, and James Rumbaugh, *UML User Guide*, Addison Wesley, 1998.

Walker Royce, *Software Project Management*, Addison Wesley, 1998.

Ben Shneiderman, *Designing the User Interface*, Addison Wesley, 3rd ed., 1998.

Susan Weinschenk, Pamela Jamar, and Sarah C. Yeo, *GUI Design Essentials*, Wiley, 1997.

Howard Baetjer, Jr., "Software As Capital," IEEE Computer Society, 1998.

Grady Booch, *Object Solutions: Managing the Object-Oriented Project*, Addison Wesley, 1996.

R.G.R. Cattell, *Object Data Management,* Addison-Wesley, 1991.

David A. Taylor, *Object Technology: A Manager's Guide,* Addison-Wesley, 1997.

Fowler, *UML Distilled: Applying the Standard Object Modeling Language*, Addison-Wesley, 1997.

Dwayne Phillips, "The Software Project Manager's Handbook," IEEE, 1998.

Meredith and Mantel, *Project Management, A Managerial Approach*, John Wiley and Sons, 3rd ed., 1995.

Len Bass, Paul Clements, and Rick Kazman, *Software Architecture in Practice*, Addison Wesley, 1998.

Kennedy C. Laudon and Jane P. Laudon, *Essentials of Management Information Systems, 2nd ed.*, Prentice Hall, 1997.

Jolyon E. Hallows, *Information Systems Project Management*, Amacom, 1998.

Fintan Culwin, *A Java GUI Programmer's Primer*, Prentice Hall, 1998.

Web Services Architecture (accessed at *www.w3.org/TR/2004/NOTE-ws-arch-20040211/*, 10 April 2007).

Architecture of the World Wide Web, 1st ed. (accessed at *www.w3.org/TR/2004/WD-webarch-20040816/*, 10 April 2007).

Java Community Process (accessed at *http://jcp.org/en/home/index*, 10 April 2007).

American National Standards Institute (accessed at *www.ansi.org/*, 10 April 2007).

Inderjeet Singh, Beth Stearns, Mark Johnson, et al., "Designing Enterprise Applications with the J2EE™ Platform," 2nd ed. (accessed at *http://java.sun.com/blueprints/guidelines/designing_enterprise_applications_2e/titlepage.html*, 10 April 2007).

Comparison of Integrated Development Environments (accessed at *http://en.wikipedia.org*, 11 April 2007).

Larry Dignan, "JetBlue fiasco: A database could have made a difference," ZD:Net (accessed at *http://blogs.zdnet.com/BTL/?p=4523*, 12 April 2007).

Office of Government Compliance (accessed at *www.ogc.gov.uk/* and *www.ogc.gov.uk/guidance_itil.asp*, 12 April 2007).

Scott C. Beardsley, James M. Manyika, and Roger P. Roberts, "The Next Generation of Interactions," *The McKinsey Quarterly*, no. 4, 2005.

Java EE Tutorial (accessed at *http://java.sun.com/javaee/reference/tutorials/index.jsp*, 14 April 2007).

Craig Larman, *Applying UML and Patterns*, Prentice Hall, 2005.

Kathy Schwalbe, *Information Technology Project Management*, Course Technology, 2006.

Ed Mendel, "1.2 billion in fines over child support system," *San Diego Tribune*, *www.signonsandiego.com/news/state/20070328-9999-1n28computer.html*, 28 March 2007.

Todd Weiss, "Colorado DMV Puts Brakes On $13M Registration System," *Computerworld*, 9 April 2007.

Malcolm Gladwell, *The Tipping Point, 1st ed.*, Boston: Little, Brown, 2000.

James Surowiecki, *The Wisdom of Crowds: Why the Many Are Smarter Than The Few and How Collective Wisdom Shapes Business Economies, 1st ed.*, Doubleday, 2004.

Jonathon Cagin, Craig M. Vogel, *Creating Breakthrough Products*, Prentice Hall, 2002.

"Federal Enterprise Architecture Framework," Chief Information Officers Council, 1999.

Bass, Clements, and Kazman, *Software Architecture in Practice, 2nd ed.,* Addison-Wesley, 2003.

"How Do You Define Software Architecture?" (accessed at *www.sei.cmu.edu/architecture/definitions.html*, 17 April 2007).

The Open Group, "The Open Group Architecture Framework" (accessed at *www.opengroup.org/architecture/togaf8-doc/arch/*, 25 April 2007).

"Alistair A.R. Cockburn's Resources for Writing Use Cases" (accessed at *http://alistair.cockburn. us/index.php/Resources_for_writing_use_cases*, 17 April 2007).

The Architecture Journal, Microsoft Corporation (accessed at *http://msdn2.microsoft.com/en-us/ arcjournal/default.aspx*, 17 April 2007).

Paul Krill, "IBM conference tackles system complexity," *InfoWorld, www.infoworld.com/article/07/ 04/11/HNcomplexity_1.html*, 11 April 2007.

Jason Lyman, Sandra Pelletier, Ken Scully, James Boyd, Jason Dalton, Csaba Egybazy, and Steve Tropello, "Applying the HL7 reference information model to a clinical data warehouse," Systems, Man and Cybernetics, IEEE International Conference, 2003.

"Microsoft Research: Natural Language Processing Hits High Gear" (accessed at *www.microsoft. com/presspass/features/2000/05-03nlp.mspx*, April 22, 2007).

IEEE Computer Society, "Guide to the Software Engineering Body of Knowledge, 2004 Version," SWEBOK®, *www.swebok.org/*, 2004.

Mary Hayes Weier, "SOA Is The Future For SAP, Says Company CEO," *InformationWeek, www. informationweek.com/story/showArticle.jhtml?articleID=199201125*, 24 April 2007.

Software as a Service (accessed at *http://en.wikipedia.org/wiki/Software_as_a_Service*, 25 April 2007).

Frederick Chong and Gianpaolo Carraro, "Architecture Strategies for Catching the Long Tail," Microsoft Corporation (accessed at *http://msdn2.microsoft.com/en-us/library/aa479069. aspx*, 25 April 2007).

Michael Platt, "Microsoft Architecture Overview," Microsoft Corporation, *http://msdn2.microsoft. com/en-us/library/ms978007.aspx*, July 2002.

Roger Sessions, "A Better Path to Enterprise Architectures," Microsoft Corporation (accessed at *http://msdn2.microsoft.com/en-us/library/aa479371.aspx*, 25 April 2007).

Architecture, IBM (accessed at *www-128.ibm.com/developerworks/architecture*, 25 April 2007).

Christopher Rhoads and Li Yuan, "How Motorola Fell A Giant Step Behind," *Wall Street Journal*, 27 April 2007.

Tom Koehler, "Standards Pay," Boeing Frontiers Online (accessed at *www.boeing.com/news/ frontiers/archive/2006/ december/ts_challenge.html*, 28 April 2007).

Don Tapscott and Anthony D. Williams, *Wikinomics*, Portfolio Books, 2006.

Laurie Orlov, *The Three Archetypes of IT*, Forrester Research, 22 March 2006.

Frank Davies, "'Innovation agenda' is advancing in Congress," *San Jose Mercury News*, 1 May 2007.

Evan Schuman, "Wal-Mart to Add RFID to 400 More Stores," *Baseline Magazine*, 2 May 2007.

Grady Booch, *The Irrelevance of Architecture*, IEEE Software, May–June 2007.

Grady Booch Online (accessed at *www.booch.com/architecture/index.jsp*, 28 April 2007).

Veterans Administration Enterprise Centers, Technical Architecture (accessed at *www.aac.va.gov/ technicals.htm*, 8 May 2007).

Office of Enterprise Architecture Management (accessed at *www.va.gov/OIT/EAM/default.asp*, 8 May 2007).

Office of Information & Technology (accessed at *www.va.gov/OIT/CIO/default.asp*, 8 May 2007).

"GM's Cure for Complexity," *CIO, www.cio.com.au/index.php/id;1706983620;fp;;fpid;;pf;*, 1 October 2004.

Christopher Koch, "A New Blueprint For The Enterprise," *CIO*, 1 March 2005.

Information Technology Infrastructure Library (accessed at *www.itil.co.uk/*, 10 May 2007).

Information Technology Infrastructure Library (accessed at *http://itil.technorealism.org/index.php?page=Introduction_To_ITIL*, 10 May 2007).

SAP, Enterprise Architecture (accessed at *www.sap.com/platform/esoa/index.epx*, 10 May 2007).

Christopher Koch, "A New Blueprint For the Enterprise," *CIO*, 8 April 2005.

Zachman Institute for Framework Advancement (accessed at *www.zifa.com/*, 10 May 2007).

Federal Enterprise Architecture (accessed at *www.whitehouse.gov/omb/egov/a-1-fea.html*, 10 May 2007).

Galen Gruman, "The Four Stages of Enterprise Architecture," *CIO*, 7 February 2007.

Center for Information Systems Research (accessed at *http://mitsloan.mit.edu/cisr/*, 10 May 2007).

Enterprise Architecture As Strategy: Creating a Foundation for Business Execution (accessed at *www.architectureasstrategy.com/book/eas/about.htm#*, 10 May 2007).

Adrian Grigoriu, *An Enterprise Architecture Development Framework*, Trafford, 2006.

Scott A. Bernard, *An Introduction to Enterprise Architecture, 2nd ed.,* Authorhouse, 2005.

Tom Davenport, "Managing Customer Knowledge," *CIO*, 9 May 2007.

Capability Maturity Model Integration (CMMI), Software Engineering Institute (accessed at *www.sei.cmu.edu/* and *www.sei.cmu.edu/managing/*, 11 May 2007).

Diann Daniel, "The Rising Importance of the Enterprise Architect," *CIO*, *www.cio.com/article/print/101401*, 31 March 2007.

Cliff Peale," Not just floating soap anymore," *Cincinnati Enquirer*, 13 May 2007.

Jena McGregor, "The World's Most Innovative Companies," *BusinessWeek*, 4 May 2007.

Juris Kaza, "Have Cell Phone, Will Travel," *Computerworld*, 14 May 2007.

Kim S. Nash and Deborah Gage, "We Really Did Screw Up," *Baseline Online* (accessed at *www.baselinemag.com/article2/0,1540,2131032,00.asp?kc=CIOMINUTE051607CIO1*, 16 May 2007).

Jared T. Howerton, "Service-Oriented Architecture and Web 2.0," *IT Professional*, IEEE Computer Society, Volume 9, no. 3, *www.computer.org/portal/cms_docs_itpro/itpro/homepage/2007/may_june/f3062.pdf*, May/June 2007.

Duffie Brunson and Sid Frank, "The Partnership of Six Sigma and Data Certification," *www.b-eye-network.com/view/2263*, 23 January 2006.

Duffie Brunson, "Certified Data and the Certification Process for Financial Institutions," *www.b-eye-network.com/view/2081*, 6 December 2005.

Sid Frank, "The Importance of Data Quality in Service-Oriented Architectures," *www.b-eye-network.com/view/4086*, 6 March 2007.

Meredith Levinson, "ABC: An Introduction to KM," *www.cio.com/article/40343/ABC_An_Introduction_to_KM*, 7 March 2007.

"ABC: An Introduction to SOA,"*CIO*, *www.cio.com/article/40941*, 7 March 2007.

Mark Cooper and Paul Patterson, "ABC: An Introduction to BPM," *www.cio.com/article/106609/4*, 27 April 2007.

Ben Worthen, "ABC: An Introduction to SCM," *www.cio.com/article/40940*, 27 April 2007.

Christopher Koch, "ABC: An Introduction to ERP," *www.cio.com/article/40323*, 7 March 2007.

Thomas Wailgum, "ABC: An Introduction to CRM," *www.cio.com/article/40295*, 6 March 2007.

Robinson College of Business, Georgia State University, "Innovation: The DNA of UPS" (accessed at *www.robinson.gsu.edu/magazine/fall2004/UPS.html*, 27 April 2007).

John McCormick, "6 Keys to SOA Success," *Baseline Magazine* (accessed at *www.baselinemag.com/article2/0,1540,2129603,00.asp*, 31 May 2007).

Yefim V. Natis, "Applied SOA: Transforming Fundamental Principles Into Best Practices," Gartner Research, 4 April 2007.

Boeing Dreamliner Web site, "Boeing 787 Dreamliner Will Provide New Solutions for Airlines, Passengers" (accessed at *www.boeing.com/commercial/787family/background.html*, 31 May 2007).

Sharon Gaudin, "Social Security Administration Worker Charged In Identity Theft Scheme," *www.informationweek.com/shared/printableArticle.jhtml?articleID=199000813*, April 2007.

Security Breaches, "Statement by Ohio State University on recent data breaches," accessed at *www.osu.edu/news/newsitem1673*, 1 June 2007).

Edward Prewitt, "Disruption is Good, Ignoring it is Bad," *CIO*, *www.cio.com.au/index.php?id=1918308937*, 7 May 2001.

Thomas A. Stewart, "Intellectual Capital: The New Wealth of Organizations" (accessed at *http://members.aol.com/thosstew/forward.html*, 21 May 2007).

Paul Sloane, "Ten Great Ways to Crush Creativity," *Innovative Leader*, Volume 12, no. 7, *www.winstonbrill.com/bril001/html/article_index/articles/551-600/article581_body.html*, July 2003.

Howard Baejter Jr., *Software as Capital: An Economic Perspective on Software Engineering*, The Institute of Electrical and Electronics Engineers, Inc., 1998.

Addept Solutions, "Capitalising on Knowledge" (accessed at *www.addept.com/km.aspx?CGID=60*, 21 April 2007).

Kelly Spors, "States That Foster 'New Economy' Growth," *Wall Street Journal* (accessed at *www.startupjournal.com/howto/management/20070301-memos.html*, 26 April 2007).

The Kauffman Foundation, "The 2007 State New Economy Index" (accessed at *www.kauffman.org/pdf/2007_State_Index.pdf*, 26 April 2007).

Fangqi Xu, Ginny McDonnell, and William R. Nash, "A Survey of Creativity Courses at Universities in Principal Countries," *The Journal of Creative Behavior*, The Creative Education Foundation, Inc., Volume 39, no. 2, *www.creativeeducationfoundation.org/univ_creativity.shtml*, Second Quarter, 2005.

Dan Saffer, "The Cult of Innovation," *BusinessWeek*, 5 March 2007.

Jim Collins, *Good to Great: Why Some Companies Make the Leap... and Others Don't*, Harper Collins, 2001.

"Why they don't buy what you sell," *www.marketingweb.co.za/marketingweb/view/marketingweb/en/page73590?oid=80206&sn=Marketingweb%20detail*, 7 March 2007.

Thomas H. Davenport, Laurence Prusak, "Working Knowledge: How Organizations Manage What They Know" (accessed at *www.acm.org/ubiquity/book/t_davenport_1.html*, 21 April 2007).

Laurence Prusak, "Where did knowledge management come from?" IBM Systems Journal, Volume 40, no. 4, www.research.ibm.com/journal/sj/404/prusak.html, 2001.

Tom Davenport, "Managing Customer Knowledge," *CIO*, 9 May 2007.

Michael Polanyi, *The Tacit Dimension*, Doubleday & Co., Inc., Garden City, NY, 1967.

Todd Zwillich, "82 Million in U.S. Without Health Insurance," WebMD Medical News, *www.webmd.com/skin-problems-and-treatments/news/20040617/millions-in-us-without-health-insurance*, 16 June 2004.

Robert K. Merton, "On Social Structure and Science," The University of Chicago Press, *www.compilerpress.atfreeweb.com/Anno%20Merton%20Unintended.htm*, 1996.

Eliyahu Goldratt, *Critical Chain*, The North River Press, 1997.

Eliyahu Goldratt, *The Goal: A Process of Ongoing Improvement, 2nd ed.,* The North River Press, 1992.

Tim O'Reilly, "Open Source Paradigm Shift," *http://tim.oreilly.com/articles/paradigmshift_0504.html*, May 2004.

Thomas Kuhn, *The Structure of Scientific Revolutions, 3rd ed.,* University of Chicago Press, 1962, 1970, 1996.

Tim O'Reilly, "The Network Really Is the Computer," *www.oreillynet.com/pub/a/251*, 8 June 2000.

Andrew Lavallee, "At Some Schools, Facebook Evolves From Time Waster to Academic Study," *Wall Street Journal, http://online.wsj.com/article/SB117917799574302391.html*, 29 May 2007.

Random Walk Diagram (accessed at *www.chemistrydaily.com/chemistry/Random_walk*, 2 June 2007).

Thomas J. Peters and Robert H. Waterman, *In Search of Excellence: Lessons from America's Best-Run Companies*, Harper & Row (New York), 1982.

"How to Be a Smart Innovator," WSJ Online, *http://online.wsj.com/article/SB115755363514155116.html? mod=2_1241_2*, 11 September 2006.

Enterprise Architecture (accessed at *http://en.wikipedia.org/wiki/Enterprise_architecture*, 3 June 2007).

Institute For Enterprise Architecture Developments (accessed at *www.enterprise-architecture.info/*, 3 June 2007).

Enterprise Architecture Portal (accessed at *www.cioindex.com/eap.asp*, 3 June 2007).

U.S. Department of Housing and Urban Development (HUD) Enterprise Architecture, accessed at *www.hud.gov/offices/cio/ea/newea/index.cfm*, 3 June 2007).

John Edwards, "On-Demand Software: Software as a Service Appeal," *CIO, www.cio.com/article/29093/On_Demand_Software_Software_as_a_Service_Appeal*, 1 March 2007.

Meredith Levinson, "ABC: An Introduction to Software as a Service," *www.cio.com/article/109704/ABC_An_Introduction_to_Software_as_a_Service*, 15 May 2007.

Galen Gruman, "Get Smart About SaaS," *CIO*, 1 June 2007.

Dan Tynan, "The 50 Greatest Gadgets of the Past 50 Years," *PCWorld*, 24 December 2005.

Christopher Rhoads, "Motorola to Slash 4,000 Additional Jobs," *http://online.wsj.com/article/SB118055898779719029.html?mod=djemalert*, 31 May 2007.

Robert L. Scheier, "Storage 2.0—Web-based storage is coming," *Computerworld*, 4 June 2007.

Doug Bartholomew, Mel Duval, "Does GE Have the Best I.T.?", *Baseline, www.baselinemag.com/article2/0,1540,2142230,00.asp*, 14 June 2007.

Jim Rapoza, "Weaving the Semantic Web," *http://etech.eweek.com/content/web_technology/spinning_the_semantic_web.html*, 30 May 2007.

Antony Adshead, "New routes with enterprise mashups," *ComputerWeekly, www.computerweekly.com/Articles/2007/05/18/223929/ new-routes-with-enterprise-mashups.htm*, 18 May 2007.

Christopher Alexander, *A Pattern Language*, Oxford University Press, 1977.

Sun One Architecture Guide (accessed at *www.sun.com/software/sunone/docs/arch/*, 12 June 2007).

eProject (accessed at *www.eproject.com* and *www.eproject.com/products/software_as_a_service.htm*, 12 June 2007).

Bill Rosser, "Creating a Business Architecture: Where Does It Lead You?" Gartner Research, 30 November 2006.

Jay DiMare, "Service-oriented architecture: A practical guide to measuring return on that investment," IBM Web site, *www-935.ibm.com/services/us/index.wss/ibvstudy/bcs/a1025716?ca=rss_bcs*, 12 October 2006.

Antone Gonsalves, "Intel Drives Itanium Road Map Toward 32 Nanometers," *InformationWeek,* *www.informationweek.com/news/showArticle.jhtml?articleID=199904627*, 15 June 2007.

Scott Ferguson, "Data Center Power Consumption on the Rise, Report Shows," *eWeek*, *www.eweek.com/article2/0,1895,2095409,00.asp*, 15 February 2007.

Jim Gray, "A Conversation with Werner Vogels," ACM Queue, Web Services, Volume 4, no. 4, *www.acmqueue.com/modules.php?name=Content&pa=showpage&pid=388*, May 2006.

J. Lynn Lunsford, "Boeing Plans a Grand Unveiling For Dreamliner—but Can It Fly?" *Wall Street Journal* (accessed at *http://online.wsj.com/article_print/SB118375713162359544.html*, 7 July 2007).

Evan Schuman, "At Wal-Mart, World's Largest Retail Data Warehouse Gets Even Larger," Ziff Davis Internet, *www.eweek.com/article2/0,1895,1675960,00.asp*, 13 October 2004.

The Dartmouth Atlas of Health Care (accessed at *www.dartmouthatlas.org/*, 21 July 2007).

Center for Disease Control and Prevention Web site (accessed at *www.cdc.gov/*, 21 July 2007).

Organisation for Economic Co-operation and Development (accessed at *www.oecd.org/home/0,2987,en_2649_201185_1_1_1_1_1,00.html*, 21 July 2007).

U.S. Census Report (accessed at *www.census.gov/Press-Release/www/releases/archives/income_wealth/005647.html*, 21 July 2007).

Human Genome Project (accessed at *www.ornl.gov/sci/techresources/Human_Genome/home.shtml*, 28 July 2007).

Roche Pharmaceuticals (accessed at *www.roche.com/home.html*, 28 July 2007).

Timothy Redman, *Data Quality*, Digital Press, pages 3, 47–49, 51–67, 78–79, 2001.

Timothy Redman, *Data Quality for the Information Age*, Digital Press, 1996.

Ahmed Elfatatry, "Dealing With Change: Components Versus Services," Communications of the ACM, Volume 50, no. 8, August 2007.

John Naisbitt and Patricia Aburdeen, *Megatrends*, Warner Books, New York, 1982.

330

CHAPTER 12

ETHICAL, PRIVACY, AND SECURITY ISSUES

THE PERVASIVENESS OF COMPUTER VIRUSES

In view of all the deadly viruses that have been spreading lately, Weekend Update would like to remind you: when you link up to another computer, you're linking up to every computer that that computer has ever linked up to.

— Dennis Miller, Saturday Night Live, U.S. television show

HANNAFORD BROTHERS ILLUSTRATES WHY MANAGERS MUST UNDERSTAND THE ETHICAL, PRIVACY, AND SECURITY ISSUES RELATING TO IT

Hannaford Brothers is a supermarket chain that employs 27,000 workers with 167 stores in 5 northeastern states and Florida.[1] In December 2007, a security breach began at Hannaford involving customer credit and debit card data. It took three months before the breach was uncovered by customers complaining to their banks about fraudulent transactions on their cards. The breach was finally contained two weeks later.[2]

The data was captured illegally as the cards were swiped at the check-out line. To its credit, Hannaford Brothers met the payment card industry (PCI) standards for data protection, and the company did not use wireless technology to transmit unencrypted data. (These two factors have played a part in other customer data breaches.) The PCI standards, however, do not require that card data be encrypted at the instant the card is swiped. At Hannaford, the unencrypted card data traveled over the store's private network before reaching a server where it was encrypted and routed to the credit card company to complete the approval process.

While the investigation is continuing, one probable scenario is that an employee with administrative network access was involved. Malicious software was planted on servers in each of Hannaford's stores; the software captured the unencrypted card data from customers and transferred it to an accomplice located overseas.[3] Unfortunately, many businesses have spent considerable money to implement the current PCI data protection standards, which now appear to be inadequate.[4]

Hannaford has cooperated with credit and debit card issuers to ensure that customers whose data was stolen are protected. The firm also notified law enforcement authorities and is working with them to track down those who are responsible.

Just a few days after Hannaford Brothers announced the data breach, multiple class action lawsuits were filed against the company alleging it was negligent for failing to maintain adequate computer data security for customer credit and debit card data.[5] At the time the initial class action suit was filed, there had already been 1800 cases of reported credit and debit card fraud arising from the breach. Hannaford is likely facing years of litigation; tens of millions of dollars in legal fees, settlement costs, and customer credit monitoring services; and a reduction in sales revenue due to loss of customer goodwill.

LEARNING OBJECTIVES

As you read this chapter, ask yourself:

- What are some of the ethical issues raised by the use of information technology?
- What privacy issues are raised by the use of information technology, and how do organizations deal with them?
- What are some common information technology security issues, and how can organizations minimize their potential negative impact?

This chapter will identify some of the ethical and social issues associated with the use of information technology, point out some of the potential negative impacts, and provide guidance to help minimize these. But first we begin with a definition of ethics and a discussion of some of the measures organizations are taking to ensure that their employees act in an ethical manner.

WHAT IS ETHICS?

Ethics is a set of beliefs about right and wrong behavior. Ethical behavior conforms to generally accepted social norms—many of which are almost universally accepted. Doing what is ethical can be difficult in certain situations. For example, although nearly everyone would agree that lying and cheating are unethical, some people might consider it acceptable to tell a lie to protect someone's feelings or to keep a friend from getting into trouble.

Making ethical decisions in the area of information technology is really no different than in other areas, although the specific issues may be different. Is it okay to download copyrighted material without paying a fee? Should you point out to a supplier that their accounting system consistently under-bills your firm, or should you take advantage of the error to save your firm some money? Can you cut some corners on a software implementation project to meet a tight deadline?

The next section outlines actions that many organizations are taking to improve their ethics and suggests a model of ethical decision making.

Improving Corporate Ethics

In recent years, we have seen the failure of major corporations like Enron and WorldCom due to accounting scandals. We also have seen the collapse of many financial institutions due to unwise and unethical decision making regarding the approval of mortgages and lines of credit to unqualified individuals and organizations. Clearly such unethical behavior has led to serious negative consequences that have had a global impact. We also have witnessed an increasing number of corporate officers and senior managers sentenced to prison terms for their unethical behavior. Many organizations today recognize the need to take action to ensure that their employees operate in an ethical manner when using technology and in the general course of business. The following sections will summarize the key actions organizations are taking to improve business ethics.

Appointing a Corporate Ethics Officer

Corporate ethics can be defined broadly to include ethical conduct, legal compliance, and corporate social responsibility. The primary functions of a corporate ethics policy are setting standards, building awareness, and handling internal reports—tasks that are neither consolidated nor handled well in many organizations. Some organizations are choosing to pull these functions together under a corporate officer to ensure that they receive sufficient emphasis and cohesive treatment.

The **corporate ethics officer** is a senior-level manager who provides vision and direction in the area of business conduct. The role includes "integrating their organization's ethics and values initiatives, compliance activities, and business conduct practices into the decision-making processes at all levels of the organization."[6] The ethics officer tries to establish an environment that encourages ethical decision making. Specific responsibilities might include "complete oversight of the ethics function, collecting and analyzing data, developing and interpreting ethics policy, developing and administering ethics education and training, and overseeing ethics investigations."[7] The presence of a corporate ethics officer has become increasingly common. Often a corporation will place a higher emphasis on ethics policies following a major scandal within the organization, as illustrated in the following example.

Former Hewlett Packard Chairwoman Patricia Dunn and former Compliance Officer Kevin Hunsaker were involved in an internal investigation of HP board members suspected of leaking information about ongoing board room disputes to the news media. Three detectives involved in the investigation allegedly engaged in pretexting (the use of false pretenses) to gain access to the telephone records of HP directors, certain employees, and nine journalists. The detectives allegedly obtained and used the targeted individuals' Social Security numbers to impersonate those individuals in calls to the phone company with the goal of obtaining private phone records.

The state of California charged that such pretexting practices are illegal as they represent an invasion of privacy and involve gaining personal information under false pretenses.[8] Eventually, the state settled a civil complaint against the company under which HP paid $14.5 million to cover fines and legal costs. The settlement did not involve any admission or conclusion of guilt on the part of HP. Dunn and Hunsaker resigned as a result of the scandal.[9] In the aftermath of this scandal, HP appointed Jon Hoak, a former legal counsel for NCR Corporation, to be its Ethics and Compliance Officer, reporting directly to CEO, President, and Chairman Mark Hurd. Hoak is responsible for HP's adherence to its Standards of Business Conduct and performs an independent assessment of HP's investigative practices and develops future best practices.

Ethical Standards Set by Board of Directors

The board of directors is responsible for the careful and responsible management of an organization. In a for-profit corporation, the board's primary objective is to oversee the organization's business activities and management for the benefit of all stakeholders, including shareholders, customers, employees, suppliers, and the community. In a nonprofit corporation, the board reports to a different set of stakeholders, in particular, the local communities that the nonprofit serves.

The board fulfills some of its responsibilities directly and assigns others to various committees. The board is not normally responsible for day-to-day management and operations; these responsibilities are delegated to the organization's management team. The board, however, is responsible for supervising the management team.

Directors of the company are expected to conduct themselves according to the highest standards of personal and professional integrity. Directors also are expected to set the standard for company-wide ethical conduct and ensure compliance with laws and regulations.

Establishing a Corporate Code of Ethics

A **code of ethics** highlights an organization's key ethical issues and identifies the over-arching values and principles that are important to the organization. The code frequently includes a set of formal, written statements about the purpose of the organization, its values, and the principles that guide its employees' actions. An organization's code of ethics applies to its directors, officers, and employees. The code of ethics should focus employees on areas of ethical risk relating to their role in the organization. It should also provide guidance to help them recognize and deal with ethical issues, provide mechanisms for reporting unethical conduct, and foster a culture of honesty and accountability in an organization. The code of ethics helps ensure that employees abide by the law, follow necessary regulations, and behave in an ethical manner.

A code of ethics cannot gain company-wide acceptance unless it is developed with employee participation and fully endorsed by the organization's leadership. It also must be easily accessible by employees, shareholders, business partners, and the public. The code of ethics must continually be applied to a company's decision making and emphasized as an important part of its culture. Breaches in the code of ethics must be identified and treated appropriately so that its relevance is not undermined. Establishing a code of ethics is an important step for any company, and most large organizations have developed such a code.

In March 2007, *Business Ethics* magazine rated publicly held U.S. companies based on a statistical analysis of corporate service to seven stakeholder groups—employees, customers, community, minorities and women, shareholders, the environment, and non-U.S. stakeholders. The top IT company, based on performance between 2000 and 2007, was Intel Corporation, the world's largest computer chip maker. A summary of Intel's code of ethics is shown in Figure 12-1. A more detailed version is presented in a 22-page document (Intel Code of Conduct May 2007 found at **www.intel.com/intel/finance/docs/code-of-conduct.pdf**), which offers employees guidelines designed to deter wrongdoing, encourage honest and ethical conduct, and promote behavior that complies with applicable laws and regulations. Intel's code of ethics also expresses its policies regarding the environment, health and safety, intellectual property, diversity, nondiscrimination, supplier expectations, privacy, and business continuity.

- Intel conducts business with honesty and integrity.
- Intel follows the letter and spirit of the law.
- Intel employees treat each other fairly.
- Intel employees act in the best interests of Intel and avoid conflicts of interest.
- Intel employees protect the company's assets and reputation.

Source: Intel, accessed at *www.intel.com/intel/finance/docs/code-of-conduct.pdf*

FIGURE 12-1   Intel's Five Principles of Conduct

Requiring Employees to Take Ethics Training

The ancient Greek philosophers believed that personal convictions about right and wrong behavior could be improved through education. Today, most psychologists agree with them. Lawrence Kohlberg, the late Harvard psychologist, found that many factors stimulate a person's moral development, but one of the most crucial is education. Other researchers have repeatedly supported these findings—people can continue their moral development through further education that involves critical thinking and examining contemporary issues.

Thus, a company's code of ethics must be promoted and continually communicated within the organization, from top to bottom. Organizations should show employees examples of how to apply the code of ethics in real life. One approach is through a comprehensive ethics education program that encourages employees to act responsibly and ethically. Such programs are often presented in small workshop formats in which employees apply the organization's code of ethics to hypothetical but realistic case studies relating to the use of technology, interactions with vendors, and a variety of other topics. Not only do these courses make employees more aware of a company's code of ethics and how to apply it, they demonstrate that a company intends to operate in an ethical manner. The existence of formal training programs also can reduce a company's liability in the event of legal action.

Including Ethical Criteria in Employee Appraisals

Employees are increasingly evaluated on their demonstration of qualities and characteristics that are highlighted in the corporate code of ethics. For example, many companies base a portion of their employee performance evaluations on treating others fairly and with respect; operating effectively in a multicultural environment; accepting personal accountability for meeting business needs; and operating openly and honestly with suppliers, customers, and other employees. These factors are considered along with more traditional criteria used in performance appraisals, such as an employee's overall contribution to moving the business ahead, successful completion of projects, and maintenance of good customer relations.

PRIVACY

Often the use of information about people (employees, customers, business partners, etc.) in business requires balancing the needs of those who use the information against the rights and desires of the people whose information may be used.

On the one hand, information about people is gathered, stored, analyzed, and reported because organizations can use it to make better decisions. Some of these decisions can affect people's lives profoundly—whether or not to extend credit to a new customer, to hire one job candidate or another, to offer a scholarship or not. In addition, increased competitiveness in the global marketplace has intensified the need to understand consumers' purchasing habits and financial condition. Companies use this information to target marketing efforts to consumers who are most likely to buy their products and services. Organizations also need basic information about existing customers in order to serve them better. It is hard to imagine an organization having a relationship with its customers without having data about them. Thus, organizations implement customer relationship management systems that collect and store key data from every interaction they have with a customer.

On the other hand, many people object to the data collection policies of government and other organizations on the basis that they strip people of the power to control their own personal information. Many individuals are also concerned about the number of data breaches in which personal data stored by an organization falls into the hands of criminals. For many, the existing hodgepodge of privacy laws and practices fails to provide adequate protection and fuels a sense of distrust and skepticism, as illustrated by this chapter's opening vignette.

As a result of the frequency of data breaches and the reluctance of many organizations to report them, various states have passed laws that, in effect, require any agency, person, or business conducting business in the state to disclose any breach of security to any resident whose data is believed to have been compromised. Table 12-1 identifies the largest U.S. data breaches since 2003.

TABLE 12-1   Ten largest recent data breaches

Organization	Date	Number of individuals impacted
Data Processors International	March 6, 2003	5 million
America Online	June 24, 2004	30 million
Citigroup	June 6, 2005	20 million
Visa, MasterCard, American Express	June 19, 2005	40 million
U.S. Department of Veteran Affairs	May 22, 2006	26 million
TJX Companies, Inc	January 17, 2007	94 million
Dal Printing	March 12, 2007	9 million
Fidelity National Information Services	July 3, 2007	9 million
TD Ameritrade	September 14, 2007	6 million
HM Revenue and Customs	November 20, 2007	20 million
Best Western	August 23, 2008	8 million

Source: Attrition.org Data Loss Archive and Database at http://attrition.org/dataloss accessed January 31, 2009.

A combination of approaches—new laws, technical solutions, and privacy policies—is required to balance the scales. Reasonable limits must be set on government and business access to personal information; new information and communication technologies must be designed to protect rather than diminish privacy; and appropriate corporate policies must be developed to set baseline standards for people's privacy. Education and communication are essential as well.

Right to Privacy

This section will help you understand the right to privacy. Then we will cover information technology developments that affect the privacy of personal information.

First, it is important to gain a historical perspective on the right to privacy. When the U.S. Constitution took effect in 1789, the drafters were concerned that a powerful government would intrude on the privacy of individual citizens. As a result, they added the Bill of

Rights. So, although the Constitution does not contain the word "privacy," the U.S. Supreme Court has ruled that the right to privacy is protected by a number of amendments in the Bill of Rights. For example, the Supreme Court has stated that American citizens are protected by the Fourth Amendment when there is a "reasonable expectation of privacy." The Fourth Amendment is as follows:

"The right of the people to be secure in their persons, houses, papers, and effects, against unreasonable searches and seizures, shall not be violated, and no Warrants shall issue, but upon probable cause, supported by Oath or affirmation, and particularly describing the place to be searched, and the persons or things to be seized."

The next two sections will address two key privacy issues—treating customer data responsibly and workplace monitoring.

Treating Customer Data Responsibly

When dealing with customer data, strong measures are required to avoid customer relationship problems. One widely accepted approach to treating customer data responsibly is for a company to adopt the Code of Fair Information Practices and the 1980 Organization for Economic Cooperation and Development (OECD) privacy guidelines. The code of Fair Information Practices defines five widely accepted core principles concerning fair information practices of privacy protection: (1) Notice/Awareness; (2) Choice/Consent; (3) Access/Participation; (4) Integrity/Security; and (5) Enforcement/Redress. The 1980 Organization for Economic Cooperation and Development (OECD) privacy guidelines continue to represent the international consensus on general guidance concerning the collection and management of personal information. Under these two guidelines, an organization collects only personal information that is necessary to deliver its product or service. The organization ensures that the information is protected carefully and accessible only by those with a need to know, and it provides a process for consumers to review their own data and make corrections. The company informs customers if it intends to use customer information for research or marketing, and it provides a means for them to opt out of the data collection process.

The European Union Data Protection Directive prohibits the transfer of personal data to non-European Union nations that do not meet the European adequacy standard for privacy protection. Some of these standards require the creation of government data protection agencies, registration of databases with those agencies, and in certain cases, approval before personal data processing can begin. The United States does not meet these standards. The U.S. Department of Commerce together with the European Commission developed a "safe harbor" framework to ensure that U.S. companies don't experience interruptions in their dealings with countries in the European Union. U.S. organizations that can verify their policies and practices are compliant with the safe harbor's requirements will be recognized as meeting the European adequate standard privacy for privacy protection.

Organizations should appoint an executive (often called a Chief Privacy Officer or CPO) to define, implement, and oversee a set of data privacy policies. This individual must ensure that the organization avoids violating state and federal government regulations. If an organization works with European customers and organizations, the CPO also must ensure that the organization meets the safe harbor requirements regarding the collection and use of customer and employee data. This individual should be briefed on *planned* marketing programs, information systems, or databases that involve the collection or dissemination of consumer data and, importantly, be given the power to modify or stop initiatives that violate established data privacy policies. The rationale for early involvement in such initiatives is to ensure that potential problems can be identified in the earliest stages, when it is easier and less expensive to correct them.

There are several tasks critical to establishing an effective data privacy program, including:

- Conduct a thorough assessment to document what sensitive information your organization is collecting, where it is stored, how long it is kept, who has access to it, and how your organization is using this data.
- Define a comprehensive data privacy program that encompasses the development of a set of data privacy policies that meet or exceed industry and government requirements; addresses ongoing employee education and compliance; and provides for regular updates to suppliers, customers, contractors, and employees.
- Assign a high level executive to implement and monitor the data privacy program.
- Develop a data breach response plan to be implemented in the event of such an incident.
- Track ongoing changes to regulatory and legal requirements and make necessary changes to your data privacy program.

Some organizations fail to address privacy issues early on, and it takes a negative experience to make them appoint an executive to define, implement, and manage data privacy policies. For example, U.S. Bancorp, a bank that in early 2009 had $247 billion in assets, appointed a CPO, but only after spending $3 million to settle a lawsuit that accused the bank of selling confidential customer financial information to telemarketers.[10] This was one of the first of what turned out to be many lawsuits against banks alleging violations of customer privacy.

Many organizations that operate a Web site place a cookie—a small file containing a string of characters that uniquely identifies a customer's browser—on the computer hard drive of visitors to the organization's site. For each visit to the Web site, data about user preferences and activity is captured and stored under that cookie on the company's Web server. Additional information that a customer submits, such as name, address, and credit card information, as well as information gleaned from third parties, also is associated with the cookie and added to the customer's file on the server. In this manner, it is possible for the operator of the Web site to gain a fairly complete and accurate picture of their customers. The Web site usually has a privacy policy that states what sort of information about customers is captured and how that information may be used by the capturing organization.

The world's largest online store, Amazon.com, captures a lot of data on its more than 60 million active customers. For example, it uses data about previous purchases by its customers to make recommendations to them for future purchases. So if one of your recent Amazon.com purchases was a book by suspense author Dean Koontz, the next time you visit Amazon.com, you are likely to see a recommendation to purchase books by other authors of this same genre, such as Stephen King. While some people appreciate this "service," others are concerned over just how much Amazon.com knows about them and what it is doing with this knowledge.

Workplace Monitoring

Many organizations have developed a policy on the use of information technology to protect against employee abuses that reduce worker productivity or that could expose the employer to harassment lawsuits. The institution and communication of such an IT usage policy establishes boundaries of acceptable behavior and enables management to take action against violators.

The potential for decreased productivity, coupled with increased legal liabilities, have forced many employers to monitor workers to ensure compliance with the corporate IT usage policy. More than 80 percent of major U.S. firms find it necessary to record and review employee communications and activities on the job, including e-mail, Web surfing, and phone usage (see Table 12-2). Some are even videotaping employees on the job. In addition, some companies employ psychological testing and random drug testing. With few exceptions, these increasingly common (and many would say intrusive) practices are legal.

TABLE 12-2 Extent of workplace monitoring

Subject of Workplace Monitoring	Percent of Employers that Monitor Workers	Percent of Companies that Have Fired Employees for Abuse or Violation of Company Policy
E-mail	43%	28%
Web surfing	66%	30%
Time spent on the phone as well as phone numbers called	45%	6%

Source: "2007 Electronic Monitoring & Surveillance Survey," American Management Press Room, February 28, 2008, http://press.amanet.org, accessed November 24, 2008.

The Fourth Amendment of the Constitution protects citizens from unreasonable searches by the government and is often used to protect the privacy of government employees. The Fourth Amendment cannot be used to control how a private employer treats its employees, however, because such actions are not taken by the government. As a result, public-sector employees have far greater privacy rights than those in private industry. Although private-sector employees can seek legal protection against an invasive employer under various state statutes, the degree of protection varies widely by state. Furthermore, state privacy statutes tend to favor employers over employees. For example, for employees to successfully sue an organization for violation of their privacy rights, the employees must prove that they were in a work environment where they had a reasonable expectation of privacy. As a result, courts typically rule against employees who file privacy claims for being monitored while using company equipment. A private organization can defeat a privacy claim simply by proving that an employee had been given explicit notice that e-mail, Internet, and phone usage were not private and that their use might be monitored. When an employer engages in workplace monitoring, though, it must ensure that it treats all types of workers equally. For example, a company could get into legal trouble for punishing an hourly employee more seriously for visiting inappropriate Web sites than it punished a salaried employee.

Society is struggling to define the extent to which employers should be able to monitor the work-related activities of employees. On the one hand, employers want to be able to guarantee a work environment that is comfortable for all workers, ensures a high level of worker productivity, and limits the costs of defending against "frivolous" privacy violation lawsuits filed by disgruntled employees. On the other hand, privacy advocates want federal legislation that keeps employers from infringing upon the privacy rights of employees. Such legislation would require prior notification to all employees of the existence and location of all electronic monitoring devices. Privacy advocates also want restrictions on the types of information collected and the extent to which an employer may use electronic monitoring. As a result, many laws are being introduced and debated at both the state and federal level. As the laws governing employee privacy and monitoring continue to evolve, business managers must stay informed in order to avoid enforcing outdated usage policies. Organizations with global operations face an even bigger challenge because the legislative bodies of other countries also debate these issues.

A MANAGER TAKES INAPPROPRIATE ACTION

City of Ontario, California

The city of Ontario, California contracted with Arch Wireless to provide wireless text-messaging services. The city received 22 alphanumeric pagers, which it distributed to its employees. Jeff Quon, a member of the Ontario Police Department (OPD) SWAT team, received one of these pagers and used it in the normal course of his duties. He also used his pager to transmit sexually-explicit messages to two other workers in the police department and to his wife. None of the recipients complained about the messages.

    The city has a general computer usage, Internet, and e-mail policy. While the policy does not specifically address the use of pagers, it does state that the use of city-owned computers and all associated equipment, software programs, networks, Internet, e-mail, and other systems operating on these computers is limited to city of Ontario related business. The use of these tools for personal benefit is a significant violation of city of Ontario policy. The policy also states:

> Access to all sites on the Internet is recorded and will be periodically reviewed by the city. The city of Ontario reserves the right to monitor and log all network activity including e-mail and Internet use, with or without notice. Users should have no expectation of privacy or confidentiality when using these resources.

> Access to the Internet and the e-mail system is not confidential; and information produced in either hard copy or electronic form is considered city property. As such, these systems should not be used for personal or confidential communications. Deletion of e-mail or other electronic information may not fully delete the information from the system.

> The use of inappropriate, derogatory, obscene, suggestive, defamatory, or harassing language in the e-mail system will not be tolerated.

continued

A year before the city acquired the pagers, Sgt. Quon signed an "Employment Acknowledgement" that borrowed language from the general policy, indicating he had read and fully understood the city of Ontario's computer usage, Internet, and e-mail policy. A year after the city acquired the pagers, Sgt. Quon attended a meeting during which all people present were informed that pager messages were considered e-mail and those messages would fall under the city's policy as public information and were eligible for auditing.

Under the city's contract with Arch Wireless, each pager was allowed 25,000 characters per month. Beyond this limit, the city was required to pay overage charges. An informal policy was that if there was an overage, the employee would pay the additional charges. Lt. Duke with the Police Department was in charge of the purchasing contract with Arch Wireless and was responsible for collecting the overage charges from employees.

Quon exceeded his monthly allowance three or four times, but he paid all overage charges. According to Sgt. Quon, Lt. Duke told him that the city would not monitor the content of his messages unless he exceeded the budgeted monthly usage and failed to pay the overage fee.

Lt. Duke let it be known that he was tired of being a bill collector. Chief Scharf then asked Lt. Duke to request the transcripts of those officers who had exceeded their limit to determine if the messages were exclusively work-related or if employees were using the pagers for personal matters. Because the Ontario Police Department was unable to access the message directly, Lt. Duke requested that Arch Wireless provide the transcripts. One of the officers whose transcripts he requested was Sgt. Quon. Following a review of the transcripts by Lt. Duke and an investigation by Internal Affairs, it was determined that Sgt. Quon had exceeded his monthly allotment by more than 15,000 characters and that many of the messages were personal in nature and were often sexually explicit.

Quon, his wife, and the two employees with whom he had exchanged the sexually explicit messages filed a lawsuit alleging that "1) the pager service provider had violated the federal Stored Communications Act by releasing transcripts of Quon's messages to the City and 2) the City, and others, had violated their rights under the Fourth Amendment to the United States Constitution and Article I, Section 1 of the California Constitution."[11]

The Stored Communications Act (SCA) was enacted as part of the Electronic Communications Privacy Act in 1986 and is an attempt to address a number of potential privacy issues not addressed by the Fourth Amendment. The statute defines an electronic communications service (ECS) as any service that provides its users with the ability to send and receive wire or electronic communications. The U.S. Court of Appeals for the Ninth Circuit ruled that Arch Wireless was an electronic communications service and had violated the SCA when it provided transcripts of Quon's messages to the OPD in the absence of a court order or consent of sender or intended recipients.

The court also ruled that the OPD, as a *public* employer, had violated Quon's Fourth Amendment and California Privacy rights. For this to be true, the court had to agree that Quon had a reasonable expectation of privacy and that the Ontario Police Department had conducted an unreasonable search. While the OPD had an acceptable use policy that informed employees they had no reasonable expectation of privacy, OPD employees were told their texts would not be audited as long as they paid any overage charges. The court ruled that the search was unreasonable because it was too broad in scope—the OPD did not need to review the contents of the text messages to determine if their text message quota was too low.

continued

CYBERCRIME AND COMPUTER SECURITY

Cybercrime refers to criminal activity in which a computer or a computer network is used as a tool to commit a crime or is the target of criminal activity. Examples include gaining unauthorized access to data stored on a computer, illegal interception of non-public communications or data transmissions, and interfering with the functioning of a computer system. **Electronic fraud** is a broad class of cybercrime that involves the use of computer hardware, software, or networks to misrepresent facts for the purpose of causing someone to do or refrain from doing something that causes loss. An example would be altering the transactions entered into an information system, or altering or deleting stored data. According to the 2007 Computer Crime and Security Survey, electronic fraud, followed by virus attacks, is the leading cause of financial loss from computer incidents.[12]

343

No one really knows the extent of cybercrime as many crimes go unreported. Most companies that have been the victim of cybercrime simply won't talk to the press, although, as mentioned earlier, many states have passed Data Disclosure Laws that require disclosure to those affected. The concern of companies who are victims of cybercrime is loss of public trust and image—not to mention the fear of encouraging copycat hackers. In 2007, the FBI received 206,844 complaints of cybercrime committed over the Internet, with losses estimated at $240 million.[13] The actual cost of cybercrime is certainly much higher because not all crimes are reported and not all the costs (legal fees, loss of revenue, etc.) to companies affected by data breaches can be accurately estimated. The cost of the TJX data breach for example, is estimated to have cost the firm over $256 million in loss of business and legal fees.[14]

The following sections discuss the most frequent types of computer attacks, identify the various types of computer crime perpetrators, and provide action steps that managers can take to protect their organization from computer crime.

Types of Attacks

Security incidents can take many forms, but one of the most frequent is an attack on a networked computer from an outside source. Numerous types of attacks exist, and new types are being invented all the time. Some of the more common attacks involve a virus, worm, or distributed denial-of-service attack. Many computer attacks take advantage of some sort of vulnerability associated with the computer's operating system or a software application. As software manufacturers become aware of these vulnerabilities, they issue software patches to address them. Thus, it is important for organizations and individual users to continually update their system with software patches.

Viruses

"Computer virus" has become an umbrella term for many types of malicious code. Technically, a **virus** is a piece of programming code, usually disguised as something innocuous that causes some unexpected and usually undesirable event. Often, a virus is attached to a file so that when the infected file is opened, the virus executes. Other viruses sit in a computer's memory and infect files as the computer opens, modifies, or creates the files. Thus, viruses are said to be self-replicating. (The name virus derives from the analogous behavior of biological viruses that insert copies of themselves into living cells.) Most viruses deliver a "payload" or malicious act. For example, the virus may be programmed to display a certain message on the computer's display screen, delete or modify a certain document, or reformat the hard drive.

Viruses do not spread themselves from computer to computer; they are not self-propagating. To propagate to other machines, a virus must be passed on to other users through infected e-mail document attachments, programs on storage devices, or shared files. In other words, it takes action by the computer user to spread a virus.

Macro viruses are easily created and have become the most common type of virus. They use an application macro language (such as Visual Basic or VBScript) to create programs that infect documents and templates. After an infected document is opened, the virus executes and infects the user's application templates. Macros can wreak all sorts of havoc—including inserting unwanted words, numbers, or phrases into documents and altering command functions. More seriously, macro viruses can delete and change files, automatically run scripts, and overwrite standard application macros so that they can be spread to other machines. After a macro virus infects a user's application, it can embed itself in all future documents created with the application.

Worms

Worms are harmful computer programs that reside in the active memory of the computer. They differ from viruses in that they can propagate over a network without human intervention, sending copies of themselves to other computers by e-mail or Internet Relay Chat (IRC). Thus, they are self-propagating. The harm caused by a worm depends on the code written into the worm. Some worms do damage by consuming large amounts of system resources as they self-propagate; others erase data or execute instructions that install **malware** (malicious software) on a computer without the user's knowledge.

The negative impact of a virus or worm attack on an organization's computers can be considerable—lost data and programs, lost productivity because workers cannot use their computers, additional lost productivity as workers attempt to recover data and programs, and lots of effort for IT workers to clean up the mess and restore systems. The cost to repair the damage done by each of the Code Red, SirCam, Melissa, and ILOVEYOU worms was estimated to exceed $1 billion.

In late 2008, the Koobface worm began spreading rapidly through the social networking site Facebook. Targeted users received a message in their Facebook Inbox with a subject line of "You look funny in this new video," or something similar. Recipients were instructed to click on a provided link to view the video. Once on the video site, a message displayed indicating that an update of Flash was needed before the video could be displayed. The viewer was prompted to open a file called flash_player.exe. If the user opened the file, the Koobface worm downloaded malicious code to the user's computer. The worm then attempted to spread itself by sending similar infected messages to the user's Facebook friends. The code also was able to redirect future user searches on Google or Yahoo! to lesser known search sites. Of more concern is the worm's ability to install other malicious code at a later time.

Distributed Denial-of-Service Attack (DDOS)

A **distributed denial-of-service attack** is one in which a malicious hacker takes over computers connected to the Internet and causes them to flood a target site with demands for data and other small tasks (see Figure 12-2). A distributed denial-of-service attack does not involve taking over the targeted system. Instead, it keeps the target site so busy responding to a stream of automated requests that legitimate users cannot get in—the Internet equivalent of dialing a phone number repeatedly so that all other callers hear a busy signal.

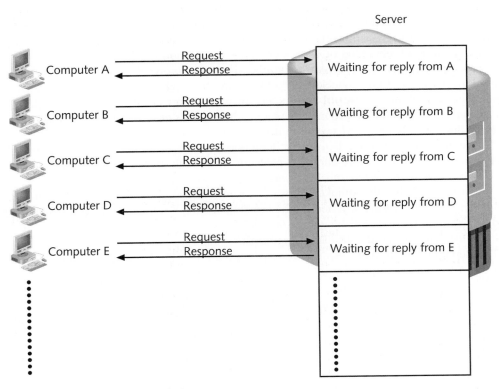

FIGURE 12-2   Distributed Denial-of-Service Attack

Software to initiate a distributed denial-of-service attack is simple to use, and many versions of such software can be found on the Web. A tiny program is downloaded surreptitiously from the attacker's computer to dozens, hundreds, or even thousands of computers all over the world. Based on a command by the attacker, or at a preset time, the malware loaded onto these computers go into action, each sending a simple request for access to the target site, again and again and again—dozens of times per second. A compromised computer is called a **zombie**. The term **botnet** is generally used to refer to a group of zombie computers running software that is being remotely controlled without the knowledge or consent of the owners of the compromised computers. Depending on the code planted on the zombie computers, a botnet also can be created for other purposes, such as sending out large quantities of spam e-mail.

Most zombies are home-based computers, and their owners are unaware of their compromise. It is estimated that there are millions of active botnet computers.[15] Arbor Networks, a network traffic analysis company, estimates that one to three percent of all Internet traffic is made up of packets of data used in denial-of-service attacks designed to knock Web sites offline.[16]

The zombies are often programmed to put false return addresses on the packets they send out (a practice known as **spoofing**) so that the sources of the attack are obscured and cannot be identified and turned off. Spoofing actually provides an opportunity to prevent distributed denial-of-service attacks. Internet service providers (ISPs) can prevent incoming packets with false IP addresses from being passed on by a process called **ingress filtering**. Corporations can use **egress filtering** to ensure that spoofed packets do not leave their corporate network. Such checking of addresses takes a tremendous amount of Internet router processing power, however. As the number of packets increases, more and more processing capacity is required to check the IP address on each packet. Companies would have to deploy faster and more powerful routers and switches to maintain the same level of performance, which would be expensive. As a result, few ISPs or corporations perform this checking. Such capabilities may be built into the next generation of network equipment.

The zombies involved in a distributed denial-of-service attack are often compromised seriously and are left with more enduring problems than their target. As a result, a user who discovers that his or her machine was compromised needs to have the computer inspected to ensure that the attacker software is removed completely from the system. In addition, system software will need to be reinstalled from a reliable backup to reestablish the system's integrity, and an upgrade or patch must be implemented to eliminate the vulnerability that allowed the attacker to enter the system.

The Republic of Estonia is a small country (population 1.4 million) in the Baltic region of northern Europe. Occupied by the Soviet Union following World War II, it gained its independence in 1991. In April and May of 2007, a global botnet of compromised home computers was used to launch hundreds of distributed denial-of-service attacks, which disrupted the Web sites of numerous Estonian government agencies, financial institutions, and media outlets. Pro-Russian activists led the attacks in retaliation for the Estonian government's decision to move a Soviet World War II memorial.[17]

Perpetrators

There are many types of computer criminals, and each type of perpetrator has different objectives as shown in Table 12-3.

TABLE 12-3 Classification of perpetrators of computer crime

Type of Perpetrator	Typical Objectives
Hacker	Test limits of system and/or gain publicity
Cracker	Cause problems, steal data, and corrupt systems
Insider	Gain financially and/or disrupt company's information systems
Industrial spy	Capture trade secrets and gain competitive advantage
Cybercriminal	Gain financially
Hacktivist	Promote political ideology
Cyberterrorist	Destroy infrastructure components of financial institutions, utilities, and emergency response units

Defensive Measures

The security of any system or network is a combination of technology, policy, and people, and it requires a wide range of activities to be effective. In addition to elements designed to prevent, detect, and respond to security incidents, a strong security program must include preliminary defensive measures, such as an overall security assessment. Assessment includes evaluating threats to the organization's computers and network, examining those threats in relation to the organization's ability to meet key business objectives, taking actions to address the most serious threats in a cost-effective manner, and educating end users about the risks and the actions they must take to help prevent a security incident. The IT security group must lead the effort to prevent security breaches by implementing security policies and procedures as well as effectively employing available hardware and software tools. Business managers must take the lead in assessing the potential impact of various threats on meeting key business objectives. Together with IT, managers must weigh the cost and potential benefits of additional security measures. No security system is perfect, however, so systems and procedures must be monitored to detect a possible intrusion. If an intrusion occurs, there must be a clear action plan that addresses notification, evidence protection, activity log maintenance, containment, eradication, recovery, and incident follow up.

Risk Assessment

A **risk assessment** is an organization's review of potential threats to its computers and networks along with an analysis of the probability that these will occur and prevent the organization from meeting key business objectives. The goal of risk assessment is to identify which investments of time and resources will best protect the organization from its most likely and serious threats. No amount of resources can guarantee a perfect security system, so organizations frequently have to balance the risk of a security breach with the cost of preventing one. The concept of **reasonable assurance** recognizes that managers must use their judgment to ensure that the cost of control does not exceed the system's benefits or the risks involved. Table 12-4 illustrates a risk assessment for a hypothetical organization.

TABLE 12-4 Risk assessment for hypothetical company

Risk	Business Objective Threatened	Estimated Probability of Such an Event Occurring	Estimated Cost of a Successful Attack	Probability × Cost = Expected Cost	Assessment of Current Level of Protection	Relative Priority to Be Fixed
Distributed denial-of-service attack	24 x 7 operation of B2C Web site	40%	$500,000	$200,000	Poor	1
E-mail attachment with harmful worm	Rapid and reliable communications among employees and suppliers	70%	$200,000	$140,000	Poor	2
Harmful virus	Employees' use of personal productivity software	90%	$50,000	$45,000	Good	3
Invoice and payment fraud	Reliable cash flow	10%	$200,000	$20,000	Excellent	4

A completed risk assessment identifies the most dangerous threats to a company and helps focus security efforts on the areas of highest payoff. For each risk area, the estimated probability of an attack occurring is multiplied by the estimated cost of a successful attack. The result is the expected cost impact for that risk area. Organizations can then assess the current level of protection against that event occurring—poor, good, or excellent. The risk areas with the highest estimated cost and the poorest level of protection are where security measures need to be improved.

Establishing a Security Policy

A **security policy** defines an organization's security requirements as well as the controls and sanctions needed to meet those requirements. A good security policy delineates responsibilities and the behavior expected of members of the organization. A security policy outlines *what* needs to be done, but not *how* to do it. The details of how to accomplish the goals of the policy are provided in separate documents and procedure guidelines.[18] In a recent survey of over 500 security professionals, 68 percent of the respondents said that their organizations had a formal information security policy, while 18 percent said they were developing such a policy.[19]

The National Institute of Standards and Technology (NIST) is a non-regulatory federal agency within the U.S. Department of Commerce. Its Computer Security Division develops security standards and technology against threats to the confidentiality, integrity, and availability of information and services.[20] The Computer Security Division has published the

NIST SP 800 series of documents, which provides useful definitions, policies, standards, and guidelines related to computer security. These may be found at the Computer Security Division Computer Security Resource Center Web site at **http://csrc.nist.gov**.

Whenever possible, automated system rules should mirror an organization's written policies. Automated system policies often can be put into practice using the configuration options in a software program. For example, if a written policy states that passwords must be changed every 30 days, then all systems should be configured to enforce this policy automatically. When applying system security restrictions, there are some trade-offs between ease of use and increased security; however, when a decision is made to favor ease of use, security incidents sometimes increase. As security techniques continue to advance in sophistication, they become more transparent to end users.

The use of e-mail attachments is a critical security issue. Sophisticated attackers can try to penetrate a network via e-mail attachments, regardless of the existence of a firewall and other security measures. As a result, some companies have chosen to block any incoming mail that has a file attachment. This greatly reduces their vulnerability. Some companies allow employees to receive and open e-mail with attachments, but only if the e-mail is expected and from someone known by the recipient. Such a policy can be risky, however, because worms often use the address book of their victims to generate e-mails to a target audience.

Another growing area of concern is the use of wireless devices to access corporate e-mail, store confidential data, and run critical applications such as inventory management and sales force automation. The primary security threat for mobile devices continues to be loss or theft of the device. However, mobile devices such as smartphones can be susceptible to viruses and worms. Wary companies have begun to include special security requirements for mobile devices as a part of their security policies. In some cases, users of laptops and mobile devices must use a virtual private network to gain access to their corporate network. A **virtual private network (VPN)** works by using the Internet to relay communications, but maintains privacy through security procedures and tunneling protocols, which encrypt data at the sending end and decrypt it at the receiving end. An additional level of security involves encrypting the originating and receiving network addresses. Because of the ease of loss or theft, it also is vital to encrypt all sensitive corporate data stored on handhelds and laptops. Unfortunately, it is hard to apply a single, simple approach to securing all handheld devices because so many manufacturers and models exist.[21]

Educating Employees, Contractors, and Part-Time Workers

According to a recent survey, one of the major security problems for U.S. companies in 2007 was creating and enhancing user awareness of security policies.[22] Employees, contractors, and part-time workers must be educated about the importance of security, so they will be motivated to understand and follow the security policies. Often, this can be accomplished by discussing recent security incidents that affected the organization. Users must understand that they are a key part of the security system and that they have certain responsibilities. For example, users must help protect an organization's information systems and data by doing the following:

- Guarding their passwords to protect against unauthorized access to their accounts
- Prohibiting others from using their passwords

- Applying strict access controls (file and directory permissions) to protect data from disclosure or destruction
- Reporting all unusual activity to the organization's IT security group

Prevention

No organization can ever be completely secure from attack. The key is to implement a layered security solution to make computer break-ins so difficult that an attacker eventually gives up. In a layered solution, if an attacker breaks through one layer of security, there is another layer to overcome. These layers of protective measures are explained in more detail in the following sections.

Installing a Corporate Firewall

Installation of a corporate firewall is the most common security precaution taken by businesses. A **firewall** stands guard between your organization's internal network and the Internet, and limits network access based on the organization's access policy (Figure 12-3). Firewalls can be established through the use of software, hardware, or a combination of both. Any Internet traffic that is not permitted explicitly into the internal network is denied entry. Similarly, most firewalls can be configured so that internal network users can be blocked from gaining access to certain Web sites based on content such as sex, violence, and so on. Most firewalls also can be configured to block instant messaging, access to newsgroups, and other Internet activities.

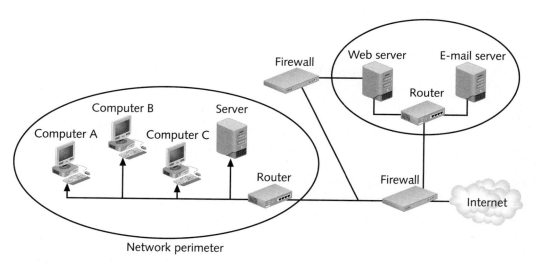

FIGURE 12-3   Firewall

Installing a firewall can lead to another serious security issue—complacency. For example, a firewall cannot prevent a worm from entering the network as an e-mail attachment. Most firewalls are configured to allow e-mail and benign-looking attachments to reach their intended recipient.

Table 12-5 lists some of the top-rated firewall software used to protect home personal computers. Typically, the software sells for $30 to $60 for a single user license.

TABLE 12-5 Popular firewall software for personal computers

Software	Vendor
Norton Personal Firewall	Symantec
Comodo	Comodo Security Solutions, Inc
Online Armor	Tall Emu Pty Ltd
ZoneAlarm Pro	Zone Labs
Personal Firewall	McAfee

Intrusion Prevention Systems

Intrusion prevention systems (IPSs) work to prevent an attack by blocking viruses, malformed packets, and other threats from getting into the company network. The IPS sits directly behind the firewall and examines all the traffic passing through it. A firewall and a network IPS are complementary. Most firewalls can be configured to block everything except what you explicitly allow through; most IPSs can be configured to let everything through except what it is told to block.

Installing Antivirus Software on Personal Computers

Antivirus software should be installed on each user's personal computer to scan a computer's memory and disk drives regularly for viruses. Antivirus software scans for a specific sequence of bytes, known as a **virus signature**. If it finds a virus, the antivirus software informs the user and may clean, delete, or quarantine any files, directories, or disks affected by the malicious code. Good antivirus software checks vital system files when the system is booted up, monitors the system continuously for virus-like activity, scans disks, scans memory when a program is run, checks programs when they are downloaded, and scans e-mail attachments before they are opened. Two of the most widely used antivirus software products are Norton Antivirus from Symantec and Personal Firewall from McAfee.

The United States Computer Emergency Response Team (US-CERT) is a partnership between the Department of Homeland Security and the public and private sectors. It was established in 2003 to protect the nation's Internet infrastructure against cyber attacks. US-CERT has long served as a clearinghouse for information on new viruses, worms, and other computer security topics. According to US-CERT, most of the virus and worm attacks that the team analyzes use already known programs. Thus, it is crucial that antivirus software be updated continually with the latest virus detection information, called **virus definitions**. In most corporations, the network administrator is responsible for monitoring network security Web sites frequently and downloading updated antivirus software as needed. Many antivirus vendors recommend, and provide for, automatic, frequent updates.

Implementing Safeguards Against Attacks by Malicious Insiders

User accounts that remain active after employees leave the company are potential security risks. To reduce the threat of attack by malicious insiders, IT staff must delete promptly the computer accounts, login IDs, and passwords of departing employees.

Organizations also need to define carefully employee roles and to separate key responsibilities properly, so that a single person is not responsible for accomplishing a task that has high security implications. For example, it would not make sense to allow an employee to initiate as well as approve purchase orders. That would allow an employee to input large invoices on behalf of a "friendly vendor," approve the invoices for payment, and then disappear from the company to split the money with the vendor. In addition to separating duties, many organizations frequently rotate people in sensitive positions to prevent potential insider crimes.

Another important safeguard is to create roles and user accounts so that users have the authority to perform their responsibilities and no more. For example, members of the Finance Department should have different authorizations from members of Human Resources. An accountant should not be able to review the pay and attendance records of an employee, and a member of Human Resources should not know how much was spent to modernize a piece of equipment. Even within one department, not all members should be given the same capabilities. Within the Finance Department, for example, some users may be able to approve invoices for payment, but others may only be able to enter them. An effective administrator will identify the similarities among users and create profiles associated with these groups.

Addressing the Most Critical Internet Security Threats

The overwhelming majority of successful computer attacks are made possible by taking advantage of well-known vulnerabilities. Computer attackers know that many organizations are slow to fix problems, which makes scanning the Internet for vulnerable systems an effective attack strategy. "The easy and destructive spread of worms, such as Blaster, Slammer, and Code Red, can be traced directly to exploitation of unpatched vulnerabilities."[23]

Both the SANS (System Administration, Networking, and Security) Institute and US-CERT regularly update a summary of the most frequent, high-impact vulnerabilities being reported to them. You can read these summaries at *www.sans.org/top20* and *www.us-cert.gov/current*, respectively. The actions required to address these issues include installing a known patch to the software, and keeping applications and operating systems up-to-date. Those responsible for computer security must make it a priority to prevent attacks using these vulnerabilities.

Conducting Periodic IT Security Audits

Another important prevention tool is a security audit that evaluates whether an organization has a well-considered security policy in place and if it is being followed. For example, if a policy says that all users must change their passwords every 30 days, the audit must check how well the policy is being implemented. The audit also should review who has access to particular systems and data and what level of authority each user has. It is not unusual for an audit to reveal that too many people have access to critical data and that many people have capabilities beyond those needed to perform their jobs. One result of a good audit is a list of items that need to be addressed in order to ensure that the security policy is being met.

A thorough security audit also should test system safeguards to ensure that they are operating as intended. Such tests might include trying the default system passwords that are active when software is first received from the vendor. The goal of such a test is to ensure that all such "known" passwords have been changed.

Some organizations will also perform a penetration test of their defenses. This entails assigning individuals to try to break through the measures and identify vulnerabilities that still need to be addressed. The individuals used for this test are often contractors rather than employees. The contractors may possess special skills or knowledge and are likely to take unique approaches to test the security measures.

An example of an organization that employs security audits is the U.S. government. U.S. government agencies must maintain security for their information systems and data to prevent data tampering, disruptions in critical operations, fraud, and the inappropriate disclosure of sensitive information. Title III of the E-Government Act, entitled the Federal Information Security Management Act (FISMA), "requires each federal agency to develop, document, and implement an agency-wide program to provide security for the information and information systems that support the operations and assets of the agency, including those provided or managed by another agency, contractor, or other source."[24]

The annual Federal Computer Security Report Card is based on evaluations defined in FISMA and compiled by the House Government Reform Committee from information provided by each agency's inspector general. These results for selected agencies are shown in Table 12-6.[25, 26] The important thing to note is that significant improvements in security can require years and do not come easy. The overall security of federal government computer systems earned only a C average in the 2007 security report card.

TABLE 12-6   Selected federal agencies' computer security report card for 2004 to 2007

Federal Agency	2007	2006	2005	2004
Department of Homeland Security	B+	D	F	F
Department of Justice	A+	A-	D	B-
Nuclear Regulatory Commission	F	F	D-	B+
Department of State	C	F	F	D+
Department of Treasury	F	F	D-	D+
Department of Defense	D-	F	F	D
NASA	C	D-	B-	D-
Department of Energy	B+	C-	F	F
Government Wide Grade	C	C-	D+	D+

Detection

Even when preventive measures are implemented, no organization is completely secure from a determined attack. Thus, organizations should implement detection systems to catch intruders in the act. Organizations often employ an intrusion detection system to minimize the impact of intruders.

Intrusion Detection Systems

An **intrusion detection system** is software and/or hardware that monitors system and network resources and activities, and notifies network security personnel when it identifies possible intrusions from outside the organization or misuse from within the organization. Two fundamentally different approaches to intrusion detection are knowledge-based approaches and behavior-based approaches. Knowledge-based intrusion detection systems contain information about specific attacks and system vulnerabilities and watch for attempts to exploit these vulnerabilities, such as repeated failed login attempts or recurring attempts to download a program to a server. When such an attempt is detected, an alarm is triggered. A behavior-based intrusion detection system models normal behavior of a system and its users from reference information collected by various means. The intrusion detection system compares current activity to this model and generates an alarm if it finds a deviation. Examples include unusual traffic at odd hours or a user in the Human Resources Department who accesses an accounting program that she has never used.

Response

An organization should be prepared for the worst—a successful attack that defeats all or some of a system's defenses and damages data and information systems. A response plan should be developed well in advance of any incident and be approved by both the organization's legal department and senior management. A well-developed response plan helps keep an incident under technical and emotional control.

In a security incident, the primary goal must be to regain control and limit damage, not to attempt to monitor or catch an intruder. Sometimes system administrators take the discovery of an intruder as a personal challenge and lose valuable time that should be used to restore data and information systems to normal.

Incident Notification

A key element of any response plan is to define who to notify and who *not* to notify. Within the company, who needs to be notified, and what information does each person need to have? Under what conditions should the company contact major customers and suppliers? How does the company inform them of a disruption in business without unnecessarily alarming them? When should local authorities or the FBI be contacted?

Most security experts recommend against giving out specific information about a compromise in public forums, such as news reports, conferences, professional meetings, and online discussion groups. All parties working on the problem need to be kept informed and up-to-date, without using systems connected to the compromised system. The intruder may be monitoring these systems and e-mail to learn what is known about the security breach.

Protecting Evidence and Activity Logs

An organization should document all details of a security incident as it works to resolve the incident. Documentation captures valuable evidence for a future prosecution and provides data to help during the incident eradication and follow-up phases. It is especially important to capture all system events, the specific actions taken (what, when, and who), and all external conversations (what, when and who) in a log book. Because this may become court evidence, an organization should establish a set of document handling procedures using the legal department as a resource.

Incident Containment

Often, it is necessary to act quickly to contain an attack and to keep a bad situation from becoming even worse. The response plan should define clearly the process for deciding if an attack is dangerous enough to warrant shutting down or disconnecting critical systems from the network. How such decisions are made, how fast they are made, and who makes them are all elements of an effective response plan.

Eradication

Before the IT security group begins the eradication effort, it must collect and log all possible criminal evidence from the system, and then verify that all necessary backups are current, complete, and free of any virus. Creating a forensic disk image of each compromised system on write-only media for later study, and as evidence, can be very useful. After virus eradication, the group must create a new backup. Throughout this process, a log should be kept of all actions taken. This will prove helpful during the follow-up phase and ensure that the problem does not recur. It is imperative to back up critical applications and data regularly. Many organizations, however, have implemented inadequate backup processes and found they could not restore original data fully after a security incident. All backups should be created with enough frequency to enable a full and quick restoration of data if an attack destroys the original. This process should be tested to confirm that it works.

Incident Follow-up

Of course, an essential part of follow-up is to determine how the organization's security was compromised so that it does not happen again. Often, the fix is as simple as getting a software patch from a product vendor. It is important to look deeper than the immediate fix and discover why the incident occurred, however. If a simple software fix could have prevented the incident, then why wasn't the fix installed *before* the incident occurred?

A review should be conducted after an incident to determine exactly what happened and to evaluate how the organization responded. One approach is to write a formal incident report that includes a detailed chronology of events and the impact of the incident. This report should identify any mistakes so that they are not repeated in the future. The experience from this incident should be used to update and revise the security incident response plan.

Creating a detailed chronology of all events also will document the incident for later prosecution. To this end, it is critical to develop an estimate of the monetary damage. Potential costs include loss of revenue, loss in productivity, and the salaries of people working to address the incident, along with the cost to replace data, software, and hardware.

Another important issue is the amount of effort that should be put into capturing the perpetrator. If a Web site simply was defaced, it is easy to fix or restore the site's HTML (Hypertext Markup Language, the code that describes to your browser how a Web page should look). What if the intruders inflicted more serious damage, however, such as erasing proprietary program source code or the contents of key corporate databases? What if they stole company trade secrets? Expert crackers can conceal their identity and tracking them down can take a long time as well as a tremendous amount of corporate resources.

The potential for negative publicity also must be considered. Discussing security attacks through public trials and the associated publicity not only has enormous potential costs in public relations, but real monetary costs. For example, a brokerage firm might lose many customers who learn of an attack and then think their money or records aren't secure. Even if a company decides that the negative publicity risk is worth it and goes after the perpetrator, documents containing proprietary information that must be provided to the court could cause even greater security threats in the future. On the other hand, does an organization have an ethical or legal duty to inform customers or clients of a cyberattack that may have put their personal data or financial resources at risk?

Table 12-7 recommends a set of actions an organization can take to implement a successful IT security initiative. The appropriate answer to each question is "yes."

TABLE 12-7   A manager's checklist

Recommended Management Actions	Yes	No
Has a risk assessment been performed to identify investments of time and resources that can protect the organization from its most likely and most serious threats?		
Has a security policy been formulated and shared broadly throughout the organization?		
Is there an effective security education program for employees, contractors, and part-time employees?		
Has a layered security solution been implemented to prevent computer incidents?		
Has a comprehensive incident response plan been developed?		
Does the organization have a written data privacy policy that is followed?		
Have you identified a person who has full responsibility for implementing your data policy and dealing with consumer data issues?		
Have you developed and communicated an acceptable computer usage policy?		

Chapter Summary

- Ethics is a set of beliefs about right and wrong behavior.

- Key actions that many organizations are taking to improve business ethics include appointing a corporate ethics officer, setting of ethical standards by the board of directors, establishing a corporate code of ethics, requiring employees to take ethics training, and including ethical criteria in employee appraisals.

- The Supreme Court has ruled that citizens are protected by the Fourth Amendment from unreasonable searches and seizures by the government when there is a reasonable expectation of privacy.

- There are few laws that provide individuals with privacy protection from private industry.

- An organization can treat customer data responsibly by collecting only personal information necessary to deliver its product or service, ensuring that the data is protected carefully and accessible only by those with a need to know, and providing a process for consumers to review their own data and make corrections.

- Organizations should appoint an executive to define, implement, and oversee a set of data privacy policies.

- Many Web sites use cookies to capture data about visitors and their activity while at the Web site. These Web sites typically have a privacy policy that states what sort of information is captured and how that information may be used.

- Many organizations have an information technology usage policy to protect against employee abuses that reduce worker productivity or that could expose the employer to harassment lawsuits. Such a policy establishes boundaries of acceptable behavior and enables management to take action against violators.

- Laws governing employee privacy and monitoring continue to develop as society struggles to define the extent to which employers should be able to monitor the work-related activities of their employees.

- Cybercrime includes a wide range of activities but electronic fraud, followed by virus attacks, is the leading cause of financial loss from computer incidents.

- A botnet can be used to initiate a distributed denial-of-service attack, generate volumes of spam, and perform other disruptive acts.

- An organization's security program should begin by assessing threats to the organization's computers and network, identifying actions that address the most serious vulnerabilities, and educating users about security risks and the actions they must take to prevent a security incident.

- While no organization can ever be completely secure from attack, implementation of a layered security solution can make computer break-in extremely difficult. The layers of security should include a firewall, antivirus software, safeguards against attacks by malicious insiders, addressing the most critical Internet security threats, verifying backup processes, and conducting security audits.

- The use of intrusion detection systems can reduce the impact of intruders.

An organization should be prepared with a response plan in the unfortunate event that a successful attack defeats all defenses and damages data and information systems. This plan should address incident notification, protection of evidence and activity logs, incident containment, incident eradication, and incident follow-up.

Discussion Questions

1. How would you define ethical? How would you define legal? Provide an example of an action that is legal but not ethical.

2. What is the role of the board of directors in setting the ethical standards of the organization?

3. What is a code of ethics? Can you find a code of ethics for your school, university, or place of employment?

4. Imagine that you must develop ethics training for a small group of fellow employees or students. What do you think should be the primary objective(s) of such training? What topics do you think should be covered?

5. What is meant by "reasonable expectation of privacy"? Provide an example of a situation where an individual has such an expectation. Provide an example of a situation where an individual should not have such an expectation.

6. Briefly define virus, worm, and botnet.

7. Which of the various types of perpetrators of computer crime has the greatest potential to cause serious harm to an organization? Why?

8. What is a risk assessment? What is the concept of reasonable assurance as it applies to the implementation of computer security measures?

9. What is an organization's computer security policy?

10. Why is it important to implement a multiple-layered computer security defense?

Action Memos

1. You just received an e-mail request from your friend who is the vice president of Human Resources. She is taking an informal survey of a few close confidants on the topic of adding ethical criteria and evaluations to the organization's employee appraisals process. She has asked you to provide your opinion in a brief e-mail to her by the end of the day. How would you respond?

2. You are the new CIO of a market research and consulting firm. During a discussion last week with the CEO and her direct reports, you learned that the firm has no Consumer Data Privacy Policy. Prepare a one-page set of talking points you can use with the other executives to convince them why the creation of such a policy should be a priority.

Web-Based Case

Find and read the privacy policies for Web sites that you frequently visit. What questions do these policies raise about the collection and use of your personal data? With which policy do you feel most comfortable? Why? What changes would you like to see made with this policy to ease any concerns you may have?

Case Study

Trading Scandal at Société Générale

In January 2008, Société Générale (SocGen), France's second largest banking establishment, was a victim of internal fraud carried out by an employee, Jérôme Kerviel. SocGen bank lost €4.9 billion (euros) as an immediate result of the fraud. (At the time of this incident, the euro was worth approximately $1.45 dollars.)

In 2007, SocGen was rated the best equity derivatives operation in the world by *Risk* magazine. Its internal control system of checks and balances was world renown. For example, its trading room has five levels of hierarchy. Each of those levels has a clear set of trading limits and controls, which are checked daily by a small army of compliance officers.[27] In addition, "the bank also has a shock team of internal auditors who descend on a corner of the bank without warning and pull apart its operations to ensure they conform to bank rules."[28]

During the summer of 2000, Kerviel began employment in the bank, ironically, in its compliance department. Five years later, he was promoted to a junior trader in the arbitrage desk, which deals in program trading, exchange traded funds, swaps, stock index futures trading, and quantitative trading. Kerviel was responsible for generating profits for the bank and its customers by betting on the market's future performance. His first major win came in 2005 when he shorted stock of German insurer Allianz and earned the bank €55,000.

Thanks to his years of experience in the compliance department, Kerviel was an expert in the proprietary information system SocGen used to book trades. He knew that while the risk-control department monitored the bank's overall positions very closely, it did not verify the data that individual traders entered into the system. Kerviel also knew the timing of the nightly reconciliation of the day's trades, so that he was able to delete and then re-enter unauthorized transactions without getting caught.

On November 7, 2007, SocGen received an e-mail alert from a surveillance officer at Eurex (one of Europe's largest exchanges). The message stated that Kerviel had engaged in several transactions that had set off alarms at the exchange over the past seven months. A SocGen risk-control expert responded two weeks later that there was nothing irregular about the transactions. A week later, Eurex sent a second e-mail alert stating that they were not satisfied with Soc-Gen's explanation and demanding more details. Following another two-week delay, SocGen provided further details, and both Eurex and SocGen let the matter drop. The compliance officer who made both replies to Eurex used accounts provided by Kerviel and his supervisor as well as a compliance officer at a SocGen subsidiary. Kerviel's supervisor stated that there was no anomaly whatsoever.

Following the Eurex warnings, Kerviel took additional steps to cover his tracks by manipulating portions of the internal risk-control system with which he was unfamiliar. This ultimately led to the discovery of his alleged fraud.[29] On January 18, 2008 Keviel executed trades, which set off another alarm. This time, upon a more thorough investigation, a major problem became apparent. As SocGen risk-control experts reviewed Kerviel's latest transactions carefully, they were shocked to discover that they had resulted in a position of €50 billion (obviously far beyond Kerviel's trading limit) which, when finally cleared, resulted in a loss of more than €4.9 billion!

As of this writing, Kerviel is still under investigation and involved in litigation charging him of using his insider knowledge to falsify records and commit computer fraud. Prosecutors suspect his motivation was to boost his income by making successful trades far beyond his trading limits, thus earning large bonuses (his total salary and bonus for 2007 was a relatively modest €94,000). Kerviel spent five weeks in jail but is currently free on bond. He was hired in February 2008 as a computer consultant by the French firm Lemaire Consultants & Associates, however, he is said to be "traumatized" by his new-found infamy.

Kerviel admits he took trading positions beyond his authorized limit to make transactions involving European index futures. Kerviel told prosecutors "the techniques I used aren't at all sophisticated and any control that's properly carried out should have caught it."[30] He insists he did no wrong and that the bank was fully aware of his transactions. Kerviel has said he refuses to be made a scapegoat for the bank's lapses in oversight. He argues that his superiors tacitly approved his activities—as long as they were generating a profit. Kerviel had earned a profit for the bank of nearly €1.5 billion in 2007 by exceeding his trade limit and executing similar, but successful, trades. The bank meanwhile said the fraud was based on simple transactions, but concealed by "sophisticated and varied techniques."[31] If convicted, Kerviel faces up to five years in jail and fines for as much as €300,000.[32]

The sterling reputation of SocGen was tarnished badly and the market value of the firm dropped 50 percent over the course of just a few months. The bank's highly respected CEO and Chairman of the board, Daniel Bouton, was put under enormous pressure to step down; this included requests for his resignation from French President Nicholas Sarkozy. Bouton eventually resigned as CEO in May 2008, but he remains chairman of the board.[33] In December 2008, European hedge fund GLG Partners entered into an agreement to acquire the bank in the second half of 2009.[34]

Several internal and external investigations of the bank's operating procedures and internal controls have been completed. The French banking regulator stated there were "grave deficiencies" in the bank's internal controls and fined it €4 million. The Banking Commission said SocGen did not focus sufficiently on fraud weaknesses and there were "significant weaknesses" in the bank's IT security systems. Another report pointed out that Kerviel's direct supervisor was inexperienced and received insufficient support to do his job properly. It also stated that Kerviel's fraudulent transactions were entered by an unnamed assistant trader thus raising the issue of collusion and indicating even more widespread weaknesses in internal controls.

Pascal Decque, a financial analyst who covers SocGen for Natixis (a leading player in corporate and investment banking) commented, "SocGen was brilliant in their achievement, they were the world leader in derivatives. Maybe when you are that good, you think you will never fail."[35]

Discussion Questions

1. Peter Gumble, European editor for *Fortune* magazine comments: "Kerviel is a stunning example of a trader breaking the rules, but he is by no means alone. One of the dirty little secrets of trading floors around the world is that every so often, somebody is caught concealing a position and is quickly—and quietly—dismissed. Traders do this not infrequently, and the question is how quickly compliance systems pick it up." [This] "might be shocking for people unfamiliar with the high-risk, high-reward culture of most trading floors, but consider this: the only way banks can tell who will turn into a good trader and who won't is by giving every youngster it hires a chance to show his mettle. This means allowing even the most junior traders to take aggressive positions. The leeway is supposed to be matched

by careful controls, but clearly they aren't foolproof."[36] What is your reaction to this statement by Mr. Gumble?

2. What explanations can there be for the failure of SocGen's internal control system to detect Kerveil's transactions while Eurtex detected many suspicious transactions?

3. Should banks and investment firms permit members of their compliance departments to become traders?

4. Do research on the Web to find out if Kerveil was found guilty and punished. What other outcomes resulted from this incident?

Endnotes

[1] "About Hannaford," Hannaford Web site at *http://hannaford.com*, accessed November 21, 2008.

[2] "Hannaford Bros Supermarkets Hit By Big Data Breach," *http://wbztv.com, March 17, 2008.*

[3] Bill Brenner, "Hannaford Breach Details Indicate Inside Job," *http://searchsecurity.techtarget. com/news/article/0,289142,sid14_gci1307486,00.html*, March 28, 2008.

[4] Ed Dickson, "Hannaford Brothers Data Breach Might Reveal Current Security Standards Are Outdated," Blogger News Network, *www.bloggernews.net/114589*, March 19, 2008.

[5] "Hannaford Bros.Faces Class Action Over Data Breach," ConsumerAffairs.com, *www.consumeraffairs.com/news04/2008/03/hannaford_data2.html*, March 21, 2008.

[6] "What is an Ethics Officer?" Web site of Ethics Officer Association, *www.eoa.org*, accessed November 22, 2008.

[7] Patricia Harned, "A Word from the President: Ethics Offices and Officers," *Ethics Today Online*, *www.ethics.org*, Volume 3, Issue 2, October 2004.

[8] Robert Mullins, "HP Hires Ethics and Compliance Officer," *Computerworld*, October 17, 2006.

[9] K.C. Jones, "Calif. Attorney General Attempting Deal Between HP, Pretext Victims," *Computerworld*, December 8, 2006.

[10] "Chief Privacy Officers: Forces or Figureheads?" *Computerworld*, March 24, 2001.

[11] David Herron, Scott H. Dunham, Linda Kwak, and Shannon Gibson, "Ninth Circuit Court Addresses Privacy Rights for Employer-Provided Text-Messaging Capabilities," *O'Melveny & Myers LLP Employment Law Newsletter*, October 3, 2008.

[12] "CSI Survey 2007," *GoCSI.com*, accessed June 27, 2008.

[13] Keith Regan, "Web Crime Spikes in 2007, Losses Near $240 M," *Electronic Commerce Times*, April 4, 2008.

[14] Ross Kerber, "Cost of Data Breach at TJX Soars to Over $256 M," *Boston Globe*, August 15, 2007.

[15] "Botnet," *SearchSecurity.com*, accessed December 8, 2008.

[16] Robert McMillan, "Internet has A Trash Problem, Researcher Says," *Network World*, April 1, 2008.

[17] Carolyn Duffy Marsan, "How Close is World War 3.0," *Network World*, August 22, 2007.

[18] Marc Gartenberg, "How to Develop an Enterprise Security Policy," *Computerworld*, *www. computerworld.com*, January 13, 2005.

[19] Robert Richardson, "2008 CSI Computer Crime & Security Survey," accessed at *www.gocsi. com/forms/csi_survey.jhtml*, January 12, 2009.

361

20 "2007 Computer Security Division Report," National Institute of Standards and Technology, accessed at *http://csrc.nist.gov/publications/nistir/ir7442/NIST-IR-7442_2007CSDAnnualReport.pdf*, February 2, 2009.

21 Jaikumar Vijayan, "Handheld Risks Prompt Push for Usage Policy," *Computerworld*, *www.computerworld.com*, February 21, 2005.

22 Larry Greenemeier, "The Threat Within: Employees Pose the Biggest Security Risk," *InformationWeek*, July16, 2007.

23 "The SANS Top 20 Internet Vulnerabilities," *www.sans.org/top20*, August 22, 2005.

24 "Background, FISMA Implementation Project," *http://csrc.nist.gov/sec-cert/ca-background.html*, August 13, 2005.

25 May 2008 Federal Security Report Card, accessed at *http://republicans.oversight.house.gov/media/PDFs/Reports/FY2007FISMAReportCard.pdf* on December 9, 2008.

26 FISMA Grades 2005 at *http://republicans.oversight.house.gov/FISMA/* accessed December 9, 2008.

27 Peter Gumbel, "4 Things I Learned at Société Générale," *Fortune*, February 1, 2008.

28 Peter Gumbel, "4 Things I Learned at Société Générale," *Fortune*, February 1, 2008.

29 Nelson D. Schwartz and Katrin Bennhold, "Société Générale Scandal: 'A Suspicion That This Was Inevitable,'" *International Herald Tribune*, February 5, 2008.

30 Peter Gumbel, "4 Things I Learned at Société Générale," *Fortune*, February 1, 2008.

31 "Rogue Trader to Cost SocGen $7b," *BBS News*, January 24, 2008.

32 Nicola Clark and James Kanter, "Decision Delayed on Releasing ex-Trader at Center of Société Générale Inquiry," *International Herald Tribune*, March 14, 2008.

33 "Société Boss Burton to Step Down," *BBC News*, April 17, 2008.

34 "GLG to Acquire SocGen Long Only Operation," *Hedge Funds Review*, December 24, 2008.

35 Nelson D. Schwartz and Katrin Bennhold, "Société GénéraleScandal: 'A Suspicion That This Was Inevitable,'" *International Herald Tribune*, February 5, 2008.

36 Peter Gumbel, "4 Things I Learned at Société Générale," *Fortune*, February 1, 2008.

bandwidth The range of frequencies that an electronic signal occupies on a given transmission media.

best practice The most efficient and effective way of accomplishing a task, based on procedures that have proven themselves repeatedly over an extended period of time.

blog A Web site in which contributors ("bloggers") provide ongoing commentary on a particular subject.

botnet A group of zombie computers running software that is being remotely controlled without the knowledge or consent of the owners of the compromised computers.

business continuity plan A plan that defines the people and procedures required to ensure timely and orderly resumption of an organization's essential processes with minimal interruption.

business intelligence (BI) A wide range of applications, practices, and technologies used for the extraction, translation, integration, analysis, and presentation of data to support improved decision making.

business performance management (BPM) An application of BI that enables the continuous and real-time analysis of operational data to measure actual performance and forecast future performance.

business rule management system (BRMS) Software used to define, execute, monitor, and maintain the decision logic used by the operational systems to run the organization.

business-to-business (B2B) e-business The exchange of goods and services between businesses via computer networks.

business-to-consumer (B2C) e-business The exchange of goods and services between businesses and individual consumers via computer networks.

calendaring software Software that allows people to capture and record scheduled meetings and events.

centralized architecture A type of software architecture based on the use of a mainframe computer that supports a variety of local and remote devices, such as printers, terminals, and workstations.

client/server A type of distributed architecture where clients request services and resources over the network and servers provide those services and resources.

code of ethics A written statement that highlights an organization's key ethical issues and identifies the overarching values and principles that are important to the organization and its decision making.

cohesion A measure of how strongly related and focused the various responsibilities of a software or hardware component are.

communications channel A path that carries a signal from sender to receiver.

communications management An area of project management that involves generating, collecting, disseminating, and storing project information in a timely and effective manner.

community of practice (CoP) A group whose members share a common set of goals and interests and regularly engage in sharing and learning as they strive to meet those goals.

consumer-to-consumer (C2C) e-business The exchange of goods and services among individuals, typically facilitated by a third party, via computer networks.

Control OBjectives for Information and Related Technology (COBIT) A set of guidelines whose goal is to align IT resources and processes with business objectives, quality standards, monetary controls, and security needs.

core business process A business process which provides valuable customer benefits and typically has a direct impact on the organization's customers, is a major costs driver, or is essential for providing services.

core competency An activity that an organization performs well and leverages widely to many products and markets; a core competency provides value to customers and is hard for competitors to imitate.

corporate ethics officer A senior-level manager who provides vision and direction in the area of business conduct.

cost management An area of project management that involves developing and managing a project budget.

cost-reimbursable contract A contract that requires paying the provider an amount that covers the provider's actual costs plus an additional amount or percentage for profit.

coupling A measure of the degree to which each software and hardware component relies on other modules to perform its function.

customer relationship management (CRM) system An enterprise system that supports the processes performed by all the entities involved in creating or increasing the demand for an organization's products and services.

customer service Increasing customer satisfaction and improving the customer experience by, for example, dealing with problems caused by over (customer receives more of a particular item than he expected), short (customer receives less of a particular item than he expected), and damaged shipments.

cybercrime Criminal activity in which a computer or a computer network is used as a tool to commit a crime or is the target of criminal activity.

data cube A subset of a database built to support OLAP processing. Data cubes contain numeric facts called measures, which are categorized by dimensions such as time and geography.

data mart A smaller version of a data warehouse—scaled down to meet the specific needs of a business unit.

data warehouse A database that stores large amounts of historical data in a form that readily supports analysis and management decision making.

decision support system (DSS) An information system that employs models and analytic tools to help users gain insights into data, draw conclusions from the data, and make recommendations.

demand planning Determining the demand for products taking into account all the factors that can affect that demand—general economic conditions, actions by competitors, your own pricing, and promotion and advertising activities.

desktop sharing A method of collaborating electronically that includes a number of technologies and products that allow remote access and collaboration.

disaster recovery plan A subset of the business continuity plan that focuses on keeping components of the IT infrastructure functioning during a disaster or recovering them quickly afterward.

distributed applications A software architecture style that involves sharing the processing, formatting, presentation, and storage functions across clients and servers.

distributed denial-of-service attack A type of computer attack in which a malicious hacker takes over computers connected to the Internet and causes them to flood a targeted site with demands for data and other small tasks.

drill-down analysis The interactive examination of high level, summary data in increasing detail to gain insight into certain elements.

due diligence The effort made by an ordinarily prudent or reasonable party to avoid harm to another party.

e-business The transformation of key business processes though the use of Internet technologies.

e-government (e-gov) The use of information technology (such as Wide Area Networks, the Internet, and mobile computing) by government agencies to transform relations between the government and citizens (G2C), the government and businesses (G2B), and among various branches of the government (G2G).

egress filtering A computer security technique in which an organization ensures that spoofed data packets do not leave its network.

e-learning systems A range of computer-enhanced learning techniques, including computer-based simulations, multimedia CD-ROMs, Web-based learning materials, hypermedia, podcasts, and Webcasts.

electronic bulletin board A collaboration tool that allows users to leave messages or read public messages that provide information or announce upcoming events.

electronic corporate directory An electronic directory used in a large organization to find the right person with whom to collaborate on an issue or opportunity.

electronic data interchange (EDI) An interorganizational system that supports the direct, computer-to-computer transfer of information in the form of predefined electronic documents.

electronic fraud A broad class of cybercrime that involves the use of computer hardware, software, or networks to misrepresent facts for the purpose of causing someone to do or refrain from doing something which causes loss.

end users The people most directly affected by a project. To complete their work, end users probably will have to learn new work processes and tools created by a project.

Enhanced Data Rates for Global Evolution (EDGE) A type of wireless network connection that provides faster data transfer rates than GPRS over a similar-sized area.

enterprise architecture A set of models that describe the technical implementation of an organization's business strategy and business processes.

enterprise IT Information systems used by organizations to define interactions among their own employees or with external customers, suppliers, and other business partners.

enterprise resource planning (ERP) system A set of core software modules that enable organizations to share data across the entire enterprise through the use of a common database and management reporting tools.

enterprise search The application of search technology to find information within an organization.

e-procurement software Software that allows a company to create an electronic catalog with search capability.

ethics A set of beliefs about right and wrong behavior.

extract-transform-load (ETL) Process used to pull data from disparate data sources to populate and maintain a data warehouse.

firewall   A system of software, hardware, or a combination of both, that stands guard between an internal network and the Internet; a firewall also limits network access based on access policy.

fixed-price contract   A contract in which the buyer and provider agree to a total fixed price for a well-defined product or service.

Forming-Storming-Norming-Performing model   A model first proposed by Bruce Tuckman to describe how teams develop and evolve.

function IT   Information systems that improve the productivity of individual users in performing stand-alone tasks.

general packet radio service (GPRS)   A type of mobile data service available to users of GSM mobile phones; it provides fast data transfers over a very large area.

global service providers (GSP)   Outsourcing firms that evaluate all aspects of an organization's business activities to take advantage of an outsourcer's best practices, business contacts, capabilities, experience, intellectual property, global infrastructure, or geographic presence by tapping resources and providing capabilities anywhere around the globe.

global system for mobile communications service (GSM)   The most widely adopted digital cellular technology in use today; it uses a time and frequency division technique to optimize the call-carrying capacity of a wireless network.

goal   A specific result that must be achieved to reach an objective.

growth-share matrix   A model used to allocate resources among various business units; it enables managers to divide their organization's collection of business units and products into four distinct groups and offers advice for each group.

hertz (Hz)   The measure of frequency at which a signal is transmitted (cycles per second).

hot spot   The area covered by one or more interconnected wireless access points.

human resource management   An area of project management that involves making the most effective use of the people involved with a project. It includes organizational planning, staff acquisition, and team development.

industry consortia-sponsored marketplace   An electronic marketplace set up by several different companies in a particular industry that join forces to gain the advantages of a private company marketplace.

information systems   Systems that enable a firm to meet fundamental objectives, such as increasing revenue, reducing costs, improving decision making, enhancing customer relationships, and speeding up products' time to market.

information technology (IT)   All the tools that capture, store, process, exchange, and use information, including software, hardware, and networks.

ingress filtering   A computer security technique in which Internet service providers (ISPs) prevent incoming data packets from being passed on with false IP addresses.

instant messaging (IM)   Real-time, informal communications based on the often rapid exchange of typed messages.

intangible benefit   A benefit that cannot be measured directly nor quantified easily in monetary terms.

internal control   The process established by an organization's board of directors, managers, and IT systems to provide reasonable assurance for effective and efficient operations, reliable financial reporting, and compliance with applicable laws and regulations.

interorganizational information systems   An IT system that supports the flow of data among organizations to achieve shared goals.

intrusion detection system   A network security mechanism that monitors system and network resources and activities, and notifies network security personnel when it identifies possible intrusions from outside the organization or misuse from within the organization.

intrusion prevention system   A network security mechanism that works to prevent an attack by blocking viruses, malformed packets, and other threats from getting into the company network.

IT governance   A decision-making process relating to investments in IT.

IT infrastructure   An organization's set of IT hardware, software, and networks.

IT Infrastructure Library (ITIL)   A set of guidelines initially formulated by the UK government in the late 1980s and widely used today throughout Europe and the United States to standardize, integrate, and manage IT service delivery.

IT support organization   The group of employees within an organization that plans, implements, operates, and supports IT.

key performance indicators (KPIs)   Metrics that track progress in executing chosen strategies in terms of direction, measure, target, and time frame.

knowledge management (KM)   The practice of increasing awareness, fostering learning, speeding collaboration and innovation, and exchanging insights in an organization.

logistics   The process of establishing a network of warehouses for storing products, choosing carriers for product delivery, scheduling carrier pick-ups, and invoicing the customer.

lump-sum contract   A contract in which the buyer and provider agree to a total fixed price for a well-defined product or service.

make-or-buy decision   The process of comparing the pros and cons of in-house production versus outsourcing of a given product or service.

malware   Malicious software, usually installed without a computer owner's knowledge.

manufacturing   Producing, testing, packaging, and preparing products for delivery.

market options matrices   A decision-making model that identifies an organization's product and market options.

measures   Metrics that track progress in executing chosen strategies to attain an organization's objectives and goals.

Michael Porter's Five Forces Model   A model used to assess the nature of industry competition; it identifies fundamental factors that determine the level of competition and long-term profitability of an industry.

mobile commerce (m-commerce)   The buying and selling of goods and services via mobile devices such as cell phones, smartphones, PDAs, and other such devices.

network IT   Information systems that improve communications and support collaboration among members of a workgroup.

objective   A statement of a compelling business need that an organization must meet to achieve its vision and mission.

offshore outsourcing   An arrangement in which a company contracts with another organization, whose workers are located in a foreign country, to provide services that could be provided by company employees.

Online Analytical Processing (OLAP)   A method to analyze multidimensional data from many different perspectives.

organic list   A type of search engine result in which users are given a listing of potential Web sites based on their content and keyword relevancy.

outsourcing   An arrangement in which a company contracts with another organization to provide services that could be provided by the company's employees.

paid listings   Search engine results that appear because the owners of certain sites have paid fees to the search engine firm.

Payment Card Industry (PCI) data security standard   A multifaceted security standard that requires retailers to implement a set of security management policies, procedures, network architecture, software design, and other critical protective measures to safeguard cardholder data.

podcast   A digital media file distributed over the Internet using syndication feeds; it is designed to be played on portable media players and personal computers.

private company marketplace   A Web site set up by a large manufacturer to manage its purchasing functions.

private store   A Web site that functions as a private store for each of an organization's major customers with access provided through a company identification code and password enabling purchases from a selection of products at pre-negotiated prices.

process efficiency monitoring BPM software   Software that connects with and monitors each system used by a company to support a particular process in order to identify bottle necks and inefficiencies.

procurement management   An area of project management that involves acquiring goods or services for a project from sources outside the performing organization.

project   A temporary endeavor undertaken to create a unique product, service, or result.

project champion   A senior-level executive who is a strong advocate for a project.

project management   The application of knowledge, skills, and techniques to project activities in order to meet project requirements.

project risk   An uncertain event or condition that, if it occurs, has an effect on a project objective.

project scope   A definition of the work included and not included in a project.

project sponsor   A senior manager in an organization who will be most affected by a project's implementation.

project stakeholders   The people involved in a project or those affected by its outcome.

quality   The degree to which a project meets the needs of its users.

quality assurance   The ongoing evaluation of a project to ensure that it meets the identified quality standards.

quality control   The process of checking project results to ensure that they meet identified quality standards.

quality management   An area of project management that involves ensuring that a project will meet the needs for which it was undertaken.

quality planning   The process of determining which quality standards are relevant to a project and determining how they will be met.

Really Simple Syndication (RSS)   A family of data formats that allows end users to automatically receive feeds anytime there are new postings to their favorite blog sites, updated news headlines, or new information posted at specified Web sites.

reasonable assurance   A concept in computer security that recognizes that managers must use their judgment to ensure that the cost of control does not exceed the system's benefits or the risks involved.

recovery time objective   The time within which a business function must be recovered before an organization suffers serious damage.

reporting and insight BPM software   Software that gathers data from a business process and provide reports and dashboards to create actionable information to decision makers.

risk assessment   An organization's review of potential threats to its computers and networks along with an analysis of the probability that these will occur in such a way as to prevent the organization from meeting key business objectives.

risk management   An area of project management that involves identifying, analyzing, and managing project risks.

risk owner   Person responsible for developing a risk management strategy and monitoring the project to determine if the risk is about to occur or has occurred.

scope management   An area of project management that involves defining the work that must be done as part of a project and then controlling the work to stay within the agreed upon scope.

search engine optimization   The process of ensuring that a Web site appears at or near the top of the search engine results whenever someone enters search terms that relate to a company's products or services.

Secure Sockets Layer (SSL)   A protocol used to verify that the Web site to which a consumer is connected is what it purports to be. SSL also encrypts and decrypts the information flowing between the Web site and the consumer's computer.

security policy   A written statement that defines an organization's security requirements as well as the controls and sanctions needed to meet those requirements.

service-oriented architecture (SOA)   A software application development approach based on building user applications out of software services.

shared workspace   An area on a Web server in which project members and colleagues can share documents, models, photos, and other forms of information to keep each other current on the status of projects or topics of common interest.

smart card   A card, similar to a credit card in size and shape, that contains an embedded microchip to process instructions and store data for use in various applications such as telephone calling, electronic cash payments, storage of patient information, and security access.

smart sourcing   An approach to analyzing outsourcing needs based on the work to be done, its associated processes, and the level of effectiveness and resources required.

social network analysis (SNA)   A method of documenting and measuring flows of information between individuals, workgroups, organizations, computers, Web sites, and other information sources.

sourcing   The process of choosing suppliers and establishing the contract terms in order to provide and deliver a product's raw materials to the manufacturing locations.

spoofing   Providing false return addresses on packets of data sent over the Internet.

strategic planning   A process that helps managers to identify desired outcomes and formulate feasible plans to achieve their objectives by using available resources and capabilities.

strategy   Specific actions that an organization will take to achieve its vision/mission, objectives, and goals.

strengths, weaknesses, opportunities, threats (SWOT) matrix   A model used for the analysis of the internal and external environment; it illustrates what the firm is doing well, where it can improve, what opportunities are available, and what environmental factors threaten the future of the organization.

supply chain   The flow of materials, information, and dollars from supplier to manufacturer to wholesaler to retailer to supplier.

Supply chain management (SCM)   Planning, executing, monitoring, and controlling the set of processes in the supply chain.

tangible benefit   A benefit that is measured directly and assigned a monetary value.

time and material contracts   A contract in which the buyer pays the provider for both the time and materials required to complete the contracted work.

time management   An area of project management that involves estimating a reasonable completion date, developing a workable project schedule, and ensuring the timely completion of the project.

transaction processing system (TPS)   An information system that captures data from company transactions and other key events, and updates the firm's records, which are maintained in electronic files or databases.

transmission media   Media used to propagate a communication signal; it may be guided, in which case the signal travels along a solid medium, or wireless, in which case the signal is broadcast over airwaves as a form of electro-magnetic radiation.

Unified Modeling Language (UML)   A language for specifying, constructing, visualizing, and documenting the artifacts of a software-intensive system.

value proposition   A clear statement of the tangible benefits that a customer obtains from using a company's products or services.

virtual private network (VPN)   A computer network that uses the Internet to relay communications, but which maintains privacy through security procedures and tunneling protocols that encrypt data at the sending end and decrypt it at the receiving end.

virus   A piece of programming code, usually disguised as something innocuous, which causes some unexpected and usually undesirable event.

virus definitions   A compilation of the latest virus detection information.

virus signature   A specific sequence of bytes in a virus.

vision/mission statement   A document that communicates an organization's overarching aspirations, which form a foundation for making decisions and taking action.

Web 2.0   A term describing changes in technology and Web site design to enhance information sharing, collaboration, and functionality on the Web.

Web conference   A way to conduct live meetings or presentations over the Internet.

Wi-Fi   A wireless communications technology brand owned by the Wi-Fi Alliance, which includes more than 300 technology companies.

wiki   A collaborative Web site, which allows users to edit and change its content easily and quickly.

work breakdown structure (WBS)   An outline of the work to be done to complete a project; it is critical to effective time management.

workflow designer BPM software   Software that enables business managers and analysts to design a business process complete with all of the associated forms, business rules, role definitions, and integration to other systems involved in the process.

Worldwide Interoperability for Microwave Access (WiMAX)   The common name for a set of 802.16 wireless metropolitan-area network standards that support different types of communications access.

worms   Harmful computer programs that reside in the active memory of the computer; worms can propagate over a network without human intervention.

zombie   A computer that has been compromised by a virus, worm, or some other type of malware.

INDEX

Index